W9-CEI-138

Environmentally Conscious Mechanical Design

Edited by
Myer Kutz

John Wiley & Sons, Inc.

This book is printed on acid-free paper. ♾

Published by John Wiley & Sons, Inc., Hoboken, New Jersey
Published simultaneously in Canada

Wiley Bicentennial Logo: Richard J. Pacifico

Library of Congress Cataloging-in-Publication Data:

Environmentally conscious mechanical design, volume 1 / edited by Myer Kutz.
p. cm.
ISBN-13 978-0-471-72636-4
1. Machine design. 2. Sustainable engineering. I. Kutz, Myer.
TJ233.E58 2007
621.8′15—dc22

2006029315

10 9 8 7 6 5 4 3 2 1

To David and Evelyn and to Eric and Raffie, my *dear friends.*

[illegible] and Raffy, my pet friends.

Contents

Contributors

Berdinus A Bras
Georgia Institute of Technology
Atlanta, Georgia

Michael Castellani
University at Buffalo–SUNY
Buffalo, New York

Abigail Clarke
Michigan Technological University
Houghton, Michigan

N. De Silva
University of Kentucky
Lexington, Kentucky

B. S. Dhillon
University of Ottawa
Ottawa ON

O. W. Dillon Jr.
University of Kentucky
Lexington, Kentucky

Daniel Fitzgerald
University of Maryland
College Park, Maryland

John K. Gershenson
Michigan Technological University
Houghton, Michigan

Thornton H. Gogoll
Black & Decker Inc.
Towson, Maryland

Jeffrey W. Herrmann
University of Maryland
College Park, Maryland

Dr. Sridhar Idapalapati
Nanyang Technological University
Singapore

I. H. Jaafar
University of Kentucky
Lexington, Kentucky

I. S. Jawahir
University of Kentucky
Lexington, Kentucky

K. Joshi
University of Kentucky
Lexington, Kentucky

Hartmut Kaebernick
The University of New South Wales
Sydney nsw
Australia

Sami Kara
The University of New South Wales
Sydney NSW
Australia

R. Alan Kemerling
Ethicon Endo-Surgery, Inc.
Cincinnati, Ohio

Kemper E. Lewis
University at Buffalo–SUNY
Buffalo, New York

O. Geoffrey Okogbaa
University of South Florida
Tampa, Florida

Wilkistar Otieno
University of South Florida
Tampa, Florida

William Regli
Drexel University
Philadelphia, Pennsylvania

K. E. Rouch
University of Kentucky
Lexington, Kentucky

Peter Sandborn
University of Maryland
College Park, Maryland

Linda Schmidt
University of Maryland
College Park, Maryland

Timothy W. Simpson
Penn State University
University Park, Pennsylvania

Robert B. Stone
University of Missouri–Rolla
Rolla, Missouri

A. C. Ungureanu
University of Kentucky

A. Venkatachalam
Lexington, Kentucky

William H. Wood
United States Naval Academy
Annapolis, Maryland

Preface

Many readers will approach this series of books in environmentally conscious engineering with some degree of familiarity with, knowledge about, or even expertise in one or more of a range of environmental issues, such as climate change, pollution, and waste. Such capabilities may be useful for readers of this series, but they aren't strictly necessary. The purpose of this series is not to help engineering practitioners and managers deal with the *effects* of man-induced environmental change. Nor is it to argue about whether such effects degrade the environment only marginally or to such an extent that civilization as we know it is in peril, or that any effects are nothing more than a scientific-establishment-and-media-driven hoax and can be safely ignored. (Other authors, fiction and nonfiction, have already weighed in on these matters.) By contrast, this series of engineering books takes as a given that the overwhelming majority in the scientific community is correct, and that the future of civilization depends on minimizing environmental damage from industrial, as well as personal, activities. However, the series goes beyond advocating solutions that emphasize only curtailing or cutting back on these activities. Instead, its purpose is to exhort and enable engineering practitioners and managers to reduce environmental impacts—to engage, in other words, in *environmentally conscious engineering*, a catalog of practical technologies and techniques that can improve or modify just about anything engineers do, whether they are involved in designing something, making something, obtaining or manufacturing materials and chemicals with which to make something, generating power, or transporting people and freight.

Increasingly, engineering practitioners and managers need to know how to respond to challenges of integrating environmentally conscious technologies, techniques, strategies, and objectives into their daily work, and, thereby, find opportunities to lower costs and increase profits while managing to limit environmental impacts. Engineering practitioners and managers also increasingly face challenges in complying with changing environmental laws. So companies seeking a competitive advantage and better bottom lines are employing environmentally responsible design and production methods to meet the demands of their stakeholders, who now include not only owners and stockholders, but also customers, regulators, employees, and the larger, even worldwide community.

Engineering professionals need references that go far beyond traditional primers that cover only regulatory compliance. They need integrated approaches centered on innovative methods and trends in design and manufacturing that help them focus on using environmentally friendly processes and creating green products. They need resources that help them participate in strategies for designing environmentally responsible products and methods, resources that provide a foundation for understanding and implementing principles of environmentally conscious engineering.

To help engineering practitioners and managers meet these needs, I envisioned a flexibly connected series of edited handbooks, each devoted to a broad topic under the umbrella of environmentally conscious engineering, starting with three volumes that are closely linked—environmentally conscious mechanical design, environmentally conscious manufacturing, and environmentally conscious materials and chemicals processing.

The intended audience for the series is practicing engineers and upper-level students in a number of areas—mechanical, chemical, industrial, manufacturing, plant, and environmental—as well as engineering managers. This audience is broad and multidisciplinary. Some of the practitioners who make up this audience are concerned with design, some with manufacturing, and others with materials and chemicals processing, and these practitioners work in a wide variety of organizations, including institutions of higher learning, design, manufacturing, and consulting firms, as well as federal, state, and local government agencies. So what made sense in my mind was a series of relatively short handbooks, rather than a single, enormous handbook, even though the topics in each of the smaller volumes have linkages and some of the topics (*design for environment, DfE,* comes to mind) might be suitably contained in more than one freestanding volume. In this way, each volume is targeted at a particular segment of the broader audience. At the same time, a linked series is appropriate because every practitioner, researcher, and bureaucrat can't be an expert on every topic, especially in so broad and multidisciplinary a field, and may need to read an authoritative summary on a professional level of a subject that he or she is not intimately familiar with but may need to know about for a number of different reasons.

The Environmentally Conscious Engineering series is composed of practical references for engineers who are seeking to answer a question, solve a problem, reduce a cost, or improve a system or facility. These handbooks are not research monographs. The purpose is to show readers what options are available in a particular situation and which option they might choose to solve problems at hand. I want these handbooks to serve as a source of practical advice to readers. I would like them to be the first information resource a practicing engineer reaches for when faced with a new problem or opportunity—a place to turn to even before turning to other print sources, even any officially sanctioned ones, or to sites on the Internet. So the handbooks have to be more than references or collections of background readings. In each chapter, readers should feel that they are in the hands of an experienced consultant who is providing sensible advice that can lead to beneficial action and results.

The first volume in the series, the **Handbook of Environmentally Conscious Mechanical Design**, seeks to minimize the overall environmental impact throughout the entire life cycle of products, by addressing environment impacts in the design phase of products, because the benefits of environmentally conscious design increase as one moves upstream in a life cycle. The handbook provides a foundation for understanding and implementing environmentally conscious design principles. Contributors discuss the important practical and analytical techniques—Design for Environment (DFE), life cycle design, reverse engineering, design for reliability, design for maintainability, reuse and recycling technologies, design for remanufacturing, materials selection for green design, and Total Quality Management (TQM) - that constitute the practices for designing industrial, business, and consumer products that are environmentally friendly and meet environmental regulations.

I asked the contributors, located not only in North America, but also across the Pacific Ocean, to provide short statements about the contents of their chapters and why the chapters are important. Here are their responses:

Daniel Fitzgerald (University of Maryland in College Park, Maryland), who along with Jeffrey W. Herrmann, Peter Sandborn, Linda Schmidt, and Thornton H. Gogoll, contributed the chapter on **Design for Environment (DFE) Strategies, Practices, Guidelines, Methods, and Tools**, writes, "This chapter provides an overview of the design for environment (DfE) concept and examples of DfE strategies, practices, guidelines, methods, and tools in product development. Given the variety of techniques available, it is important to have a systematic approach for selecting the most appropriate ones. Therefore, the chapter also describes how a product development organization can develop an effective DfE process."

I.S. Jawahir (University of Kentucky in Lexington), who together with I. H. Jaafar, A. Venkatachalam, K. Joshi, A. C. Ungureanu, N. De Silva, K. E. Rouch, and O. W. Dillon Jr. contributed a chapter on **Total Life-cycle Considerations in Production Design for Sustainability: A New Assessment Methodology and Case Studies**, writes, "This chapter deals with principles of product design for sustainability by considering the four life-cycle stages (premanufacturing, manufacturing, use and postuse) of products and three components of sustainability (environment, economy, and society). A framework for comprehensive evaluation of product sustainability has been developed and presented, along with case studies from the automotive and consumer electronic products industries. The significance of this chapter is in its consideration of a postuse life-cycle stage, which lays the foundation for multi life-cycle evaluation of products by considering the 6Rs: recover, recycle, reduce, redesign, remanufacture and reuse."

John K. Gershenson (Michigan Technological University in Houghton), who contributed a chapter on **Life-cycle Design** with Abigail Clarke, writes that "Life-cycle design is inspired by history and influenced by markets and politics. It catalyzes essential product improvement. This chapter on life-cycle design details the principles, methods, and tools that characterize the goals of life-cycle design, as well as the means to achieve these goals and make design decisions impacting the environment. A few illustrative examples show the possibilities of successful life-cycle design and its impact on a product and its environment."

Kemper Lewis (SUNY Buffalo in Buffalo, New York), who wrote the chapter on **Reverse Engineering** with Tim Simpson, Bill Regli, Rob Stone, and Bill Wood, writes that "Reverse engineering has moved well beyond its origin as a benchmarking exercise and is now a field marked by challenging research areas, strategic product development life-cycle issues, innovative university-level pedagogical initiatives, and integration with state-of-the-art digital technologies and tools. Reverse engineering at its core examines the existing form of an object to infer its function. In this chapter, we describe the basic reverse engineering process and discuss the motivations behind implementing reverse engineering techniques and principles, including product development strategies, environmental impacts, and digital knowledge capture.

B.S. Dhillon (University of Ottawa in Ottawa, Ontario, Canada), who contributed the chapter on **Design for Reliability**, writes, "Reliability is increasingly becoming important during the design of engineering systems, as our daily lives and schedules are more dependent than ever before on the satisfactory functioning of these systems. Some examples of these systems are computers, trains, automobiles, aircraft, and space satellites. Some of the specific factors that are playing a key role in increasing the importance of reliability in designed systems include system complexity and sophistication, competition, increasing number of reliability/safety/quality-related lawsuits, public pressures, high acquisition cost, the past well-publicized system failures, and loss of prestige."

O. Geoffrey Okogbaa (University of South Florida in Tampa), who contributed the chapter on **Design for Maintainability** with Wilkistar Otieno, writes, "Diminishing resources mean that we can longer afford the luxury of the ineffective intervention strategy depicted by the fail-fix-fail-fix cycle. Thus, the motivation for design for maintainability is to provide a 'predict and prevent' centric framework that eventually leads to autonomous and self-correcting systems. Such systems need minimal intervention and hence lead to reduced resources."

Hartmut Kaebernick (The University of New South Wales in Sydney, Australia), who contributed the chapter on **Reuse and Recycling Technologies** with Sami Kara, writes, "Reuse and recycling techniques have been developed over the last decade as part of a major strategy to save energy and material resources. Some of the techniques have reached a high technology standard whereas others, especially in the reuse area, are still at the beginning of their development. Both reuse and recycling form a central part of an overall approach towards sustainability, which could hardly be achieved without these techniques."

Berdinus A Bras (Georgia Institute of Technology in Atlanta, Georgia), who contributed the chapter on **Design for Remanufacturing Processes**, writes, "Remanufacturing is often considered a superior strategy over recycling because it maintains the geometric shape of the product and thus preserves energy and money spent in the manufacturing phase of the life cycle. Increasingly, businesses are starting to look at remanufacturing as a way of retaining product ownership and a venue into new (and more profitable) business strategies focused on selling function rather than form per se. In this chapter, product design guidelines will be discussed that foster remanufacturing. Furthermore, an overview of the industry, including typical facility-level processes, will be given in order to understand the rationale for specific design guidelines."

Sridhar Idapalapati (Nanyang Technological University in Singapore), who contributed the chapter on **Materials Selection for Green Design**, writes, "Currently, the number of materials available to a design engineer is 100,000 or so, and this number is growing day by day. The problem of screening and selecting materials to satisfy competing design requirements requires understanding of their intrinsic and extrinsic properties and development of systematic approaches. In the detailed design stage of a product it is essential to have accurate material information for proper estimation of its performance and optimization."

Finally, R. Alan Kemerling (Ethicon Endo-Surgery, Inc. in Cincinnati, Ohio), who contributed the chapter on **Employing TQM in Environmentally Conscious Design**, writes, "Total Quality Management contains proven tools and processes to enable design for the environment. When the design team employs the tools and processes of TQM, the requirements of customers will be met or exceeded and the environment becomes a customer of the process. This is accomplished because TQM allows the design team to optimize the design and supporting processes, achieving higher utilization and less waste of resources."

That ends the contributors' comments. I would like to express my heartfelt thanks to all of them for having taken the opportunity to work on this book. Their lives are terribly busy, and it is wonderful that they found the time to write thoughtful and complex chapters. I developed the handbook because I believed it could have a meaningful impact on the way many engineers approach their daily work, and I am gratified that the contributors thought enough of the idea that they were willing to participate in the project. Thanks also to my editor, Bob Argentieri, for his faith in the project from the outset. And a special note of thanks to my wife Arlene, whose constant support keeps me going.

Myer Kutz
Delmar, New York

CHAPTER 1

DESIGN FOR ENVIRONMENT (DfE): STRATEGIES, PRACTICES, GUIDELINES, METHODS, AND TOOLS

Daniel P. Fitzgerald
Department of Mechanical Engineering and Institute for Systems Research University of Maryland, College Park

Jeffrey W. Herrmann
Department of Mechanical Engineering and Institute for Systems Research University of Maryland, College Park

Peter A. Sandborn
Department of Mechanical Engineering and Institute for Systems Research University of Maryland, College Park

Linda C. Schmidt
Department of Mechanical Engineering and Institute for Systems Research University of Maryland, College Park

Thornton H. Gogoll
Director, Engineering Standards Black & Decker Inc. Towson, Maryland

1 INTRODUCTION

Throughout their life cycle, products generate environmental impacts (1) from extracting and processing raw materials; (2) during manufacturing, assembly, and distribution; (3) due to their packaging, use, and maintenance; and (4) at their end of their life. Environmentally benign products are products that comply with environmental regulations and may have significant features that reduce environmental impact. The ideal environmentally benign product is one that not only would be environmentally neutral to make and use but also would actually reverse whatever substandard conditions exist in its use environment. The ideal environmentally benign product would also end its life cycle by becoming a useful input for another product instead of creating waste.

In the past, manufacturing firms were concerned with meeting regulations that limited or prohibited the pollution and waste that are generated by manufacturing processes. However, regulations are now focusing on the material content of the products that are sold in an effort to control the substances that enter the waste stream.

There are many ways to minimize a product's environmental impacts. Clearly, however, the greatest opportunity occurs during the product design phases, as discussed by many authors, including Handfield et al., Fiksel, Bras, and Ashley.[1–4] Therefore, organizations that develop new products need to consider many factors related to the environmental impact of their products, including government regulations, consumer preferences, and corporate environmental objectives. Although this requires more effort than treating emissions and hazardous waste, it not only protects the environment but also reduces life-cycle costs by decreasing energy use, reducing raw material requirements, and avoiding pollution control.[5]

Design for Environment (DfE) tools, methods, and strategies have therefore become an important set of activities for product development organizations.

1.1 Design for Environment

Design for Environment (DfE) is "the systematic consideration of design performance with respect to environmental, health, and safety objectives over the full product and process life cycle."[2] DfE, like other concurrent engineering techniques, seeks to address product life-cycle concerns early in the design phase. Thus, it is similar to design for manufacturing (DFM), design for assembly (DFA), and design for production (DFP).[6] DfE combines several design-related topics: disassembly, recovery, recycling, disposal, regulatory compliance, human health and safety impact, and hazardous material minimization.

Some designers view DfE as simply calculating an environmental measurement, similar to estimating cost. This perception is due to the trend of companies implementing standalone DfE tools without explanation. As Lindahl states, "When the designers and actual users of the methods do not understand the reason why, or experience any benefits from using the DFE methods, there is a risk that they utilize the methods but do not use the results, or that they run through the method as quickly as possible."[7] Therefore, designers produce an environmental output and consider the environmental work finished. A more effective approach is to design and implement a sound DfE process, as this chapter explains.

1.2 Decision Making in New Product Development

Decision making is an important activity in new product development, and a great variety of decisions need to be made. Generally speaking, these fall into two types: design decisions and management decisions.

Design Decisions

Design decisions address the question, "What should the design be?" They determine shape, size, material, process, and components. These generate information about the product design itself and the requirements that it must satisfy.

Management Decisions

Management decisions address the issues of what should be done to make the design into a successful product. Management decisions control the progress of the design process. They affect the resources, time, and technologies available to perform development activities. They define which activities should happen, their sequence, and who should perform them. That is, what will be done, when will it be done, and who will do it. The clearest example is project management: planning, scheduling, task assignment, and purchasing.

In studying design projects, Krishnan and Ulrich provide an excellent review of the decision making in new product development, organized around topics that follow the typical decomposition of product development.[8] Herrmann and Schmidt describe the decision-making view of new product development in more detail.[9,10] Traditionally, factors such as product performance and product cost have dominated design decisions, while time to market and development cost have influenced management decisions. Of course, many decisions involve combinations of these objectives.

Considering environmental issues during decision making in new product development, while certainly more important than ever before, has been less successful for manufacturers than considering other objectives. Environmental objectives are not similar to the traditional objectives of product performance, unit cost, time to market, and development cost. All four objectives directly affect profitability and are closely monitored. Unit cost, time to market, and development cost each use a single metric that is well understood and uncomplicated. Although product performance may have multiple dimensions, these characteristics are quantifiable and clearly linked to the product design. Designers understand how changing the product design affects the product performance. Environmental objectives do not have these qualities.

1.3 Environmental Objectives

Under pressure from various stakeholders to consider environmental issues when developing new products, manufacturing firms have declared their commitment to environmentally responsible product development and have identified six relevant goals:

1. *Comply with legislation. Products* that do not comply with a nation's environmental regulations cannot be sold in that nation.
2. *Avoid liability.* Environmental damage caused by a product represents a financial liability.
3. *Satisfy customer demand.* Some consumers demand environmentally responsible products. Retailers, in turn, pass along these requirements to manufacturers.
4. *Participate in eco-labeling programs.* Products that meet requirements for eco-labeling are more marketable.
5. *Enhance profitability.* Certain environmentally friendly choices such as remanufacturing, recycling, and reducing material use make good business sense and have financial benefits.
6. *Behave ethically.* Being a good steward of the planet's resources by considering the environment during the product development process is the right thing to do.

Despite the high profile given to these objectives at the corporate level, product development teams assign a *back-burner* status to environmental issues. Environmental *objectives,* for the most part, are driven by regulations and social responsibility, and reducing environmental impact doesn't clearly increase profit. Product managers are not often willing to compromise profit, product quality, or time to market in order to create products that are more environmentally benign than required by regulations. (The exceptions are those organizations that court environmentally conscious consumers.)

Environmental *performance,* however, is measured using multiple metrics, some of which are qualitative. Moreover, these metrics may seem irrelevant to the firm's financial objectives. Measuring environmental performance, especially life-cycle analysis (LCA), can require a great deal of effort.

With environmental performance it is harder to make trade-offs. It is not clear how to select between design alternatives because there is no aggregate measure to calculate. One designer presents an excellent example:

> You have two ways of building a part. One option is based on metal. Metal is heavy (thus, it consumes more resources). It also creates waste during the actual manufacturing process (in form of sludge). However, it can be recycled when it reaches the end of its product life. In contrast, we make the product out of graphite. This part is lighter (which means it consumes less energy in use). In addition, it can be molded rather than machined (again resulting in less waste). However, when it

reaches the end of its life, it must be disposed of in a landfill since it cannot be recycled. Which of these two options results in a greener product?[1]

2 CREATING A DESIGN FOR ENVIRONMENT PROGRAM

This section describes the step necessary for creating a DfE program within a product development organization.

2.1 Identifying and Understanding the Stakeholders

The first step in creating a DfE program is to identify and understand the environmental stakeholders. A *stakeholder* is defined in the *American Heritage Dictionary* as one who has a share or an interest, as in an enterprise.[11] Stakeholders ultimately define the objectives and resulting environmental metrics of the DfE program. The following are examples of typical stakeholders for product development organizations:

- *Board members*. Internal stakeholders on the board of directors directly define corporate policies and culture.
- *Socially responsible investors*. These stakeholders invest in companies that demonstrate socially responsible values such as environmental protection and safe working conditions.
- *Non-government organizations*. Organizations such as the Global Reporting Initiative Work to advance specific environmental agendas.
- *Government organizations*. Organizations such as the EPA require meeting certain environmental regulations and provide incentives such as the Energy Star for achieving an exceptional level of environmental compliance.
- *Customers*. A customer is anyone who purchases the firm's product down the line. This could be a retailer, another product development organization, or an end user.
- *Competitors*. Competitors are other product development organizations that are in the same market. It is important to benchmark competitors to understand the environmental issues and strategies within the firm's market and to effectively position the organization in the market.
- *Community*. The community consists of people affected by the organization's products throughout their life cycle. Depending on the scope of the assessment, this can technically be everyone in the world. More realistically, it is the community that surrounds the organization's facilities and directly interacts with the products.

Each stakeholder has different environmental interests, which leaves the organization with a considerable amount of environmental demands to meet. Since product development organizations operate with limited resources, the stakeholders will need to be prioritized based on their influence on the organization. Influence generally correlates to the extent that profits will be affected if a stakeholder's demand is not met. Once the stakeholders are prioritized, the product development organization will have a good idea of which environmental demands need to be met. It is now possible to construct a DfE program with objectives and metrics that support these demands.

2.2 Creating Environmental Objectives

After a thorough analysis of the stakeholders, it is possible to create environmental objectives for the DfE program. The environmental objectives will need to align with as many of the environmental demands of the stakeholders as possible. The objectives will also need to align with the values and culture of the corporation. Klein and Sorra argue that successfully implementing an innovation (in this case, a DfE program) depends on "the extent to which targeted users perceive that use of the innovation will foster the fulfillment of their values."[12] Since it is necessary for an employee to adapt to the values of the corporation to be successful, a DfE program that aligns to corporate values will align with employee values and should be successfully implemented. When creating environmental objectives, it is important to use the correct level of specificity. The objectives should be broad enough that they do not have to be frequently updated but specific enough that they provide consistent direction for the DfE program. An environmental objective of "protect the Earth" would be too broad, while "eliminate the use of lead" would be too specific. Environmental objectives should have lower-level targets associated with them so the company can assess its progress toward objectives. For example "eliminate the use of lead" could be a lower-level target for the environmental objective "reduce the use of hazardous materials."

For example, Black & Decker created environmental objectives for its DfE process.[13] Its DfE process contains values that coincide with the organization's values. Within the Corporation's Code of Ethics and Standards of Conduct, there is a section titled Environmental Matters. It "places responsibility on every business unit for compliance with applicable laws of the country in which it is located, and ... expects all of its employees to abide by established environmental policies and procedures."[14] The objectives also meet the environmental demands of stakeholders as described next.

Practicing Environmental Stewardship

Black & Decker seeks to demonstrate environmental awareness through creating an environmental policy and publishing it on its Web site, including information about recycled content on packaging and its design for environment program. In addition, Black & Decker belongs to environmental organizations such as the World Environmental Center, which contributes to sustainable development worldwide by strengthening industrial and urban environment, health, and safety policies and practices. It is also member of the Rechargeable Battery Recycling Corporation (RBRC) and RECHARGE, which both promote the recycling of rechargeable batteries.

Complying with Environmental Regulations

As a global corporation that manufactures, purchases, and sells goods, Black & Decker must comply with all applicable regulations of countries where its products are manufactured or sold. Currently, the European Union exerts significant influence on addressing environmental issues through regulations and directives. This section lists some important U.S. and European environmental regulations.

Many U.S. regulations apply to U.S. and European workers, and these are set by both federal and state agencies. The Occupational Safety & Health Administration (OSHA) limits the concentration of certain chemicals to which workers may be exposed. The Environmental Protection Agency (EPA) regulates management of waste and emissions to the environment. Black & Decker provides employees with training on handling hazardous wastes, which is required by the Resource Conservation and Recovery Act and the Hazardous Materials Transportation Act. California's Proposition 65 requires a warning before potentially exposing a consumer to chemicals known to the State of California to cause cancer or reproductive toxicity. The legislation explicitly lists chemicals know to cause cancer and reproductive toxicity.

The European Union also regulates corporations with respect to environmental issues. The EU Battery Directive (91/157/EEC) places restrictions on the use of certain batteries. The EU Packaging Directive (Directive 2004/12/EC)[15] seeks to prevent packaging waste by requiring packaging reuse and recycling. In the future, countries in the European Union will require Black & Decker to adhere to certain laws so that the state achieves the goals of the EU Packaging Directive. Thus, Black & Decker will be interested in increasing the recyclability of its packaging. The new EU directives on waste electrical and electronic equipment (WEEE) and on the restriction of the use of certain hazardous substances in electrical and electronic components (RoHS) address issues of product take-back and bans on hazardous materials, respectively. Thus, Black & Decker must provide information about the material content of its products.

Addressing Customer Concerns

Black & Decker's retail customers are concerned about the environmental impacts of the products they sell. Examples of customer concerns are ensuring timber comes from appropriate forests; increasing the recyclability and recycled content in packaging; using cadmium in batteries; and using lead in printed wiring boards and electrical cords. More specifically, some retailers require that Black & Decker's products be free of lead-based surface coatings.

Mitigating Environmental Risks

An activity's environmental risk is the potential that the activity will adversely affect living organisms through its effluents, emissions, wastes, accidental chemical releases, energy use, and resource consumption.[16] Black & Decker seeks to mitigate environmental risks through monitoring chemical emissions from manufacturing plants; reducing waste produced by its operations; ensuring safe use of chemicals in the workplace; and ensuring proper off-site waste management.

Reducing Financial Liability

There are different types of environmental liabilities:[17]

- Compliance obligations are the costs of coming into compliance with laws and regulations.
- Remediation obligations are the costs of cleaning up pollution posing a risk to human health and the environment.
- Fines and penalties are the costs of being noncompliant.
- Compensation obligations are the costs of compensating damages suffered by individuals, their property, and businesses due to use or release of toxic substances or other pollutants.
- Punitive damages are the costs of environmental negligence.
- Natural resource damages are the costs of compensating damages to federal, state, local, foreign, or tribal land.

Not all of these environmental liabilities apply to all firms.

Reporting Environmental Performance

Black & Decker reports environmental performance to many different organizations with local, national, or global influence and authority. An example of such an organization is the Investor Responsibility Research Center (IRRC).

2.3 Metric Selection

After the environmental objectives are set, an organization needs environmental metrics to measure its progress. Many aspects need to be considered when selecting metrics for a DfE program. First, each metric should directly relate to at least one of the environmental objectives. Metrics that relate to many objectives tend to be more desirable. Second, the organization has to have the capability of measuring the metric. A metric that can be easily measured within an organization's systems ranks higher than a metric that requires costly changes and upgrades. Finally, the metrics should tailor to specific stakeholder reporting requests. An analysis of the most asked for metrics can help prioritize the metrics. Since organizations operate with limited resources, the metrics will need to be prioritized based on these aspects. It should be noted that while most metrics are quantitative, qualitative metrics such as an innovation statement do exist.

The following section briefly describes eight product-level environmental metrics developed by the authors and Black & Decker staff that product development teams can evaluate during the product development process.

Flagged Material Use in Product

This metric measures the mass of each flagged material contained in the product. A material is considered *flagged* if it is banned, restricted, or being watched with respect to regulations or customers. A consulting firm has provided Black & Decker with a list of materials that are banned, restricted, and being watched.

Total Product/Packaging Mass

This metric measures the mass of the product and packaging separately.

Flagged Material Generated in Manufacturing Process

This is a list of each flagged material generated during the manufacturing process. A material is considered flagged if it is banned, restricted, or being watched with respect to regulations or customers.

Recyclability/Disassembly Rating

This metric is the degree to which each component and subassembly in the product is recyclable. Recyclability and separability ratings can be calculated for each component based on qualitative rankings. Design engineers are provided with a list of statements that describe the degree to which a component is recyclable or separable, and a value from 1 to 6 is associated with each statement. Low ratings for both recyclability and separability facilitate disassembly and recycling. The design engineer rates the recyclability and separability of each component, subassembly, and final assembly. If both ratings for an item are less than 3, then the item is recyclable.[18]

Disassembly Time

Disassembly time is a measure of the time it will take to disassemble the product. Research has been conducted on how long it typically takes to perform certain actions. Charts with estimates for typical disassembly actions are provided to the design engineers, who can then estimate how long it would take to disassemble a product.[18]

Energy Consumption

The total expected energy usage of a product during its lifetime. This metric can be calculated by multiplying the total expected lifetime hours by the energy use per hour the product consumes. This metric needs to be calculated only for large energy consumers such as compressors, generators, and battery chargers.

Innovation Statement

A brief paragraph describes the ways a product development team reduced the negative environmental impact of their product. The product development team should write this after the product is launched. All environmental aspects considered should be included as well.

Application of DfE Approach

This binary measure (yes or no) is the answer to the following question: Did the product development team follow the DfE approach during the product development process? Following the DfE approach requires the team to review the DfE guidelines and evaluate the product-level environmental metrics. Although this list of metrics cannot completely measure every environmental impact, the metrics provide designers with a simple way to compare different designs on an environmental level. Black & Decker plans to track the trends of these metrics as the products advance through future redesigns. Furthermore, each product will have environmental targets set at the beginning of the project, and the metrics provide a way to track how well the product development team performed with respect to attaining the targets. The Corporate Environmental Affairs group will also use the metrics to respond to retailers' requests for environmental information.

2.4 Incorporating DfE into the Design Process

Incorporating a DfE process that fits into the existing product development process has significant potential to help manufacturing firms achieve their environmental objectives. By researching the organization's product development process and understanding the decision-making processes, information flow, and organizational and group values, it is possible to construct a DfE process that is customized and easy to implement. The product development process needs to be studied to ensure information availability for the desired metrics. Ideally, the DfE process should leverage existing processes in order to minimize time to market and require little extra effort from the designers.

The safety review process is an example of an existing process that most product development organizations have that can be combined with the DfE process. Most product development organizations implement a formal safety review process to ensure that the final product is safe for consumer use.[19,20] Typically, safety reviews are held at predetermined key points in the product development process. During these reviews, members from the design team and other safety specialists, such as liability and compliance representatives, meet to discuss the current product design. The meetings are run in a brainstorming format and can be guided by a checklist or company-specific agenda. One safety specialist is in charge of final decisions concerning safety. Since product safety includes qualitative measures, it is necessary to assess the issue in a meeting format where ideas and issues can be discussed with all interested parties.

There are major similarities between safety and environmental concerns. Both areas are important but are not closely linked to profitability (as are quality, cost, and time to market). Both areas involve subjective assessments on a variety of factors, many of which are qualitative. This suggests that most product-development organizations should treat environmental objectives in the same way that they treat safety concerns. It is necessary to assess a product's environmental performance at key stages in the product-development process. Furthermore, since assessing environmental performance requires information from multiple business units within the product-development organization, the organization will need to hold a meeting to discuss the issues with all interested parties. By expanding the safety review process, organizations that have similar corporate objectives for safety and for environmental issues can create a practical DfE process that should be simple to implement.

Manufacturing firms with elaborate safety evaluation and verification procedures (used in areas such as aircraft manufacturing) may not require a similarly sophisticated DfE process (unless the product has many environmental concerns, as in automobile manufacturing). However, in firms that don't explicitly consider safety during new product development, establishing a DfE process will be more work, but the need for a DfE process remains.

The safety and environmental objectives of product-development organizations vary considerably from firm to firm, and each firm uses different mechanisms for addressing these concerns. Certainly, practices that make sense in one domain may be impractical in another. This approach is based on the similarity of the safety objectives and environmental objectives. In firms where the safety objectives and environmental objectives are quite different in scope, other types of DfE processes will be more effective.

2.5 Fitting All the Pieces Together

A DfE program cannot be implemented in isolation from other programs within a product-development organization. The program needs to be integrated with other programs that fall under the corporate responsibility umbrella and carry the same weight. Typical corporate responsibility programs include giving back to the community, promoting diversity awareness, ensuring proper working conditions and benefits for employees, and environmental awareness. These programs have detailed plans and goals that are disseminated to all employees through a substantial medium such as a communications meeting. The employees then begin "living" these programs, which results in a corporate culture.

Most product-development organizations' environmental awareness initiatives are based at the manufacturing level rather than the product level. A new DfE program will most likely be integrated with this preexisting portion of environmental awareness. Upon implementation, the program objectives and specific process need to be clearly presented to employees. The commitment from the upper management within the organization should be enough to get the program rolling. If there is resistance, the organization may need to implement a system that rewards those who participate (and has consequences for those who don't). Only after seeing the organization's commitment and receiving direction can the engineers do their jobs and determine how to meet the goals.

3 IMPLEMENTING A DfE PROCESS

This section describes a new design for environment process that will be implemented at Black & Decker.[13] The company defined a DfE process that naturally integrates environmental issues into the existing product-development process with little extra effort or time. Black & Decker uses a *stage-gate* product-development process that has eight *stages*. Every stage requires certain tasks to be completed before management signs off, giving permission to proceed to the next stage. This sign-off procedure is known as the *gate*.

Currently, Black & Decker has safety reviews during stages 2, 3, 4, and 6. Safety reviews are meetings intended for reviewers to evaluate the assessment, actions, and process of the design team in addressing product safety. The DfE process adds an environmental review to the agenda of the safety reviews held during stages 2, 4, and 6. A separate environmental review will be held during stage 3, an important design stage, in order to focus specifically on the environmental issues for the particular product. The environmental reviews will require design teams to review the checklist of key requirements and to consider guidelines for reducing environmental impact. When the DfE process is first implemented, design teams will have to fill out the environmental scorecard only during stage 6 after the product design is complete. Doing this begins the process of recording environmental data and allows design teams to adapt gradually to the new process. When design teams become more familiar with the process, the scorecard will be completed two or more times during the stage-gate process in order to track design changes that affect environmental metrics during the development process.

Environmental targets will be set during stage 1 as goals for the new product. The design team will write a lessons-learned summary during stage 8 to highlight innovative environmental design changes. The lessons-learned summary will provide the innovation statement metric. Figure 1 shows the Safety Review Process and Environmental Review Process running in parallel. The following sections discuss the aforementioned environmental activities in more detail. Note that, throughout this process, many other product-development activities are occurring, causing changes to the product design.

3.1 Product Initiation Document

The product initiation document is a document that Black & Decker uses to benchmark competitors, define performance targets, and predict profitability and

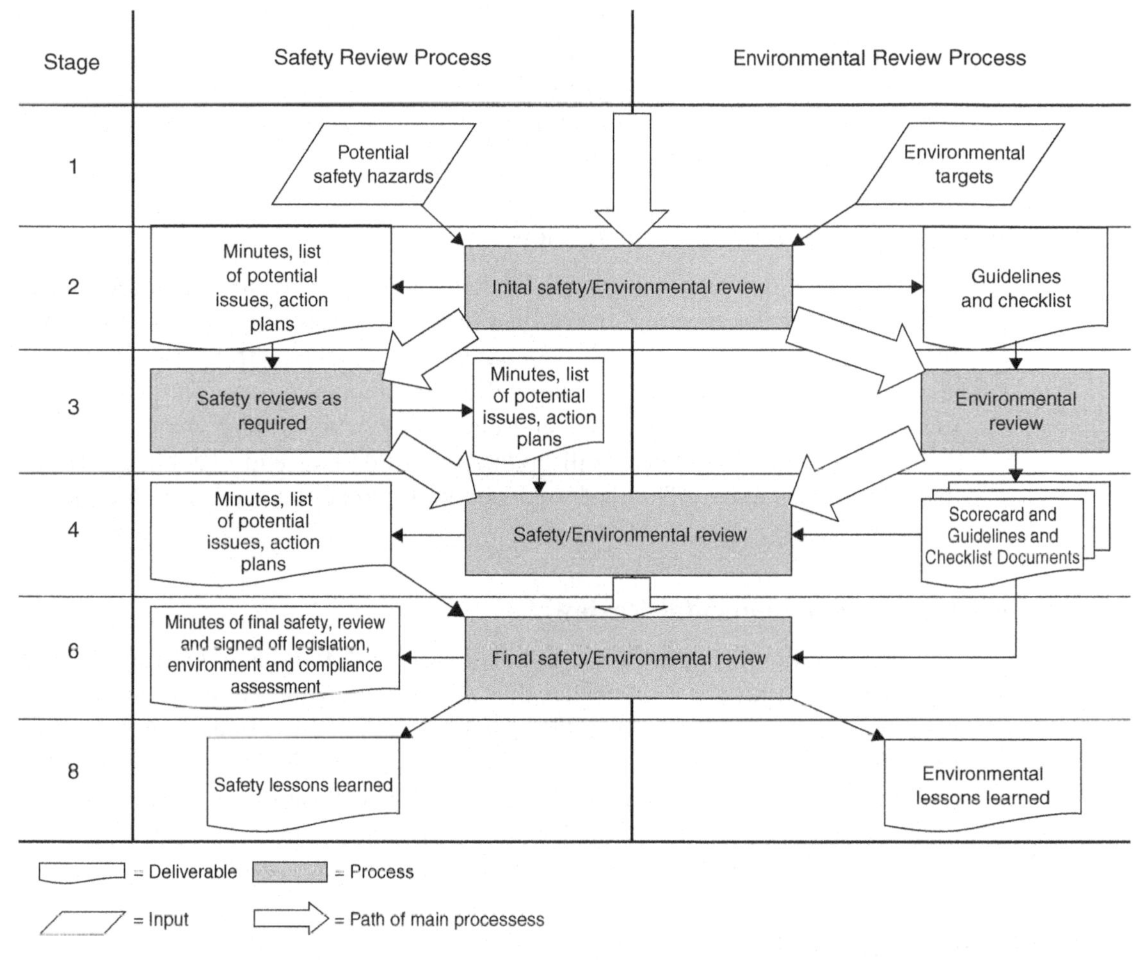

Figure 1 Combined safety and environmental review process.

market share. In addition to these issues, the product initiation document will also address environmental regulations and trends and opportunities to create environmental advantage. Targets for environmental improvement will also be included.

3.2 Initial Environmental Review

The first environmental review is coupled with a safety review. During this meeting, the design team should discuss current environmental regulations, design guidelines, and environmental metrics. A list of regulations and design guidelines can be found in the guidelines and checklist document. The environmental metrics are located in the environmental scorecard. Old lessons learned documents from similar products will be reviewed during this meeting to facilitate environmental design ideas. The result of the meeting is an initial assessment plan that includes the tests to be conducted and the analysis to be performed. The reliability representative will write the assessment plan. Also, a list of brainstormed ideas for environmental improvement and any other minutes will be included in the assessment plan.

3.3 Conceptual Design Environmental Review

The second environmental review is held separately from the safety hazard review. During this meeting, the project team will check compliance regulations, fill in the guidelines and checklist document, discuss the metrics in the scorecard, and review opportunities and additional environmental issues. The result of this meeting is an updated guidelines and checklist document and meeting minutes. The reliability representative will update the guidelines and checklist document and write the minutes. The lead engineer will update the scorecard for the next meeting.

3.4 Detailed Design Environmental Review

The third environmental review is coupled with a safety review. During this meeting, the project team should ensure that all environmental compliance issues are resolved. There should be no further changes to the design due to environmental reasons after this meeting. The result of the meeting is an updated guidelines and checklist document and meeting minutes. The reliability representative will update the guidelines and checklist document and write the minutes. The lead engineer will update the scorecard for the next meeting.

3.5 Final Environmental Review

The fourth and final environmental review is coupled with a safety review. During this meeting, all environmental compliance issues must be resolved. Optimally, no design changes due to environmental reasons would have been made between the last meeting and this meeting. The result of the meeting is a final guidelines and checklist document and meeting minutes. The reliability representative will finalize the guidelines and checklist document and write the minutes. The lead engineer will finalize the scorecard and create a Material Declaration Statement (MDS) packet for the product.

3.6 Postlaunch Review

Black & Decker includes a lessons-learned summary in their product development process. This document discusses what went well with the project, what didn't go well with the project, and reasons why the product didn't meet targets set in the trigger document. The lessons-learned summary will include environmental design innovations realized during the product development process for publicity and customer questionnaires. An example of an item to be included in the lessons learned summary is a materials selection decision. Details should include what materials were considered and the rationale of the decision. The lessons-learned summary is a very important part of the DfE process because it provides future design teams with the environmental knowledge gained by the previous designers.

3.7 Feedback Loop

The completed guidelines and checklist documents and lessons-learned summaries create a feedback loop for the DfE process. Design engineers working on similar products can use this information to make better decisions immediately, and the information is also valuable when the next generation of the product is designed years down the road. Design engineers will record what environmental decisions were made and why they were made. The decision information, scorecards, and comments on the guideline document will be archived permanently. The goal is to save the right things so the information is there in the future when more feedback activities, such as a product tear-down to verify scorecard metrics, can be introduced.

4 USING DfE TOOLS

This section will explore some general DfE tools and how they should be implemented within the product development process.

4.1 Guidelines and Checklist Document

A guidelines and checklist document is a simple DfE tool that forces designers to consider environmental issues when designing products. Integrating a guidelines/checklist document within a new DfE process is a simple and effective way to highlight environmental concerns. However, it should be noted that the guidelines/checklist document needs to be company specific and integrated systematically into the product-development process. Using an existing generic, standalone document will most likely be ineffective. First, the point of a guidelines/checklist document is to ensure that designers are taking the proper steps toward achieving specific environmental objectives. Another organization's guidelines/checklist document was designed to obtain its own objectives, which may not coincide with another company's objectives. Second, obtaining a guidelines/checklist document and simply handing it to designers will lead to confusion as to when and how to use the list. Specific procedures need to be implemented to ensure the designers are exposed to the guidelines/checklist document early in the product-development process to promote environmental design decisions. Black & Decker has systematically developed these DfE guidelines:

- Reduce the amount of flagged materials in the product by using materials not included on Black & Decker's should not use list.
- Reduce raw material used in product by eliminating or reducing components.
- Reduce the amount of flagged material released in manufacturing by choosing materials and processes that are less harmful.
- Increase the recyclability and separability of the product's components.
- Reduce the product's disassembly time.
- Reduce the amount of energy the product uses.

4.2 Product Design Matrix

The Product Design Matrix[21] is a tool that was created with the Minnesota Office of Environmental Assistance and the Minnesota Technical Assistance Program (MnTAP). The matrix helps product designers determine where the most environmental impact of their product design occurs. Two different categories are explored within the matrix, Environmental Concerns and Life Stage. The environmental concerns (Materials, Energy Use, Solid Residue, Liquid Residue, and Gaseous Residue) are listed across the top of the matrix and the Life Stages (Pre-manufacture, Product Manufacture, Distribution & Packaging, Product Use & Maintenance, and End of Life) are listed on the left side of the matrix. The matrix is shown in Figure 2 and was adapted from T. E. Graedel and B. R. Allenby.[22] Included with the matrix is a series of questions for each block. Points are associated with each question and are total for each of the 25 blocks. Then the rows and columns are totaled, providing the designers with information regarding the largest environmental concern and most environmental detrimental stage of the product life cycle. It is possible that the Product Design Matrix and accompanying questions can be varied to suit specific company needs. This tool should be used during the design review stage of the product development process so designers have an opportunity to make changes based on the results of the tool.

4.3 Environmental Effect Analysis

The environmental effect analysis was developed over time by multiple organizations, including the Swedish consulting agency HRM/Ritline, Volvo, and the University of Kalmar, Sweden. It is based on the quality assurance Failure Modes and Effects Analysis (FMEA), and the form looks much like a typical FMEA with environmental headings (Figure 3). The tool is to be used early in the product-development process by the product-development team preferably with the supervision of an environmental specialist to help with questions. First,

LIFE STAGE	Environmental Concern					Total
	1 Materials	2 Energy Use	3 Solid Residue	4 Liquid Residue	5 Gaseous Residue	
A Premanufacture	(A.1)	(A.2)	(A.3)	(A.4)	(A.5)	
B Product manufacture	(B.1)	(B.2)	(B.3)	(B.4)	(B.5)	
C Distribution, packaging	(C.1)	(C.2)	(C.3)	(C.4)	(C.5)	
D Product use, maintenance	(D.1)	(D.2)	(D.3)	(D.4)	(D.5)	
E End of life	(E.1)	(E.2)	(E.3)	(E.4)	(E.5)	
Total						

Figure 2 Product design matrix.

the team needs to identify the key activities associated with each stage of the product's lifecycle. Next, the team needs to identify the environmental aspects of the activities. Then, the team needs to identify the environmental impact associated with the environmental aspect. Some examples of environmental impacts are ozone depletion, resource depletion, and eutrophication. Next, the environmental impacts need to be evaluated to determine their significance. The evaluation technique is similar to that of the FMEA. An environmental priority number (EPN) is calculated using three variables: *S*, for controlling documents; *I*, for public image; and *O*, for environmental consequences. The variables are given a ranking from 1 to 3 based on environmental compliance, where 1 is the best possible score and 3 is the worst possible score. The EPN is calculated by adding the three scores. A fourth variable, *F*, improvement possibly, is focused on the effort in time, cost, and technical possibility of improving the product. It is based on a 1 to 9 scale, with 1 being no possibility for improvement and 9 being very large possibility for improvement. Detailed explanations of what each score means qualitatively for each variable can be found in Lindahl and Tingström.[23] After the evaluation, designers can place the results into an evaluation matrix (Figure 4) to determine what design changes should be made. Recommendations for design changes and actual design change decisions made are filled into the chart and the EPN and F are recalculated to ensure improvement is achieved. This form provides an excellent record of the aspects evaluated and design decisions made within the product development process.[23]

Environmental Effect Analysis - EEA [The SIO-Method]

Part Name	Part Number	Drawing Number	Function	Date	Issue

Project	Supplier	Info	Follow-up Date	Page No.

EEA Leader	EEA Participants

Inventory				Valuation				Actions							
Life-cycle		*Environmental Characteristics*						*Proposals for Action*		*Valuation*				*Realization*	
No.	Life-cycle Phase	Activity	Environmental Effect / Aspect	S	I	O	EPN / F	Recommended Actions	Environmental Effect / Aspect	S	I	O	EPN / F	Remarks	Responsible

Figure 3 EEA form. (From Ref. 23.)

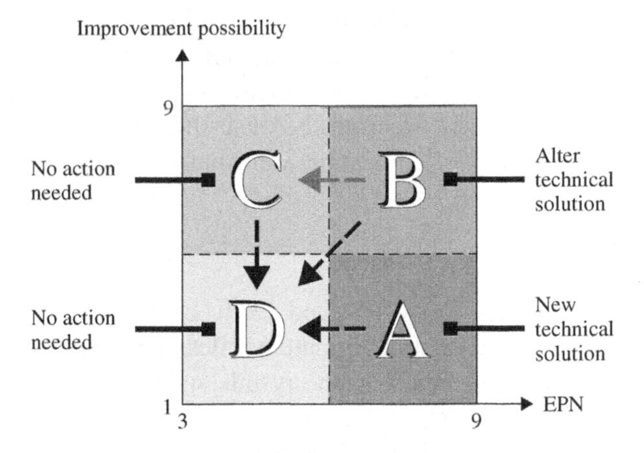

Figure 4 Evaluation matrix (From Ref. 22.)

4.4 Life-cycle Assessment

Life-cycle assessments (LCAs) are time-consuming projects that research a product's environmental impacts and conduct tests to produce environmental-impact quantities. LCAs are excellent for determining how a *current* product can be redesigned to be more environmentally benign. Unfortunately, LCAs take a long time, are very expensive, and provide information only after the design is complete. Moreover, LCAs do not help designers improve a current product's environmental impact. LCAs should not be used during the product-development process.

5 EXAMPLES OF DfE INNOVATIONS

This section provides examples of products that have been designed to reduce adverse environmental impact. Most of these products introduce increased functionality in addition to being more environmentally friendly. It is important to recognize what has been accomplished in the field of environmental design and build on this existing knowledge. By combining ideas that have been implemented in the past with their own ingenuity, designers can create new products that have minimal adverse environmental impacts, as well as adding to the environmental design knowledge base.

5.1 Forever Flashlight

The Forever Flashlight is a flashlight that does not require batteries or bulbs. Its power is generated by the user shaking the flashlight. When the user shakes the flashlight, a piece wound with copper wire moves through a magnetic field and generates power that is stored in the flashlight. Fifteen to 30 seconds of shaking can provide up to five minutes of light. Also, the Forever Flashlight uses a blue LED instead of its bulb due to its longevity. This flashlight prevents environmental harm by reducing battery usage and provides more functionality than a typical flashlight because the user will never be left in the dark due to dead batteries.[24] For more information, go to http://www.foreverflashlights.com.

5.2 Battery-Free Remote Control

The Volvo Car Corp. and Delft University of Technology created a battery-free remote control for automobiles. This was done by utilizing the *piezo effect*, the charge created when crystals such as quartz are compressed. The remote is designed with a button on top and a flexible bottom. When the user pushes the button, the top button and flexible bottom compress the crystal, creating an electrical charge that powers a circuit to unlock the car. This prevents environmental harm by reducing battery usage.[25]

5.3 Toshiba's GR-NF415GX Refrigerator

The Toshiba GR-NF415GX is an excellent example of a more environmentally benign product. It won the 2003 Grand Prize for Energy Conservation. In addition, this product example provides more insight than most because Takehisa Okamoto, an engineer who designed the refrigerator, participated in an interview discussing the design of the product.[26] Takeshisa describes the problem with previous refrigerators in this excerpt:

> To review the mechanics of earlier refrigerators, previously both the refrigerator and freezer sections were cooled by a single cooling unit. Since the refrigerator section didn't require as much cooling as the freezer section, it tended to be over-cooled. To prevent this, a damper was attached to open and close vents automatically. This would close the vents when it got too cold and open them when it got too warm. However, in the area near the vents where the cold air came out, eggs would sometimes get too cold or tofu would sometimes freeze.

Takeshisa then describes the solution to the problem and advantages of the new refrigerator:

> Then the twin cooling unit refrigerator was developed. This involved two cooling units—one in the refrigerator section and one in the freezer section—using a single compressor. This system alternates between cooling the refrigerator and cooling the freezer, which allows each section to be cooled to a more suitable temperature. While the freezer's being cooled, the frost that accumulates on the cooling unit in the refrigerator section, where coolant isn't flowing, is melted once again and returned to the refrigerator section using a fan for humidification. This prevented drying, so that cheese and ham wouldn't lose all their moisture.

> The technology makes it possible to cool the refrigerator and freezer sections simultaneously and to maintain two temperatures, with a major difference in temperatures between the temperatures of the refrigerator cooling unit (−3.5°C) and the freezer cooling unit (−24°C). Now, the refrigerator section is cooled by −2°C air to maintain a temperature of 1°C. This technology uses an ultra-low-energy freeze cycle that makes it possible to cool using cold air at temperatures close to the ideal temperature for the food.
>
> Since this is the first technology of this kind in the industry, some aspects were definitely difficult. At the same time, I think this innovation was really the key point of this development. The two-stage compressor distributes coolant compressed in two stages in two directions: to the refrigerator side and to the freezer side. For this reason, the flows of coolant to each cooling unit must be adjusted to ensure optimal flow. We achieved efficient simultaneous cooling using a pulse motor valve (PMV).

From this dialogue, one can see that there is an improvement in freshness of the food due to the accuracy of the air temperature being output into the refrigeration section. This innovation also conserves electricity because of the ultra-low-energy freeze cycle. In addition, a typical engineering solution to this problem would require two compressors to achieve the final result, while this product only required one.

5.4 Matsushita Alkaline Ion Water Purifier

The Matsushita PJ-A40MRA alkaline ion water purifier[27] has increased functionality and decreased environmental impact compared to the TK7505 alkaline ion water purifier. The new water purifier increases functionality by allowing the user to select seven kinds of water quality (as opposed to five) based on the quality a user needs in a particular situation. The new purifier also decreases environmental impact by reducing standby power from 6 watts to 0.7 watts through division of the integrated power source into two separate power sources for operation and standby.

6 CONCLUSIONS

This chapter has reviewed a large number of DfE tools, methods, and strategies that have been developed and implemented to help manufacturing firms create environmentally benign products. From this review we draw the following conclusions.

DfE tools vary widely with respect to the information they require and the analysis that they perform. Adopting a DfE tool does not automatically lead to environmentally benign products. It is important to have DfE tools that address relevant, important environmental metrics and that provide information useful to product development decision making.

Product-development organizations need DfE processes, not just DfE tools. However, a DfE process that adds a large amount of additional analysis, paperwork, and meetings (all of which add time and cost) is not desirable. Ideally, environmental objectives would be considered in every decision that occurs during new product development, like the objectives of product performance, unit cost, time to market, and development cost. However, environmental objectives are much different than these. Instead they more closely resemble safety objectives.

One possible approach to remedy this problem is for a product-development organization to create a DfE process by expanding a process that the firm may already have in place, the safety review process. In many firms, the safety review process evaluates product safety at various points during the product development process. Therefore, combining the DfE process with the safety review process would require environmental performance to be assessed multiple times during the product-development process.

This method of incorporating the DfE process into the product development process ensures environmental performance will be evaluated at key points in the design process instead of only after the design is complete.

The safety and environmental objectives of product development organizations vary considerably from firm to firm, and each firm uses different mechanisms for addressing these concerns. Certainly, practices that make sense in one domain may be impractical in another. This chapter identifies one way to create a DfE process, something that many firms are now attempting to do, and discusses the use of this approach at a power tools manufacturing firm. The chapter's analysis of this approach is based on the similarity of the safety objectives and environmental objectives. In firms where the safety objectives and environmental objectives are quite different in scope, other types of DfE processes will be more effective. Fundamentally, though, a firm still needs a DfE process, not an isolated environmental assessment tool, to achieve their environmental objectives. More generally, a DfE process must be designed to fit within the existing patterns of information flow and decision making in the product development organization, as discussed by Herrmann and Schmidt.[16]

REFERENCES

1. Robert B. Handfield, Steven A. Melnyk, Roger J. Calantone, and Sime Curkovic, "Integrating Environmental Concerns into the Design Process: The Gap between Theory and Practice," *IEEE Transactions on Engineering Management*, **48**(2), 189–208 (2001).
2. Fiksel, Joseph, *Design for Environment: Creating Eco-efficient Products and Processes*, McGraw-Hill, New York, 1996.
3. Bert Bras, "Incorporating Environmental Issues in Product Design and Realization," *UNEP Industry and Environment*, **20**(1–2), 5–13 (January–June 1997).
4. Steve Ashley, "Designing for the Environment," *Mechanical Engineering*, **115**(3), 53–55 (1993).

5. Dave Allen, Diana Bauer, Bert Bras, Tim Gutowski, Cindy Murphy, Tom Piwonka, Paul Sheng, John Sutherland, Deborah Thurston, and Egon Wolff, "Environmentally Benign Manufacturing: Trends in Europe, Japan, and the USA," DETC2001/DFM-21204, in *Proceedings of DETC'01*, the ASME 2001 Design Engineering Technical Conference and Computers and Information in Engineering Conference, Pittsburgh, Pennsylvania, September 9–12, 2001.
6. Mandar M. Chincholkar, Jeffrey W. Herrmann, and Yu-Feng Wei, "Applying Design for Production Methods for Improved Product Development, DETC2003/DFM-48133. Proceedings of the ASME 2003 International Design Engineering Technical Conferences and Computers and Information in Engineering Conference, Chicago, Illinois, September 2–6, 2003.
7. Mattias Lindahl, "Designer's Utilization of DfE Methods." Proceedings of the 1st International Workshop on "Sustainable Consumption" Tokyo, Japan: The Society of Non-Traditional Technology (SNTT) and Research Center for Life Cycle Assessment (AIST), 2003.
8. V. Krishnan, and K. T. Ulrich, "Product Development Decisions: A Review of the Literature," *Management Science* **47**(1), 1–21 (2001).
9. J. W. Herrmann and L. C. Schmidt, "Viewing Product Development as a Decision Production System, DETC2002/DTM-34030. In ASME 2002 Design Engineering Technical Conferences and Computers and Information in Engineering Conference, Montreal, Canada, September 29–October 2, 2002.
10. J. W. Herrmann and L. C. Schmidt, "Product Development and Decision Production Systems," To appear in W. Chen, K. Lewis, and L. C. Schmidt (eds.) *Decision Making in Engineering Design*, ASME Press, New York, 2006.
11. *The American Heritage® Dictionary of the English Language*, 4th ed. Copyright © 2000 by Houghton Mifflin Company. Available online at www.dictionary.com.
12. Katherine J. Klein, and Joann Speer Sorra, "The Challenge of Innovation Implementation," *Academy of Management Review*, **21**(4), 1055–1080, 1996.
13. Daniel P. Fitzgerald, J. W. Herrmann, Peter A. Sandborn, Linda C. Schmidt, and Ted Gogoll, "Beyond Tools: A Design for Environment Process," *International Journal of Performability Engineering*, **1**(2), 105–120 (2005).
14. Black & Decker Corporation Code of Ethics and Standards of Conduct. Accessed February 13, 2003, at http://www.bdk.com/governance/bdk_governance_appendix_1.pdf.
15. "Directive 2004/12/EC of the European Parliament and of the Council of 11 February 2004 Amending Directive 94/62/EC on Packaging and Packaging Waste," *Official Journal of the European Union*. Accessible online at http://www.europa.eu.int/eur-lex/pri/en/oj/dat/2004/l_047/l_04720040218en00260031.pdf.
16. Environmental Protection Agency, "Terms of Environment." Document number EPA175B97001. Accessed at http://www.epa.gov/OCEPAterms/eterms.html.
17. Environmental Protection Agency, "Valuing Potential Environmental Liabilities for Managerial Decision Making." EPA742-R-96-003 (December 1996). Accessed at http://www.epa.gov/opptintr/acctg/pubs/liabilities.pdf3.
18. Otto, Kevin N., and Kristin L. Wood. *Product Design: Techniques in Reverse Engineering and New Product Development* Prentice Hall, Upper Saddle River, NJ, 2001.
19. R. Goodden, *Product Liability Prevention—A Strategic Guide* ASQ Quality Press, Milwaukee, 2000.

20. W. F. Kitzes, "Safety Management and the Consumer Product Safety Commission," *Professional Safety*, **36**(4), 25–30 (1991).
21. Jeremy Yarwood and Patrick D. Eagan, "Design for the Environment (DfE) Toolkit," Minnesota Office of Environmental Assistance, Minnesota Technical Assistance Program (MnTAP). Accessed 2001 at http://www.moea.state.mn.us/berc/DFEtoolkit.cfm.
22. T. E. Graedel and B. R. Allenby, *Industrial Ecology* Prentice Hall, Englewood Cliffs, NJ, 1995). Copyright AT&T.
23. M. Lindahl and J. Tingström, "A Small Textbook on Environmental Effect Analysis" (shorter version of the Swedish version), Kalmar, Sweden: Dept. of Technology, University of Kalmar, 2001.
24. Hsaing-Tang Chang and Jahau Lewis Chen, "Eco-Innovative Examples for 40 TRIZ Inventive Principles," *The TRIZ Journal* (August 2003).
25. Power for Products, "Minnesota Office of Environmental Assistance. Accessed 2006 at http://www.moea.state.mn.us/p2/dfe/dfeguide/power.pdf.
26. Akane Yoshino, "Interview with Takehisa Okamoto concerning CFC-free refrigerator." Accessed January 26, 2006, at http://kagakukan.toshiba.co.jp/en/08home/newtech131.html.
27. Matsushita Group, "Alkaline Ion Water Purifier [TK7505] Factor X Calculation Data." Accessed January 26, 2006. http://panasonic.co.jp/eco/en/factor_x/m_pdf/fx_p21e.pdf.

CHAPTER 2

PRODUCT DESIGN FOR SUSTAINABILITY: A NEW ASSESSMENT METHODOLOGY AND CASE STUDIES

I. H. Jaafar, A. Venkatachalam, K. Joshi, A. C. Ungureanu, N. De Silva, K. E. Rouch, O. W. Dillon Jr., I. S. Jawahir
University of Kentucky, Lexington, Kentucky

1 INTRODUCTION

1.1 Definition

Sustainability is broadly defined by the *Brundtland Commission* as "development that meets the needs of the present without compromising the ability of future generations to meet their own needs."[1] By extending this definition to products, *sustainable products* can be defined as those products that provide environmental, societal, and economic benefits while protecting social health and welfare and maintaining the environment over their full life cycle from raw materials extraction, and use, to eventual disposal and reuse.[2]

1.2 Background

Nature's eco-balance, involving sustainable material flow, has existed on Earth for more than 3.85 billion years. The goal of sustainable products is to have zero or minimal impact on this symbiotic balance. Ideally, sustainable products would be those that are fully compatible with nature's ecosystem throughout their life cycle. It has been postulated that products that function utilizing nature's simple framework of *cyclic, solar*, and *safe* principles are most efficient in meeting this objective.[3] In this context, *cyclic* refers to the requirement for products to be made of organic materials or minerals that can be continuously recycled; *solar* refers to renewable energy utilized in the manufacture and use of products; and *safe* refers to products that release only nontoxic byproducts into the air, land, or water during their life cycle.[3,4]

Earth's water cycle is a simple sustainable product example in nature wherein these principles are efficiently utilized. Water collects into rivers, lakes, and oceans, evaporates into the air, then condenses to form clouds, and finally precipitates, falling back as rain as a "new" product to be used and reused again. The main source of energy for evaporation comes from the sun and wind—sources that, from our perspective, can be regarded as infinitely renewable and safe. Furthermore, water that collects in rivers and lakes provides a source of hydroelectric energy. Throughout the whole cycle, there is absolutely no emission of harmful byproducts. In fact, Earth's water cycle fully nourishes the environment.

In current practice, new products are developed and manufactured—and are also disposed of—at an alarmingly fast rate. This raises the concerns of overconsumption of natural resources, leading to their depletion, accumulation of waste, growing landfills, and their related adverse environmental impacts. There is also a growing concern regarding the use of hazardous materials in products that adversely affect both humans and the environment at large. Such concerns arise from documented climate change, acid rain, stratospheric ozone-layer depletion, ocean pollution, and acute health hazards such as cancer and genetic defects—all owing to the use of hazardous materials.[2]

In addition to this, common practice of traditional product life-cycle methods further exacerbates the problem. Such an approach depends on evaluating the product life cycle in only five confined stages: (1) resource extraction, (2) manufacturing operation, (3) packaging and shipping, (4) customer use, and (5) obsolescence.[5] Designing products with improvements in each of these stages would indeed enhance its life-cycle value. However, while each product is expected to satisfy its design functionality along with provisions for manufacturability, assembly requirements, reliability, safety, serviceability, and environmental compliance, its end-of-life options are usually not well thought out. Typically, economic and environmental analyses are developed almost entirely for a single life cycle of a product. Aspects such as material recovery, possible multiple reuse opportunities that are themselves associated with economic gains, and societal and environmental benefits are hardly evaluated in current manufacturing practice.

Note also that simply manufacturing and using "green" products does not wholly solve the problem. A clear distinction may have to be made at this point between *green* and *truly sustainable* products. A *green product* may only be environmentally friendly in its "use" stage of the product life cycle, where burdens still exist in the other stages of the life-cycle spectrum. *Sustainability* requires a comprehensive, multi-life-cycle approach. In this regard, industrialized countries certainly have made improvements in terms of being green in their use of materials, but waste generation still continues to increase. It is worrying that as much as 75 percent of material resources used in products and their manufacture are disposed back to the environment as waste within a year. If this trend is not curbed over the next 50 years, where demand for resources and the resulting waste may increase tenfold, the situation may even turn tragic.

In an effort to address these concerns, governments, particularly in Europe and Japan, and in the United States to a lesser extent, have developed (and are still in the process of developing) a comprehensive base of environmental laws and policies that place legal, market, and financial pressures on manufacturers to develop sustainable products. But designing and manufacturing sustainable products is not so easily done. It is considered a major and high-profile challenge involving highly complex, interdisciplinary approaches and solutions. These hurdles, however, are offset by many incentives. The motivational drivers for attaining sustainable products are discussed in the next section.

2 PRODUCT SUSTAINABILITY DRIVERS

In the twenty-first century, sustainable products have proven to increase both tangible and intangible corporate profits for a variety of reasons:

- Faster product time to market
- Fewer regulatory constraints
- High demand due to public awareness of global health and environmental benefits
- Reduced costs for raw materials and manufacturing
- Reduced liability
- Improved employee health and safety
- Delivery of value-added products to consumers.[2]

These are only some of the specific reasons to encourage the development of sustainable products. These specific *drivers of sustainability* can be grouped into three main aspects: (1) legislation, (2) economy, and (3) society.

2.1 Legislative Drivers

Legislation is one of the main motivational drivers for sustainable products. Examples of well-known legislative drivers include the following:

- The Waste Electrical and Electronic 2002/96/EC (WEEE) Directive,[6]
- The Restriction of Hazardous Substances 2002/95/EC (RoHS) Directive,[7]
- The End-of-Life Vehicles (ELVs) Directive,[8]
- The Energy Using Product (EuP) Directive[9]

These legislative drivers place responsibility of the product's conformance to specified sustainability targets throughout its life cycle squarely on producers, manufacturers, and importers. This section discusses these specific major directives.

The WEEE Directive

The WEEE Directive[6] aims both at preventing and minimizing the amount of electrical and electronic waste, at the same time maximizing the activities of reuse, recycle, and recovery of such products. The WEEE Directive requires manufacturers to be registered with the Member State Government by August 13, 2005. It applies to all electrical and electronic equipment powered at up to 1000 V AC or 1500 V DC sold in EU markets that fall within 10 identified product categories.[10] These categories are: (1) large household appliances, (2) small household appliances, (3) IT and telecommunications equipment, (4) consumer equipment, (5) lighting equipment, (6) electrical and electronic tools, (7) toys, leisure, and sports equipment, (8) medical devices, (9) monitoring and control instruments, and (10) automatic dispensers.[11] These products can be either consumer- or industry-based. Producers are responsible for financing environmentally sound methods of collection, recovery, and recycling of their products to treatment facilities that require a permit from state regulatory bodies. A report showing evidence of meeting the directive is also required. A label showing certification according to the directive's requirements and instructions are to be put on WEEE-covered products sold in the EU market. As of 2005, 13 countries had passed laws regulating the product stewardship/take-back of electrical and electronics products.[12] These regulations shift the logistic and economic responsibility of the product's end-of-life issues on the manufacturers. Table 1 gives a summary of these take-back laws for electrical and electronic products.

The RoHS Directive

The RoHS Directive[7] covers manufacturers, sellers, distributors, and recyclers of electrical and electronic equipment containing hazardous substances such as

Table 1 Breakdown of Take-back Laws for Electrical and Electronic Products, Passed in 13 Countries

Country	Law
Austria	The *Lamp Ordinance* (superseded by the *Electro Ordinance* as of August 2005) and *Batteries Take-Back Law* have been enacted. Take-back ordinance covers batteries, refrigerators, and fluorescent lamp tubes. Austria has adopted the European Union's (EU) *Waste Electrical and Electronic Equipment* (WEEE).
Belgium	WEEE (VLAREA Ordinance) applies in Flanders and the *Producer Responsibility Decree* in Wallonia and Brussels. Producers finance recovery organizations, including RECUPEL and BEBAT.
Denmark	Local authorities are responsible for the cost of collecting AV, IT, telecommunications, monitoring, medical, and lab equipment from households, commercial, and/or industrial premises for a specified fee (for businesses).
Italy	End-of-life household appliances such as refrigerators, freezers, TVs, PCs, washing machines, and dishwashers are consigned to an acknowledged collector. Mandatory deposits are required if the voluntary approach fails; producers are held responsible.
The Netherlands	Manufacturers of white and brown goods reprocess all end-of-life equipment; retailers take back components when selling replacement items, collection systems are managed by local authorities. Land filling or incineration of WEEE products is prohibited; trade of refrigeration equipment using CFCs is banned. Manufacturers and importers are charged by collectors for all electrical and electronic equipment, except computers and related telecommunications equipment.
Norway	Products that include home appliances, computers, and telecommunications and office equipment for both new and second hand products are regulated. Importers are charged collection fees.
Portugal	Producers fund disposal and are setting up a take-back organization to be called Amb3E.
Spain	Law passed as far back as 1998 deal with collection and disposal of waste products. Producers directly manage waste from their products, participate in an organized waste-management system, or fund public waste-management costs. The industry has established a number of take-back initiatives.
Sweden	Law requires collection of old refrigerators. Retailers or distributors take back all other WEEE products. Municipalities and manufacturers cover the related costs. A recovery system for ICT products is in operation.
Switzerland	Law covers almost all electrical and electronic equipment. Take-back applies to all of these products regardless of when they were purchased. Both consumers and producers cover collecting fees called *SWICO*.
Japan	*The Home Appliances Recycling Law* has been enacted since April 2001. PCs and rechargeable batteries are covered by the *Law for Promotion of Effective Utilization of Recyclable Resources*. Costs for meeting legislation are absorbed by the product. Take-back fees are covered by consumers.

Table 1 (*continued*)

Country	Law
South Korea	Consumers bring back items for free under the *EPRS Law 2003*; if not buying new, they must pay a sticker fee. Since 2005, take-back products include A/C, TVs, computers, peripherals, lamps, and cell phones. The industry either establishes a collection organization or pays a government tax.
Taiwan	Law covers computers, printers, and major appliances such as TVs, refrigerators, washing machines, and A/C. Electronics recovery has been in effect since 1998. The government requires fees on certain electronics in parts and appliances. An industry collection system has been formed.

Note: Adapted from Refs. 6, 12.

lead, mercury, cadmium, hexavalent chromium, polybrominated biphenyls, and polybrominated diphenyl ethers.[10] The RoHS Directive covers the same scope as the WEEE Directive, but excludes medical devices and monitoring and control instruments. The RoHS Directive complements the WEEE Directive by aiming at restricting the use of hazardous substances in electrical and electronic products. Under the RoHS Directive, the above mentioned parties are required to show that their products do not contain above a certain maximum level of these substances, or have a suitable nonhazardous replacement, by July 2006.[10]

By September 2005, 20 of the 25 EU member states had adopted the WEEE Directive as national law. These countries are Austria, Belgium, Cyprus, Czech Republic, Denmark, Finland, France, Germany, Greece, Hungary, Ireland, Lithuania, Luxembourg, The Netherlands, Poland, Portugal, Slovakia, Slovenia, Spain, and Sweden.[12]

The trend shows that for Western Europe and Japan, responsible treatment of consumer electronic products will continue to develop.[13] It will also continue to evolve in light of the expected change in their components and materials composition. Continued development of new technologies will allow for better end-of-life scenarios, although it must be realized that a main obstacle would include lack of economies of scale and the increasingly wide variety of products. The development of take-back legislations for consumer electronic products is expected to increase, extending to both the number of countries as well as the variety of product categories. Encouraging developments have also been seen in the industrialized and developing Asian countries such as Japan, South Korea, and Taiwan. At this time, the United States has not yet proposed any take-back legislations. Currently, no state requires manufacturers or retailers to adopt such an initiative. However, several states such as North Carolina, Wisconsin, Minnesota, and California have showed initiative geared toward the implementation of take-back schemes.[13]

The ELV Directive

In Europe, approximately 9 million *end-of-life vehicles* (ELVs) are discarded annually, with about 25 percent by weight going to landfills.[14] In 1997, the European Commission adopted a proposal for a directive that aims to make vehicle end-of-life activities such as dismantling, recovery, reuse, and recycling more environmentally friendly. This legislation was officially adopted by the European Parliament and Council in September 2000.[8,15]

The main objective of the ELV Directive is waste prevention. With this goal, the directive stipulates certain rules that vehicles, materials, and related equipment manufacturers must follow. The regulation requires these parties to take three actions:

1. Endeavor to reduce the use of hazardous substances when designing vehicles, and ensure that vehicles placed in the market after July 2003 do not contain mercury, hexavalent chromium, cadmium or lead except for certain specified applications.
2. Design and produce vehicles with a view of their dismantling, reuse, recovery, and recycling at the end of their useful life.
3. Increase the use of recycled materials in vehicle manufacture and help to develop markets for recycled materials.[8,16]

The directive also aims at improving the environmental performance of all economic operators involved in the life cycle of vehicles, especially those directly involved with the vehicles end-of-life treatment. Under the directive, EU member states are required to set up collection and transportation systems for ELVs and also for used parts to be sent to *authorized treatment facilities* (ATFs) at no cost to the previous owner. A *Certificate of Destruction* (CoD) is then issued to the previous owner upon the vehicle's deregistration at the ATF.[8]

The directive defines ELVs as follows: (1) passenger vehicles with at least four wheels, with no more than eight seats, (2) motor vehicles with at least four wheels that are used for goods transportation having a mass less than 3.75 metric tons, and (3) three-wheeled motor vehicles, excluding motor tricycles, that must comply only with collection and treatment obligations.[8] Such vehicles, which include special purpose automobiles such as motor caravans and ambulances, are exempted from the directive's recycling and recyclability requirements.

The directive set a goal of having an ELV percentage weight of 85 percent total recovery, a minimum of 80 percent recycling, up to 5 percent energy recovery, and no more than 15 percent disposal to landfills by January 2006. By January 2016, these figures are to be increased to 95 percent total recovery, a minimum of 85 percent recycling, up to 10 percent energy recovery, and no more than 5 percent waste going to the landfill.[8] However, the number and complexity of ELVs are also expected to increase over the next 20 years. Apart from having a greater amount of lightweight materials—such as aluminum and magnesium alloys and varying new composition of plastics and composites—the automobiles are also expected to have more sophisticated and complex components. These changes would require the development of new end-of-life treatment technologies. With this view in mind, the ELV Directive and specified targets are also expected to be regularly reviewed.

In the United States, end-of-life vehicle initiatives are largely an indirect result of national legislation concerning solid and hazardous-waste disposal practices. For example, the prohibition of free liquid disposal at landfills led to the practice of vehicle fluids collection recycling, and the prohibition of lead-acid battery disposal at landfills led to the practice of lead-acid battery collection for recycling.[17] Other examples of legislative development that address end-of-life vehicle practices include: (1) the state of California classifying *automobile shredder residue* (ASR) as a hazardous waste, and several other states' requirement for ASR treatment before disposal; (2) state-level restrictions on the use of mercury in auto-related devices; (3) legislations that address issues related to scrap tires; and (d) national regulatory enforcement for stormwater runoff management on dismantling operations.[17]

Sustainable Mobility Project

In the year 2000, dialogues between the world's largest energy and automotive enterprises and the *World Business Council for Sustainable Development* (WBCSD) were held to develop a vision for attaining sustainable mobility.[18] This culminated in the form of *the Sustainable Mobility project*, which from the outset sought to address sustainability issues and challenges, in both developed and developing countries—related to all modes of transportation, and emphasizing especially on road transport. The WBCSD defined *sustainable mobility* as the ability to meet society's needs "to move freely, gain access, communicate, trade, and establish relationships without sacrificing other essential human or ecological values today or in the future."[18] The term also points to addressing what is at stake in redressing the balance of costs and benefits in the transport sector. This new concept is a paradigm shift in thinking from the traditional approach, which sees transport as a derived demand and supporting framework for economic growth, toward one that considers risk assessment and acknowledges the problems associated with unconstrained/nonsustainable growth.[19]

Research and stakeholder dialogues continued throughout 2001, and this led to the "Mobility 2001" report.[20] The report highlights eight challenges in achieving sustainable mobility:[18,20]

1. Transport systems must be ensured to serve essential human needs.
2. Vehicles must be adapted to the evolving requirements on emissions, fuel consumption, capacity, and ownership structure.
3. The relationship between public and private forms of transport must be reinvented.

4. The infrastructure of mobility management and process planning must reinvented and developed.
5. Carbon emissions must be reduced.
6. The competition for infrastructure between personal and freight transport must be resolved.
7. The problem of traffic congestion must be resolved.
8. An institutional capacity must be built.

Following *Mobility 2001*, 10 *work streams* were identified. Each was categorized to concentrate on specific aspects of sustainability related issues arising from the central role played by road vehicles throughout the world. The 10 focus areas of these work streams are:

1. Methods and indicators for monitoring progress
2. Vehicle design and technology
3. Fuels
4. Infrastructure
5. Demand for personal mobility
6. Demand for commercial (goods and services) mobility
7. Policy measures
8. Urban context
9. Long-distance context[18,20]

The EuP Directive

A study headed by the European Commission found that over 80 percent of environmental impacts caused by products are influenced by decisions made at the product design stage. Based on this finding, the Commission proposed a directive to be enacted in EU states that would serve as a legislative framework for determining the ecodesign conditions to be met for *energy-using products* (EuP). The EuP Directive has now been adopted by the European Parliament and the Council.[21]

By definition, an EuP is a product that, once in the market, requires energy input in the form of electricity, fossil fuels, or renewable energy resources (like electricity) to work. It also includes products involved in the generation, transfer, and measurement of these energies, covering also parts that are dependent on energy input and intended to be incorporated into an EuP. The EuP covered in this directive refers to those placed on the market and/or put into service as individual parts for end users. The directive also specifies that the EuP must be able to be environmentally assessed independently.[9] The directive covers all energy sources, especially products using fuel, gas, and electricity, while excluding means of transport.[21]

The directive has five fundamentals:

1. An integrated framework for determining the criteria in considering the environmental aspects of EuPs throughout their life cycle.
2. A protocol for the Commission to follow in creating and implementing rules for manufacturers to follow.
3. Implementing measures that are adopted by the Commission (aided by a regulatory committee).
4. The definition of ecodesign requirements to be met by products that are included in the implementing measures, including the methodologies for establishing them.
5. Implementing measures that define ecodesign requirements, conformity assessment procedures, and implementation dates.[9,21]

2.2 Economic Drivers

Apart from legislative compliance, economic factors also motivate the adoption of sustainable-product development. Studies have shown a correlation between sustainable-product manufacturing technologies and economic efficiency; a manufacturer that is first to develop a cost-effective product in compliance with anticipated take-back laws holds a competitive advantage in the marketplace, especially in terms of having a better corporate image.[22]

Environmental regulations such as the WEEE, RoHS, ELVs, and EuP directives point to the trend that it will become vital for manufacturers to incorporate sustainability in product design to compete in the global market. Such is already the case in EU member states where producers and distributors are obliged to comply with these regulations in order to place and distribute their products in these markets.

Manufacturers have realized that products that generate large amounts of waste are costly, and that a sustainable product life cycle can actually contribute to significant savings in production, waste treatment, and disposal costs. Excessive waste is an indicator of inefficiency, which increases costs. Product design that strives to eliminate wastes and the use of toxic materials will reduce costs in the long run. Product design that aims to facilitate end-of-life activities will also benefit the manufacturer, especially in markets that force manufacturers to address and participate in these activities.

The practice of considering the product life cycle to enhance sustainability facilitates planning and assessment of options where such an approach leads to sound decision making. Furthermore, this activity also reduces the risk of liabilities related to environmentally related problems such as air pollution, water contamination, and health problems. Lowering these risks also increases the potential for attracting investment. In this regard, some groups have already begun compiling data to identify companies that emphasize sustainability in their corporate agenda. These groups include the *Dow Jones Sustainability Group Index*[23] and the *Investor Responsibility Research Center*.[24] Initial research indicates a relationship between sustainability and profitability in the long run.

Activities related to developing sustainable products also pave the way for product innovation. These and related activities promote a more thorough consideration of product function/service that minimizes or even eliminates adverse environmental impacts. This approach encourages the development of new ideas, products, services, and even markets.

These incentives, in addition to increasing public awareness, have contributed largely to the growing demand for more sustainable products. To meet this demand, the manufacturing sector requires a paradigm shift in its approach to product design—a shift toward product design for sustainability.

2.3 Societal Drivers

Society has an innate desire for a future generation that is preserved, thriving in a healthy environment with a good quality of life. This inherent nature is well in line with the Brundtland Commission's definition of sustainability as "development that meets the needs of the present without compromising the ability of future generations to meet their own needs."[1] Society's compliance with this innate nature gives rise to a sustainable society—a society that places its utmost priority on meeting the needs and comfort of future generations.

Society drives sustainable development because it realizes that attainment of that goal ensures health, well being, security, and a high quality of life. Society, as a driver, is also inherently connected with both, the economy and the environment, because the government policies that nurture the latter aspects can also be viewed as starting a realization at the societal level. In other words, the realization that sustainability is essential for continued growth and preservation of the economy, environment, and society itself starts with the society. A sustainable society realizes that an economy must function within the limits of the environment, and that sustainability is achieved when the greatest good is achieved for the greatest number. It is the inequality of sharing the environmental and economic resources between members of the current society and also that of future generations that causes instability.

2.4 Interrelationship Among the Sustainability Drivers

The interrelationship among the three major components of sustainability can be represented as a gear train, shown in Figure 1. Figure 1(a) shows a scenario where economy is the driver for sustainability. In this case, the main objectives for attaining sustainability would be to attain prosperity, market development, creation of high-quality goods and services, and employment. Such a case requires the positioning of *gears*, as exemplified in Figure 1(a), where the larger "gear" drives a smaller "gear," which would provide leverage for the attainment of these

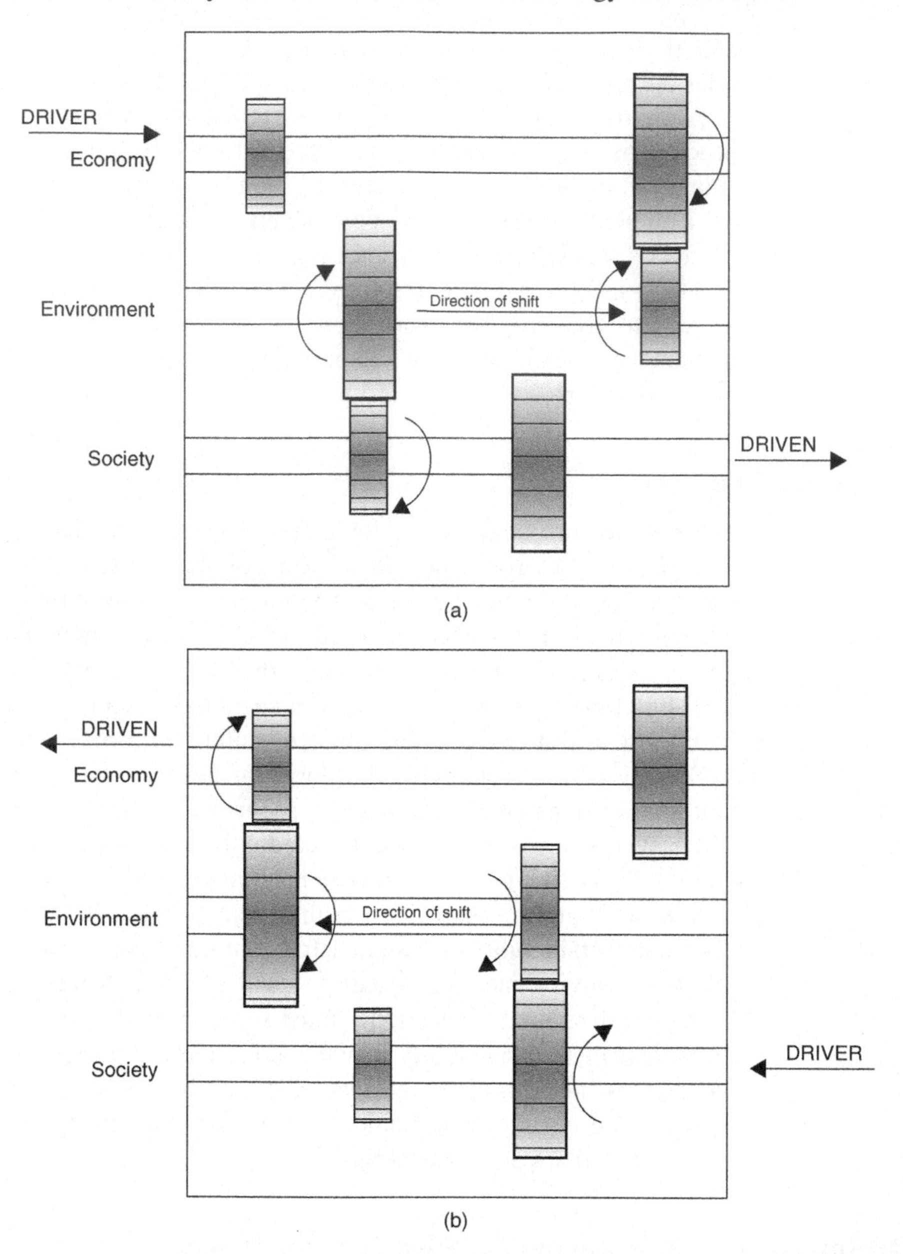

Figure 1 The gear train interrelationship representation among the main sustainability components. (a) Economy as a driver for sustainability. (b) Society as a driver for sustainability.

objectives. At the receiving end, society is driven through environment (intermediary gears), satisfying the societal needs by gaining a (mechanical) advantage and efficiency through this drive train, in mechanical sense.

Figure 1(b) shows the reverse case where society is seen as the driver for sustainability. In this case, the input objectives include the attainment of a high quality of living, health and well being, and security. This would require a paradigm shift in approach, as exemplified by the shifting of the middle shaft in the figure. In this case, the economy is at the receiving end, driven by the societal needs. In this scenario, the economic return with prosperity is determined by the societal driving power.

In both cases, the environment, which is seen as the central/intermediary shaft, plays a central role. It is the link without which the objectives cannot be met. This interrelationship among the integral sustainability elements is also represented by a gear train, because proper functioning of the whole mechanism requires proper functioning of each individual component. Moreover, these components must also fit well with each other, as depicted by the proper selection of gearing ratios, for example. Improper selections/decisions can cause unwanted outcomes.

3 PRODUCT LIFE CYCLE STAGES

This section discusses a product's life-cycle stages, where the new approach of *cradle-to-cradle* product development and assessment is primarily aimed at. Other key issues that are touched upon include the novel *6R concept* and its contribution to the naturally oriented *perpetual* material flow in a sustainable multiple product life-cycle system.

3.1 The Cradle-to-Cradle Life-Cycle Approach

A key element in designing products for sustainability is product design that goes beyond one life cycle by considering the product's end-of-life factors such as disassembly, recycling, recovery, refurbishment, and reuse. The paradigm shift that is referred to is one from the traditionally known cradle-to-grave thinking to that of cradle to cradle.[25] This product life-cycle approach is shown in Figure 2. The cradle-to-cradle approach considers the interrelationships involved between the product and the environment at all stages of its life; not only from raw materials extraction through product manufacture and use, and its eventual disposal, but also to its possible reuse or recycling to create new raw materials. Such an approach has been formally coined as *design for sustainability* (DfS) promoted through educational and training programs for the next generation of engineers and scientists.[26] It involves a systematic method for evaluating and mitigating environmental, societal, and economic concerns at the earliest possible stages of process design or product life cycle.

The product life cycle in Figure 2 exemplifies four distinct stages that interact with the environment in some way: (1) *premanufacturing*, (2) *manufacturing*, (3) *use*, and (4) *postuse* stages. Typically, the stage of *use* is much longer than the other stages, although it may not necessarily be the stage that carries the greatest environmental or societal impact. Between these stages, there may also be periods of transportation and storage. These intermediate stages of nonactivity and nonuse are usually environmentally benign.

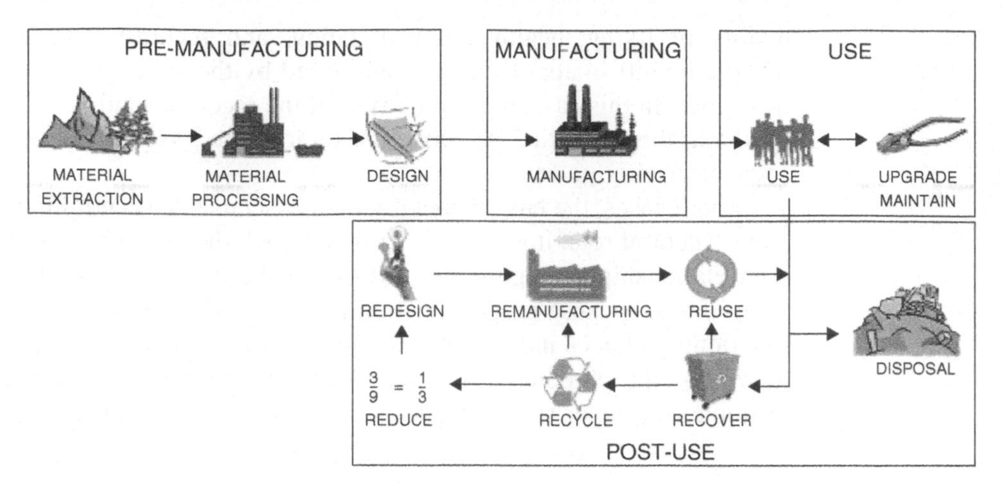

Figure 2 The closed-loop product life-cycle system showing the 6Rs for perpetual material flow.

Figure 2 also shows a feedback close loop at the *postuse* stage, representing the product's end-of-life activities. Although these activities are highly desirable, it must be realized that they may also involve processes that have nonsustainable elements (i.e., those that cause adverse environmental impacts). The inclusion of the *postuse* stage in the product life-cycle picture is the underlying basis of the cradle-to-cradle life-cycle approach.

3.2 The 6R Concept

In considering the material flow in a sustainable product life cycle, the *3Rs* (i.e., *reduce, reuse*, and *recycle*) have often been referred to as key ingredients.[27] A more comprehensive and complete depiction would include three other Rs. These are *recover, redesign,* and *remanufacture.*

Reduce involves activities that seek to simplify the current design of a given product to facilitate future *postuse* activities. Of all the end-of-life activities in the postuse stage, *reuse* may potentially be the stage incurring the lowest environmental impact, mainly because it usually involves comparatively fewer processes.[28] *Recycle* refers to activities that include shredding, smelting, and separating. *Recover* represents the activity of collecting end-of-life products for subsequent *postuse* activities. It may also refer to the disassembly and dismantling of specific components from a product at the end of its useful life. *Redesign* works in close conjunction with *reduce* in that it involves redesigning the product in view of simplifying future *postuse* processes. *Remanufacture* is similar to manufacturing. However, it is not conducted for the first time. By contrast, it is a stage conducted as a subsequent operation after other *postuse* activities. The introduction of the *6R concept* into a product's life cycle seeks to attain,

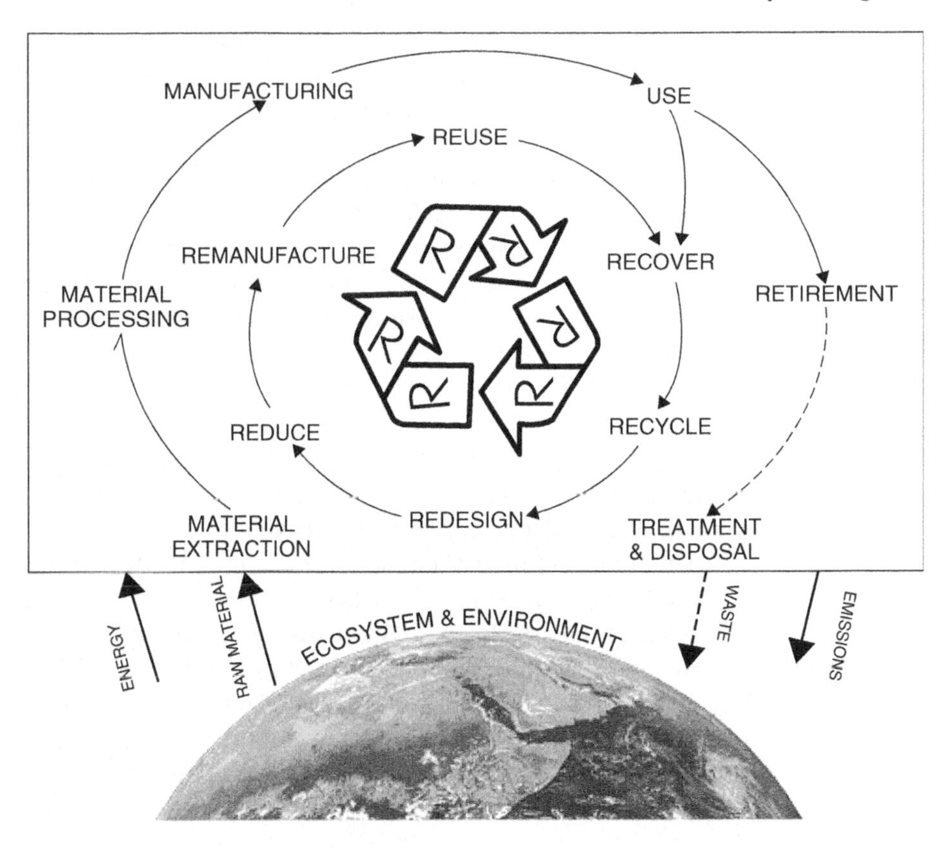

Figure 3 The closed-loop product life-cycle system.

or at least approach, the condition of a *perpetual material flow*, resulting in a minimization/elimination of that product's ecological footprint.[29,30]

3.3 Closed-loop versus Open-loop Life-cycle Systems

Figure 3 shows a closed-loop/cradle-to-cradle product life-cycle system. Conversely, in an open-loop/cradle-to-grave life-cycle system, products are consumed and disposed of at the end of their useful life. With this scenario, material resource, waste output, energy usage, and other system emissions are primarily a function of consumer demand.

To make a shift toward the closed-loop system, at least three criteria of sustainable product development must be met. These are (1) minimization of material and energy resources needed to satisfy product function and consumer demand, (2) maximization of expended resources usage, and (3) minimization/elimination of adverse impacts of wastes and emissions. A closed-loop product system as typified in Figure 3 must fulfill at least the first two criteria.[31] In this type of product system, the activities of product reuse, remanufacturing, and recycling circulate the material within the product system. These activities reduce the requirement for new material extraction to feed into the system, resulting in the reduction of the overall energy input requirement and emissions per unit of the product consumption.

It may be argued that such a closed-loop system would not be beneficial, at least from the business standpoint, to product manufacturers. This is especially for the market situation today where discrete product sales are dominant and consumers continually demand new and varied products at the lowest possible cost. Rapid technology advancement and turnover also create a situation where product life cycles shrink significantly, resulting in more and more products reaching their end-of-life period at a faster rate while they still have a considerable amount of residual (useful) life. Although on the surface, this situation may look to be going against the ideal setting for a cradle-to-cradle life-cycle implementation, it is actually ideal for implementing such a model, since there is great potential for reuse, remanufacturing, and also recycling.

A paradigm shift toward cradle-to-cradle product development would persuade manufacturers to invest more in ways to promote efficient material use and reuse. With such an approach, the value associated with the reduced energy and raw materials used are taken into account. As a result, the cradle-to-cradle paradigm shift that considers the life-cycle stages of reuse, remanufacture, and recycle not only can work toward manufacturing a more sustainable product, but also can provide realizable economic gains for the manufacturer.[31]

4 PRODUCT DESIGN FOR SUSTAINABILITY

The foregoing discussion emphasizes the importance of developing a sustainable product design by considering all stages of the product life cycle, including the stage beyond a product's useful *first life*. This importance is also underlined by the drivers for attaining such a development. The next step would be to develop a sound methodology to measure and assess product sustainability. This section presents a viable methodology for this.

4.1 Measurement and Assessment of Product Sustainability

The quantification of product sustainability is essential for a comprehensive understanding of the sustainability component in a manufactured product. It is becoming increasingly obvious that the societal appreciation, need, and even demand for such a sustainability rating would become apparent with increasing societal awareness and user value to all manufactured products. This is analogous to the current demand for comprehensive food labeling, energy efficiency levels in electrical appliances, and fuel consumption rating in automobiles. It is not surprising that a quantifiable sustainability rating would one day be required for all manufactured products through mandatory regulations.

Almost all previous research on sustainability assessment of products has produced *qualitative* results that are mostly, with the exceptions of a few recent efforts, difficult to *quantify*. Hence, these analyses are largely nonanalytical and less scientific in terms of being their perceived value of contributions. Moreover, product sustainability does not just cover a simplistic assessment of the environment as a contributing measure; it involves a comprehensive simultaneous assessment of the environmental, economic, and societal impact categories, which are all interrelated.

The three components of sustainability, each containing measurable indicators, can be used to form a meaningful sustainability matrix. Table 2 illustrates a list of quantifiable performance indicators for each of these aspects. The performance indicators listed in this table are far from exhaustive, which points to the main reason why sustainability assessment of a product is difficult: A large amount of information is required to evaluate product sustainability. Adding to this predicament would be that some of the scope of sustainability issues may also be beyond a company's power to control.[32]

Ongoing work at the University of Kentucky (UK) within the *Research Institute for Sustainability Engineering* (RISE) involves a multidisciplinary approach aimed at formulating a more comprehensive method for product sustainability assessment. A group of design, manufacturing/industrial, and materials engineers, along with social scientists, economists, and marketing specialists are actively participating in the program to establish the basic scientific principles for developing a product sustainability rating system. This includes the development of a science-based product sustainability index (PSI), which is discussed in the next section.[33]

4.2 Product Sustainability Index (PSI)

Six major sustainability elements can be identified in determining the sustainability level of a product: (1) *environmental impact*, (2) *societal impact*, (3) *functionality*, (4) *resource utilization and economy*, (5) *manufacturability*, and (6) *recyclability/remanufacturability*.[33,34,35] The requirement of having these six elements compared to the often-used three broad categories of environment, society, and economy is to incorporate the processes and systems criteria that are significant in sustainability decision making.[35] *Functionality* is a key aspect of a product where upgradeability, modularity, and maintainability all contribute to sustaining a product. *Manufacturability* deals with assembly, transportation, and packaging where new legislations are coming into effect. *Recyclability/remanufacturability* is a very extensive element focusing on aspects of waste minimization and resource preservation.

The rating of each sustainability element can be represented in several ways. For example, the rating can be given numerically between 1 and 10 (with 10 being best), or alphabetically between A and D (with A being the best). A composite

Table 2 Economic, Environmental, and Societal Performance Indicators for Sustainability Assessment[32]

Economic	Environmental	Societal
Direct • Raw material cost • Labor cost • Capital cost	**Material Consumption** • Product & packaging mass • Useful product lifetime • Hazardous materials used	**Quality of Life** • Breadth of product availability • Knowledge or skill enhancement
Potentially Hidden • Recycling revenue • Product disposition cost	**Energy Consumption** • Life-cycle energy • Power use during operation	**Peace of Mind** • Perceived risk • Complaints
Contingent • Employee injury cost • Customer warranty cost	**Local Impacts** • Product recyclability • Impact upon local streams	**Illness & Disease Reduction** • Illnesses avoided • Mortality reduction
Relationship • Loss of goodwill due to customer concerns • Business interruption due to stakeholder interventions	**Regional Impacts** • Smog creation • Acid rain precursors • Biodiversity reduction	**Accident & Injury Reduction** • Lost-time injuries • Reportable releases • Number of incidents
Externalities • Ecosystem productivity loss • Resource depletion	**Global Impacts** • CO_e emissions • Ozone depletion	**Health & Wellness** • Nutritional value provided • Food costs

Note: From Ref. 32.

index can then be calculated by averaging the sum of the sustainability element indices. Variations of this implementation are illustrated for a generic case in Figure 4.

It should be noted that the product sustainability labels shown in Figure 4 are examples for a generic case. Circles were used in the implementation to allow for flexibility in adjusting the size of the segments, based on the weighting factor of the represented sustainability element. For example, in the case of automobile as a product, a certain manufacturer may regard the environment and resource

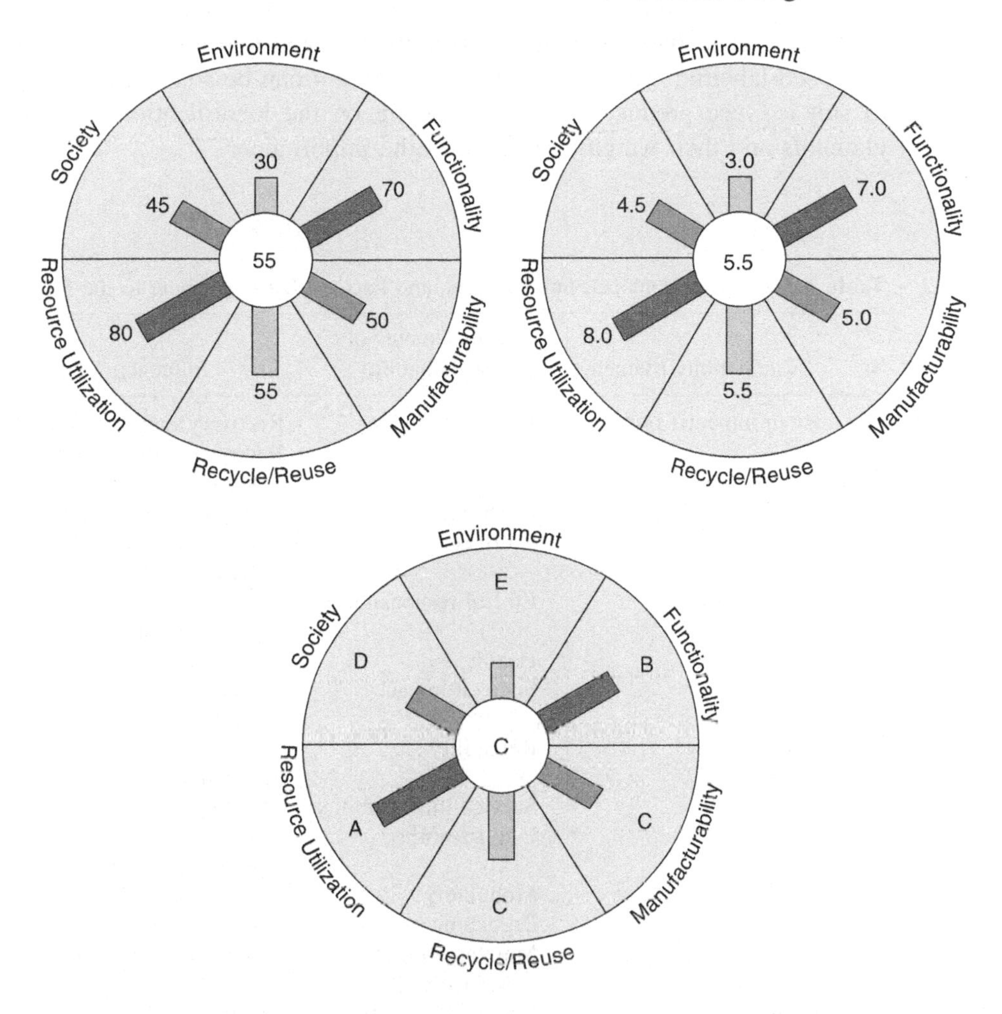

Figure 4 Variations of the proposed product label for a sustainable product. (From Ref. 33.)

utilization (fuel/energy consumption) component as having the greatest sustainability impact. In such a case, these sustainability elements may be represented by bigger pie portions. Conversely, the other impacting sustainability elements may occupy smaller portions based on their respective weighting factors assigned by the manufacturer.

Standards can be developed to establish the required minimum level of rating for each group. A sustainable product would require compliance with these acceptable levels. Although the rating of each group contributes to the product's sustainability rating, a composite rating will represent the overall PSI of the product. Each of the six sustainability elements contains a subset (subelement) of quantifiable influencing factors, which are themselves rated to arrive at an average index for that respective element group. Table 3 lists the six elements with their respective subelements. It must be noted that the generic form of the proposed labeling method illustrated in Figure 4 can be adapted and customized to suit a given product family, depending on the identification of constituting elements and their weighing to signify the importance.

Table 3 Generic Elements, Subelements, and Factors that Contribute to the PSI Score

No.	Sustainability Elements	Subelements of Sustainability	Influencing Factors
1	Environmental Impact	Life-cycle factor	Recovery rate after first life
			Recovery cost
			Potential for next life
		Environmental effects	Hazardous substances
			Emission
2	Societal Impact	Ethical responsibility	Take-back options
			Product pricing
		Health	Wellbeing
		Societal impact	Safety
			Quality of life
3	Functionality	Reliability	Type of material
			Maintenance schedule
		Service life/durability	Maintenance schedule
		Upgradeability	Ease of installation
			Option for upgrade
		Modularity	Modules available
		Ergonomics	Safety
		Maintainability/serviceability	Maintenance schedule
4	Resource Utilization and Economy	Energy efficiency	Production energy
			Energy for use
			Recycle energy
		Material utilization	Type of material
			Quantity of material
			Cost of material
		Use of renewable source of energy	Option for other energy sources
		Market value	Current market value
		Operational cost	Cost to operate
5	Manufacturability	Packaging	Take-back options
			Packaging material
			Quantity used
		Assembly	Number of parts/components
		Transportation	Cost of transportation
		Storage	Duration of storage

Table 3 *(continued)*

No.	Sustainability Elements	Subelements of Sustainability	Influencing Factors
6	Recyclability/ Remanufacturability	Recyclability	Cost of recycling
			Recycle energy
			Recycling method
			Type of material
			Separability
			Value of recycled material
		Disposability	Disposal options
		Remanufacturability	Number of recovered parts
		Disassembly	Number of parts/components
		Recovery of materials	Number of parts/components
			Type of material

4.3 A Recent Effort in Developing a Product Sustainability Index (PSI)

In line with works toward establishing an international standard for sustainability, Ford of Europe[36] also developed a PSI with indicators based partly on ISO 14040[37] (life-cycle assessment) and the ongoing work of the Society of Environmental Toxicity and Chemistry (SETAC) Europe on life-cycle costing.[38] Five key guiding principles were used in the development of Ford of Europe's PSI indicators:

1. Consideration of key attributes of environmental, social, and economic factors
2. Consideration of controllable factors mainly influenced by the product-development department
3. Use of readily available data
4. Consideration of bottom-line issues only (not going down to specific technologies such as alternative fuels, but the overall life-cycle impact)
5. Reduction to manageable amount of indicators

Based on these outlined principles, eight sustainability indicators were identified as a subset of the three main categories of environment and health, social, and economics. These indicators are listed in Table 4.[36]

It should also be noted that a product's sustainability assessment cannot be made in isolation from the manufacturing processes and systems utilized in production because these latter aspects play a catalytic role in determining the level of product sustainability. One of the main difficulties in such an assessment lies in

Table 4 Ford of Europe's PSI Indicators

	Indicator	Metric/method	Driver for inclusion
Environment and Health	Life-cycle global warming potential	Greenhouse emissions along the life cycle (vehicle production, driving 150,000 km based on EUCAR end-of-life agreement); part of an LCA according to ISO14040	Carbon intensity is the main strategic issue in the automotive industry
	Life-cycle air quality potential	Summer smog creation potential (POCP) along the same life cycle (VOCs, NO_x); part of an LCA according to ISO14040	Non-CO_2 air-quality issues have to be mentioned for trade-offs
	Sustainable materials	Recycled and natural materials related to all polymers	Resource scarcity
	Restricted substances	Vehicle interior air quality/allergy tested interior, management of substances along the supply chain (15 point rating)	Substance risk management is key
	Drive-by-exterior noise	Drive-by exterior noise = dB(a)	Main societal concern
Social	Safety	Including number of all EuroNCAP stars	Main direct impact
	Mobility capability	Mobility capacity (luggage compartment volume plus weighted number of seats) related to vehicle size	Crowded cities (future: diversity—disabled)
Economics	Life-cycle ownership costs	Vehicle price plus 3 years' fuel costs, maintenance costs, taxation minus residual value (note: for simplification reasons costs have been tracked for one selected market, life-cycle costing approach using discounting)	Customer focus, competitiveness

Key for Table 4: EUCAR = European Council for Automotive R&D; EuroNCAP = European New Car Assessment Programme; POCP = Photochemical Creation Potential
Note: Adapted from Refs. 36, 37.

consolidating the quantification of a product's sustainability, with other elements such as the material, production, and systems aspects. A complete assessment requires simultaneous consideration of both these aspects of product and process sustainability.

4.4 New Methodology for Evaluating the Product Sustainability Index over a Product's Total Life Cycle

A new comprehensive evaluation methodology to rate any given manufactured product in terms of all three components of sustainability (economy, environment and society) over its total life cycle (premanufacturing, manufacturing, use, and postuse) is presented in this section. This rating system gives the product sustainability index (PSI), which is versatile enough to be applied to a wide range of products. This system will assist product developers and manufacturers in achieving their sustainability targets. The new procedure for evaluating the PSI is given in three steps.

Step 1: Identify Potential Influencing Factors

The product developers need to identify potential influencing factors based on national/international regulations, federal and state laws, and their own perspective. The product designers should not just focus on the economic component of sustainability, but look broadly at numerous environmental and societal aspects as well. Similarly, to be effective, they should not concentrate on the product's premanufacturing and manufacturing stages, but should also consider the product's use and postuse stages. After identifying the potential influencing factors, product developers can form a three-by-four-dimensional matrix, with the three horizontal rows representing the components of sustainability (economy, environment and society) and the four vertical columns representing the four product life-cycle stages (premanufacturing, manufacturing, use, and postuse).

Step 2: Conduct a Life-cycle Assessment

Product developers will have to conduct a detailed life-cycle assessment (LCA) of all influencing factors that they have identified in the previous step to obtain the absolute values of these factors to represent the anticipated sustainability levels. Once they have obtained these absolute values for the influencing factors, they can then allocate a score/rating between 0 and 10 for each factor (with 0 being worst and 10 being best, or vice versa). Weighting can be applied to the influencing factors based on their relative importance and company priorities. Some of these factors can be nonquantifiable, in which case the designer should assign a score based on its experience and judgment.

Step 3: Determine PSI Scores

The template for the PSI matrix is shown in Table 5.

The product designers need to record the scores of the influencing factors in each box of the matrix. To evaluate the PSI, go across the matrix and sum up the scores of each influencing factors in each matrix box and calculate the percentage value using the equation (this equation represents only the first box

Table 5 Framework for a Comprehensive Total Life-cycle Evaluation Matrix for Product Sustainability (Using Fictitious Numbers)

Sustainability Components	Influencing Factors in the Product Life–cycle Stages									
	Premanufacturing	Score out of 10	**Manufacturing**	Score out of 10	**Use**	Score out of 10	**Postuse**	Score out of 10		
Environment	Material extraction	7	Production energy used	7	Emissions	9	Recyclability	7	PSI_{en} =	77.29
	Designs for environment	8	Hazardons waste produced	9	Functionality	8	Remanufacturability	8		
	Material processing	6	Renewable energy used	8	Hazardons waste generated	9	Redesign	7		
							Landfill contribution	7		
	$PSI_{(en\text{-}pm)}$ =	70	$PSI_{(en\text{-}m)}$ =	80	$PSI_{(en\text{-}u)}$ =	86.67	$PSI_{(en\text{-}pu)}$ =	72.5		
Society	Worker health	8	Work ethics	7	Product pricing	7	Take-back options	7	PSI_{so} =	73.54
	Work safety	8	Ergonomics	7	Human safety	9	Reuse	6		
	Ergonomics	7	Work safety	8	Upgradeablility	7	Recovery	7		
					Complaints	8				
	$PSI_{(so\text{-}ym)}$ =	76.67	$PSI_{(so\text{-}m)}$ =	73.33	$PSI_{(so\text{-}u)}$ =	77.5	$PSI_{(so\text{-}pu)}$ =	66.67		
Economy	Raw material cost	6	Production cost	6	Maintenance cost	7	Recycling cost	7	PSI_{ec} =	61.25
	Labor cost	3	Packaging cost	7	Repair cost	6	Disassembly cost	8		
			Energy cost	8	Consumer injury cost	8	Disposal cost	4		
			Transportation cost	5	Consumer warranty cost	7	Remanufacturing cost	7		
	$PSI_{(ec\text{-}pm)}$ =	45	$PSI_{(ec\text{-}m)}$ =	65	$PSI_{(ec\text{-}pu)}$ =	70	$PSI_{(ec\text{-}pu)}$ =	65		
	PSI_{pm} =	63.89	PSI_{m} =	72.78	PSI_{u} =	78.06	PSI_{pu} =	68.06	PSI_{TLC} =	70.69

Note: The integrated PSI in the last column and row (denoted by PSI_{TLC}) shows the computed total sustainability index with equal weighing for all elements and subelements.

of the matrix) given below:

$$PSI_{(en\_pm)} = \left\{\left[\sum_{i=1}^{n} IF_{(en\_pm)i}\right]/(n^*10)\right\}^* 100\%$$

where,

$PSI_{(en\_pm)}$ = Product Sustainability Index for Environment component of Pre-manufacturing

$IF_{(en\_pm)}$ = Influencing Factor rated on a scale of 0-10 for Environment component of Pre-manufacturing

n = Number of influencing factors considered

The PSI values for Society ($PSI_{(so\_pm)}$) and Economy ($PSI_{(ec\_pm)}$) can be calculated similarly. The product sustainability index (PSI), for a single life-cycle stage, for instance pre-manufacturing, can be evaluated by vertically adding the PSI of sustainability components over that particular life-cycle stage, as shown in the following equation:

$$PSI_{pm} = [PSI_{(en\_pm)} + PSI_{(so\_pm)} + PSI_{(ec\_pm)}]/3$$

where,

PSI_{pm} = Product Sustainability Index for Pre-manufacturing

$PSI_{(en\_pm)}$ = Product Sustainability Index for Environment component of Pre-manufacturing

$PSI_{(so\_pm)}$ = Product Sustainability Index for Society component of Pre-manufacturing

$PSI_{(ec\_pm)}$ = Product Sustainability Index for Economy component of Pre-manufacturing

The PSI values for manufacturing (PSI_m), use (PSI_u) and post-use (PSI_{pu}) stages can be obtained similarly. The Product Sustainability Index (PSI) for the Environment component of Sustainability for all four stages of product life-cycle can be calculated by horizontal summation of Product Sustainability Indices of every stage of the product life-cycle as shown below:

$$PSI_{en} = [PSI_{(en\_pm)} + PSI_{(en\_m)} + PSI_{(en\_u)} + PSI_{(en\_pu)}]/4$$

where,

$PSI_{(en\_pm)}$ = Product Sustainability Index for Environment component of Pre-manufacturing

Table 6 Visual Representation of the PSI

Symbol	Score	
	Excellent	85–90%
	Good	70–84%
	Average	50–69%
	Poor	<50%

$PSI_{(en\_m)}$ = Product Sustainability Index for Environment component of Manufacturing

$PSI_{(en\_u)}$ = Product Sustainability Index for Environment component of Use

$PSI_{(en\_pu)}$ = Product Sustainability Index for Environment component of Post-use

Similarly, the PSI for Society (PSI_{so}) and Economy (PSI_{ec}) components can be obtained. The overall product sustainability index (PSI_{TLC}) for a product over its total life-cycle can be calculated as:

$$PSI_{TLC} = PSI_{so} + PSI_{en} + PSI_{ec}$$

The PSI scores can be interpreted as shown in Table 6.

Some of the influencing factors can be subjective and company-specific, in which case, the company can still use the PSI technique for self-assessment of its products to meet its internal sustainability goals. This technique will help the product designers and manufacturers to identify opportunities to improve the performance of their product over its total life-cycle. The visual representation of the influencing factors in terms of economy, environment and society components of sustainability for a generic product is shown in Figure 5. The three concentric circles represent the three components of sustainability (Economy, Environment, and Society). The influencing factors are listed on the periphery of the outer circle. A rating scale for each influencing factor that cuts through all three sustainability component circles with markings of 0-10 indicates the impact of that particular influencing factor for the three sustainability components. It is interesting to note that in this figure, each influencing factor is represented to have possible direct or indirect impact on all the three components. For example,

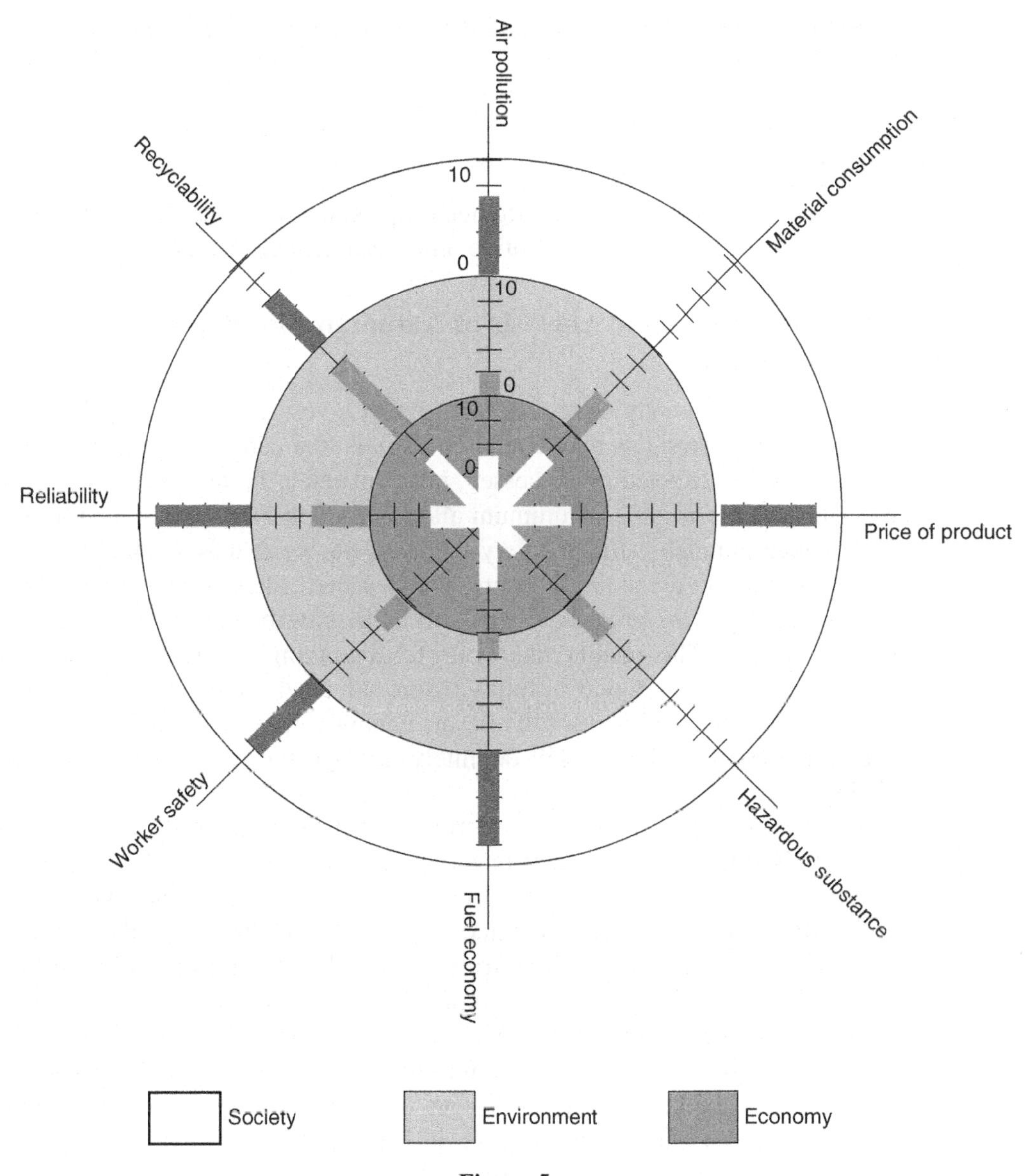

Figure 5

the *hazardous substance* content in an automobile can have a direct influence on the environment, as well as an indirect impact on the society and the economy. Disposal of hazardous substances causes unwanted additional economic burdens on the manufacturer. Seepage of these substances that leads to environmental pollution and contamination also causes unwanted environmental and societal burdens. Decontamination procedures also place unwanted economic burdens on the manufacturer. This example shows an interrelationship between all three components of sustainability caused by a single influencing factor (*hazardous substance* content of a product).

5 CASE STUDIES

This section presents two case studies on product sustainability evaluation from two industrial segments: automotive and consumer electronics.

5.1 Case Study 1: Total Life-cycle Analysis of Aluminum Products in Automotive Industry

Different products may have different impact levels at each life-cycle stage. For example, some products may contain materials that cause adverse environmental impacts when extracted or processed, but conversely be relatively safe during its use, and easy to recycle. Aluminum alloy is a good example of such a material. To produce alumina—the powdery white substance that is smelted to form aluminum—bauxite ore is initially crushed, washed, filtered, and dried to remove unwanted clay. This process, referred to as bauxite benefication, results in a 30 percent residue. This residue may collect in receiving water, resulting in reduced translucency and increased turbidity. More adversely, the settling of residue on lake and riverbed sediments will alter its chemical, physical, and biological characteristics. Such a change will definitely damage the receiving water's feeding chains.[39]

What makes aluminum very attractive from the sustainability standpoint is that not only is it the most abundant metallic element available, making up 8 percent of the Earth's crust, but it is also one of the most recyclable materials available. Aluminum can be recycled and re-recycled indefinitely without decreasing its quality or losing any of its intrinsic strength. This makes the recycling and reuse of aluminum potentially a completely closed-loop process. Hence, the high environmental price to pay for manufacturing pure aluminum is offset by the fact that it is highly recyclable and reusable. Table 7 shows data for waste accumulation for two basic life-cycle stages of aluminum in a given year, collected from a study on aluminum recycling in Europe.[40] In extracting and processing 143.3 million tonnes of bauxite to produce aluminum, as much as 52.8 tonnes of residues are generated that pollute the environment. However, in the manufacturing and use stage, compared to 27.4 million tonnes of primary aluminum entering the material flow cycle for a given year, only 8.6 million tonnes are lost. This is due to the high percentage of aluminum recycled as ingots to be reprocessed and reused.

In addition to attaining the targets as specified by the ELV Directive, other major thrusts include attaining a significant reduction in vehicle curb weight and increased fuel efficiency. This is especially the case in the United States, where an initiative was specifically established in 1993 in the form of *the Partnership*

Table 7 Material Loss and Run-off to the Environment for a Basic Aluminum Life Cycle

Losses to the Environment			
Material Extraction Stage		Manufacturing and Use Stage	
Bauxite Residues	52.8	Metal losses	1.4
		Application losses	0.8
		Landfill	3.3
		Under investigation	3.1
Total	**52.8**	**Total**	**8.6**

*All units measured in million tonnes per year.
Note: Adapted from Ref. 40.

for a New Generation of Vehicles (PNGV) involving 3 main North American automobile manufacturers, 7 government agencies, and 19 National Research Laboratories.[41] This initiative seeks to reduce the curb weight by as much as 40 percent and increase the fuel efficiency up to 80 miles/gallon.

Studies have found that the replacement of steel sheet by aluminum alloys in autobody parts can result in an approximate weight reduction by as much as 47 percent. Furthermore, secondary weight reductions can also be achieved, because the lighter overall load permits the use of lighter components in the vehicle.[42] This significant weight reduction can result in a chain effect of benefits that include: (1) greater fuel economy, (2) less energy consumption, (3) less pollution, and (4) less emission of greenhouse gases.

Aluminum is only 35 percent as dense as steel. In terms of aluminum as a replacement for steel in sheet form, the weight saved is typically more than 50 percent.[43] In addition to this, studies have shown that cutting a vehicle's weight by 10 percent results in reduced fuel consumption by as much as 8 percent (or 2.5 extra miles per gallon).[44] These findings offset the higher initial material cost of aluminum sheet that can be three to five times higher than that of steel. It should also be pointed out that although the production of new aluminum is highly energy-intensive, the recycling of aluminum consumes only up to 5 percent of that energy. Adding to this is the fact that an estimated 85 to 90 percent of post-consumer automotive aluminum scrap is recycled, amounting to at least 1 billion tonnes per year.[45]

The use of aluminum in automobiles is continuing to increase. This trend is clearly seen in Figure 6, where aluminum can amount to as high as 319 pounds per vehicle in the year 2006, compared to just 183 in 1991. The different forms of aluminum used in vehicles are 74 percent cast, 23 percent extruded, and 4 percent rolled.[46]

A recent and ongoing study at the University of Kentucky within the Research Institute for Sustainability (RISE) includes assessment and analysis of the total life cycle of aluminum and explores its potential as a substitute to steel for use in automobiles.[47] An impact and inventory matrix was formulated and completed based on research findings. This is shown in Table 8.

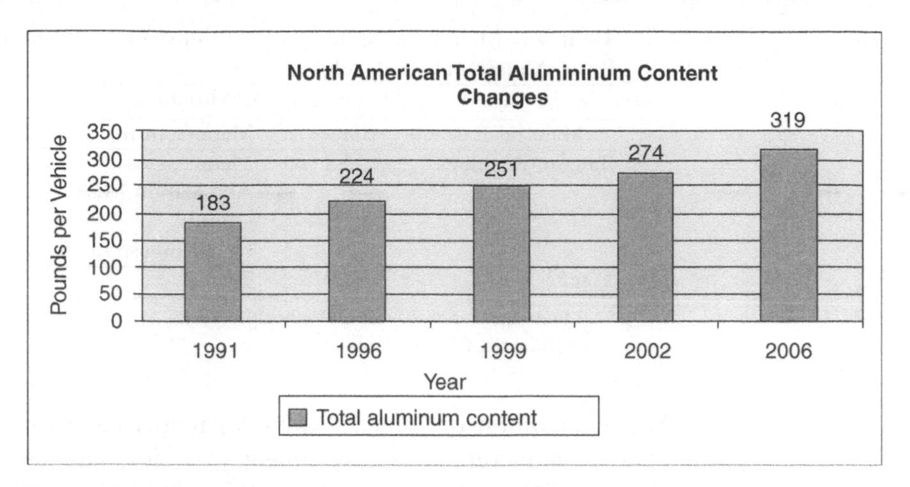

Figure 6 Change in weight of aluminum content in North American vehicles. (From Ref. 46.)

From the sustainability standpoint, the comparison shows that aluminum is a better alternative to steel at the use and postuse stage. It can also be seen that steel still holds an advantage over aluminum over the premanufacturing and manufacturing stages. However, in considering the total life cycle of both metals, aluminum is found to be the better choice from a sustainability point of view. From this example, it is clear that the sustainability of aluminum as a product material in the automotive industry can only be judged through a total life-cycle analysis. The foregoing discussion should also point to the fact that a total life-cycle assessment approach should also be applied in considering the sustainability of other materials and products as well.

5.2 Case Study 2: A New Product Scoring Method for a Consumer Electronics Product

This case study presents the development of a scoring method for assessing the product sustainability of a laser printer produced by a major consumer electronics manufacturer.[36,48] Two laser printer models, herein referred to as Product 1 and Product 2, were analyzed. For both types, a total of 10 sustainability subelements were identified as being important in the assessment: (1) energy efficiency, (2) material utilization, (3) life-cycle factor, (4) environmental effects, (5) recyclability, (6) reliability, (7) service life/durability, (8) ethical responsibility, (9) packaging, and (10) upgradeability. These 10 elements were further subdivided into *high importance* and *medium importance* categories, as listed in Table 9. The influencing factors of each subelement were also identified and included in the table.

Table 8 Impact Matrix Comparison between Aluminum and Steel for Total Life-cycle Assessment

Sustainability Measures	Body-in-White Life Cycle Stages									
	Pre-manufacturing		Manufacturing		Use		Post-use		Total	
	Aluminum	Steel	Aluminum	Steel	Aluminum	Steel	Aluminum	Steel	Aluminum	Steel
Resource Usage (energy use)	0	3	2	3	5	5	5	2	12/20	13/20
Waste Output — Solid/Liquid Residues	2	2	4	3	4	4	4	3	14/20	12/20
Air Emissions	1	3	2	3	5	2	4	3	12/20	11/20
Environmental Impact (e.g., toxicity, pollution)	2	3	3	3	5	3	5	3	15/20	12/20
Societal Impact (e.g., health, safety, life quality)	3	3	2	2	4	3	4	3	13/20	11/20
Economic Impact (e.g., costs, waste stream values)	3	4	2	3	5	3	5	3	15/20	13/20
Total	11/30	18/30	15/30	17/30	28/30	20/30	27/30	17/30	81/20	72/20

Note: From Ref. 47.

Table 9 Sustainability Elements for a Laser Printer

High Importance Subelements	Influencing Factors
Energy efficiency	Energy for use
Material utilization	Type of material Quantity of material
Life-cycle factor	Recovery cost Potential life of printer
Environmental effects	Toxic substances Emission
Recyclability	Cost of recycling Recycling method Separability Value of recycled material
Medium Importance Subelements	**Influencing Factors**
Reliability	Type of material Maintenance schedule
Service life/Durability	Maintenance schedule
Ethical responsibility	Take-back options Product pricing
Packaging	Take-back options Packaging material Quantity used
Upgradeability	Ease of installation Option for upgrade

Note: From Ref. 48.

The definitions of these subelement terminologies follow:

1. *Energy efficiency* refers to the amount of power consumption of the printer, with an average value set at 80Wh.
2. *Material utilization* was measured by grouping the materials into categories such as glass, metal, and plastics, and multiplying the corresponding eco-indicator value (pt/kg) with the weight (kg). This creates an index that indicates the environmental effects generated by the used materials.
3. *Life-cycle factor* refers to the combination of the printer's life expectancy with the recovery cost for the next life of the product cycle. It is generally known that the manufacturer involved in this case study sets the maximum number of functional years for a printer as five years, and the recovery cost at $1.00 per kg of material.

4. *Environmental effects* consider only the direct environmental effects by the use of printer, such as emissions that is measured by CO_2 output of the printer. This sub-element also includes the product evaluation for restricted hazardous material as required by the RoHS directive.
5. *Recyclability* measures all aspects related with recycling the printer at the end of its useful life. This includes the time for dismantling (time taken to separate materials, which is set at a maximum amount of 8 hours), if not shredded. For the shredding case, the index would be set to 0. The cost of recycled materials is measured against the nonrecycled virgin material for a comparison of lost value.
6. *Reliability* depends on the material type used for the printer, and the scheduled maintenance. This input is reduced to a yes/no criterion, because the laser printers referred to in this case do not require maintenance other than toner cartridge replacement.
7. *Service life/durability* includes the maintenance schedule of the product.
8. *Ethical responsibility* measures the societal commitment of the OEM by considering the take-back options at the end of life of the printer, where the input will be the economic value of the collection. The options are: (1) free collection, where both the OEM and the consumer are not charged; (2) the OEM pays the consumer for returns, such as discounts for a new product upon returning the old; and (3) the consumer pays to have the product collected or recycled. The product's price compared to the current market price is also included in the assessment.
9. *Packaging* takes into account the packaging material's recyclability, and also the quantity of packaging material that is used. Take-back option of the packaging material with the printer purchase is also included.
10. *Upgradeability* measures the options for upgrading the printer, although this is not usually practiced by the consumer. This aspect measures the options such as the availability of USB ports, connectors, extra memory slots, and so on. The difficulty of installing an upgrade is also measured.

Table 10 lists and compares quantified data on the influencing factors for the subelements of Products 1 and 2. Each influencing factor was quantified using a specific method and later represented in the form of a subelement sustainability index (energy efficiency index, material utilization index, life-cycle factor index, etc.). After all the required data for the influencing factors of each subelement were quantified and indexed, they were each given appropriate weighting based on input from the OEM. The weighting factors for each subelement were given based on their level of importance from the sustainability standpoint and also so that the amount of each category would add up to 100 percent. The given weighting factors are listed in Table 11. The products of these weighting factors

Table 10 Comparison of Compiled Data on Influencing Factors for Products 1 and 2

Comparison of Compiled Data for Laser Printer Type		
Input Parameters	Product 1	Product 2
Energy for use (Wh)	70	75
Quantity of material (kg):		
High Impact Polysterine	2.00	5.00
Polypropylene	1.00	0.20
Polyethylene LDPE	2.00	1.00
HDPE	2.00	5.00
ABS	0.10	0.10
Noryl	0.00	0.20
Sheet Metal	1.00	1.00
Zinc plated sheet steel	0.20	0.20
Steel 304	1.40	0.50
Steel 314	2.00	1.00
Aluminum 6060	0.40	0.30
Potential number of functional years (years)	4.00	3.00
Weight (kg)	6.00	5.00
Reduction in emission since 1995 (CO2) (%)	56.00	45.00
Time taken to separate/dismantle (hrs)	4.00	5.00
Percentage of separation before shredding (%)	20.00	33.00
Existence of maintenance schedule (Yes/No)	1.00	1.00
Take back options (who pays for it)	1.00	1.00
Estimated printer price ($)	189.00	200.00
"Market price of similar printer ($)	166.00	160.00
Time printer will be in market	3.00	3.00
Percentage of recycled material in packaging (%)	56.00	66.00
Number of USBs, connectors, memory slots, etc.	4.00	2.00
Ease of instillation (0 = difficult, 1 = easy)	5.00	5.00
Amount of components containing:		
Lead	1	1
Mercury	1	1
Cadmium	1	1
Hexavalent chromium	1	1
Polybrominated biphenyls (PBB)	1	1
Polybrominated diphenyl (PBDE)	1	1

Note: From Ref. 48.

with the respective subelement sustainability index were then summed to calculate the category index. Figure 7 shows an approximate flowchart describing this procedure.

The formulas for calculating the category indices follow:

High importance category index = (Energy efficiency index × 29) + (Material utilization index × 20) + (Life-cycle factor index × 13) + (Environmental effects index × 19) + (Recyclability index × 19)

Table 11 The Weighting Factors of the Sustainability Subelements

High Importance Sub elements	Weighting (%)
Energy efficiency	29
Material utilization	20
Life cycle factor	13
Environmental effects	19
Recyclability	19
Medium Importance Sub-elements	Weighting (%)
Reliability	27
Service life/Durability	22
Ethical responsibility	16
Packaging	21
Upgradeability	14

Note: From Ref. 48.

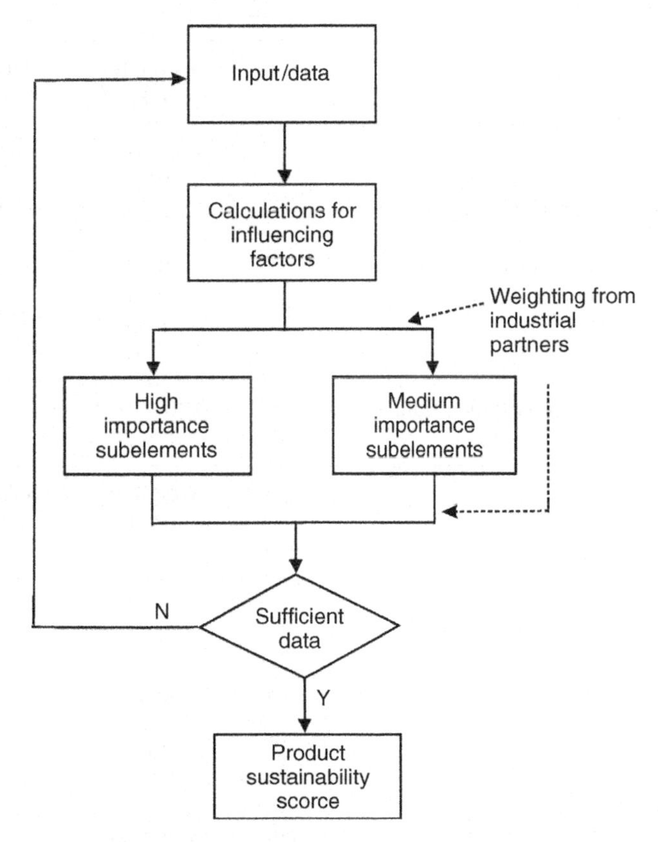

Figure 7 Flowchart showing the method used in sustainability assessment of a consumer electronics product. (From Ref. 48.)

Medium importance category index = (Reliability index × 27) + (Service life index × 22) + (Ethical responsibility index × 16) + (Packaging index × 21) + (Upgradeability index × 14)

The high importance subelement category was then assumed to contribute to 70 percent of the product's sustainability scoring and the medium importance category to 30 percent. The following formula was then used to compute the product's overall sustainability score:

Total product score = (High importance category index × 70%) + (Medium importance category index × 30%)

Data for this procedure for both Products 1 and 2 are listed in Tables 12 and 13.

Table 12 Calculations for the PSI of Product 1

Calculations for Product 1

High Importance Subelements	Weighting (%) (W)	Subelement Index (I)	Subelement Index Contribution (W × I)
Energy efficiency	0.29	0.75	0.22
Material utilization	0.20	0.12	0.02
Life-cycle factor	0.13	0.35	0.05
Environmental effects	0.19	1.56	0.30
Recyclability	0.19	0.35	0.07
Total	1.00	3.13	0.65
High Importance Category Index	0.70 × 0.65	= 0.46	

Medium Importance Subelements	Weighting (%) (W)	Subelement Index (I)	Subelement Index Contribution (W × I)
Reliability	0.27	0.56	0.15
Service life/durability	0.22	1.00	0.22
Ethical responsibility	0.16	0.94	0.15
Packaging	0.21	0.56	0.12
Upgradeability	0.14	0.20	0.03
Total	1.00	3.26	0.67
Medium Importance Category Index	0.30 × 0.67	= 0.20	
Product 1 Sustainability Index	0.46 + 0.20	= 0.66	

Note: From Ref. 48.

Table 13 Calculations for the PSI of Product 2

Calculations for Product 2

High Importance Subelements	Weighting (%) (W)	Subelement Index (I)	Subelement Index Contribution (W × I)
Energy efficiency	0.29	0.63	0.18
Material utilization	0.20	0.10	0.02
Life-cycle factor	0.13	0.26	0.03
Environmental effects	0.19	1.45	0.28
Recyclability	0.19	0.35	0.07
Total	1.00	2.79	0.58
High Importance Category Index	0.70 × 0.58	= 0.40	
Medium Importance Subelements	**Weighting (%) (W)**	**Subelement Index (I)**	**Subelement Index Contribution (W × I)**
Reliability	0.27	0.56	0.15
Service life/durability	0.22	1.00	0.22
Ethical responsibility	0.16	0.65	0.10
Packaging	0.21	0.66	0.14
Upgradeability	0.14	0.10	0.01
Total	1.00	2.97	0.63
Medium Importance Category Index	0.30 × 0.63	= 0.19	
Product 2 Sustainability Index	0.40 + 0.19	= 0.59	

Note: From Ref. 48.

Comparison between Tables 12 and 13 shows that the total PSI score of Product 1 (0.66) is slightly higher compared to Product 2 (0.59). Hence, the procedure detailed in this case study exemplifies how a product sustainability scoring method can be used to ascertain the sustainability level of a product. In particular, this method can be used to compare the built-in sustainability levels of two different product models, and to improve the sustainability of the next product designed and developed. This method also provides an opportunity for comparing competing models from different manufacturers. Although this example is specific for a certain consumer electronic product, the method could be extended to accommodate for a variety of different cases and product types.

The sustainability level of a new product can be evaluated using this method based on the integral elements and its overall impacting sustainability contents throughout its life cycle. This procedure can also be applied by design engineers as an aid in making design decisions, and also in making comparisons between similar products in terms of sustainability and environmental worthiness.

6 SUMMARY AND OUTLOOK

Current trends in product design and manufacture indicate that there is a growing awareness in designing, developing, manufacturing, and acquiring sustainable products to mitigate the increasing problems associated with the traditional and nonsustainable product life-cycle approaches. The growing consumer expectations for manufactured products and manufacturing processes include the use of nonhazardous raw materials, elimination of harmful pollutants and toxic emissions, minimized use of natural resources, reduced waste generation, increased safety requirements and personnel health conditions, and so on. Perpetual material flow in a product life cycle that mimics the self-replenishing laws of nature is an ideal target in efforts toward developing sustainable products. It has also been discussed that attainment of this ideal would benefit the economy, the environment, and society.

The need for studying the total life-cycle stages in product design requiring the economic, environmental, and societal considerations is the central part of this chapter. This comprehensive analysis, in conjunction with the recently identified 6R concept, has been shown to contribute to a naturally evolving consumer expectation for increased sustainability contents in manufactured products. Legislative, economic, and societal drivers are emerging worldwide for enforcing product sustainability requirements at various scales and levels, despite the slow acceptance by the industry.

A major starting point toward reaching the totally sustainable product is the development of a scientific method for quantifying the inherent and built-in sustainability levels of a product. This is important since the development of a systematic quantification method will also aid in gaining a better understanding of what needs to be done and achieved to meet the societal expectations, economic growth, and environmental protection without compromising the quality of life. Some initial efforts have already been set in motion toward developing a generic product sustainability evaluation method through a viable and meaningful methodology for determining the level of product sustainability, with case studies drawn from the automotive and consumer electronics industry to demonstrate the feasibility of the new sustainability rating methods that are presented in this chapter. However, a more concerted, collective effort would be required to fully develop and implement the proposed product sustainability rating method.

REFERENCES

1. "Our Common Future: From One Earth to One World, 1987." World Commission on Environment and Development, Brundtland Commission Report, Oxford University Press, pp. 22–23 IV.
2. The Institute for Market Transformation to Sustainability (MTS), Sustainable Products Corporation, Washington, D.C., Accessed October 2006 at http://MTS.sustainableproducts.com.
3. E. Datschefski, "Sustainable Products: Using Nature's Cyclic/Solar/Safe Protocol for Design, Manufacturing and Procurement." *BioThinking International, E-Monograph* (2002).
4. E. Datschefski, "Cyclic, Solar, Safe—BioDesign's Solution Requirements for Sustainability" *J. Sustainable Product Design*, 42–51 (1999).
5. T. E. Graedel, and B. R. Allenby. *Design for the Environment* (Englewood Cliffs, NJ: Prentice Hall, 1998).
6. Waste Electrical and Electronic Equipment Directive 2002/96/EC of the European Parliament and of the Council, Official Journal of the EU, L37, 13.2.2003, 24–38, January 2003.
7. Restriction of Hazardous Substances Directive 2002/95/EC of the European Parliament and of the Council, Official Journal of the EU, L37, 13.2.2003, 19–23, January 2003.
8. End-of-Life Vehicle Directive 2000/53/EC of the European Parliament and of the Council, Official Journal of the EU, L269, 21.10.2000, 32–42, September 2000.
9. Energy-using Product Directive 2005/32/EC of the European Parliament and of the Council, Official Journal of the EU, L191, 22.7.2005, 29–58, July 2005.
10. Technology International Group, Technology International Inc. http://www.techintl.com/.
11. Waste Electrical & Electronic Equipment Recycling Directory. http://www.weeedirectory.com/.
12. R. Speigel, "Take-Back Laws Taken on by 13 Countries." *Electronic News*, October 2005.
13. C. Boks, et al. "An International Comparison of Product End-of-Life Scenarios and Legislation for Consumer Electronics." Proceedings of the 1998 IEEE Int. Symposium on Electronics and the Environment, 19–24, 1998.
14. R. Lucas and D. Schwartze. "End-of-life Vehicle Regulation in Germany and Europe—Problems and Perspectives." Discussion Paper of the Project Autoteile per Mausklick, Wuppertal Institute for Climate, Environment and Energy, March 2001.
15. "End-of-life Vehicles." Europa: Gateway to the European Union. Accessed October 2006 at http://ec.europa.eu/environment/waste/elv_index.htm.
16. "Management of End-of-life Vehicles." Europa: Gateway to the European Union. Accessed October 2006 at http://europa.eu/scadplus/leg/en/lvb/l21225.htm/.
17. J. Staudlinger and G. Keoleian, *Management of End-of Life Vehicles (ELVs) in the United States* (University of Michigan: Center for Sustainable Systems, 2001).
18. "Sustainable Mobility: Overview." World Business Council for Sustainable Development (WBCSD), http://www.wbcsd.org/.
19. L. Giorgi, "Sustainable Mobility: Challenges, Opportunities and Conflicts—a Social Perspective." *International Social Science Journal*, 179–183, 2003.

20. "Mobility 2001: World Mobility at the End of the Twentieth Century and its Mobility." World Business Council for Sustainable Development (WBCSD), 2001.
21. M. Papadoyannakis, "The EuP Directive: A Framework for Continuous Improvement of the Environmental Performance of Energy-Using Products," Proceedings of 13th CIRP International Conference on Life-cycle Engineering, 1–4, 2006.
22. D. Allen, D. Bauer, B. Bras, T. Gutowski, C. Murphy, T. Piwonka, P. Sheng, J. Sutherland, J. D. Thurston, and E. Wolff. "Environmentally Benign Manufacturing: Trends in Europe, Japan, and the USA." *Journal of Manufacturing Science and Engineering*, 908–920, 2002.
23. Dow Jones Sustainability Indexes, http://www.sustainability-index.com/.
24. Investor Responsibility Research Center (IRRC), http://www.irrc.org/.
25. McDonough, W., and M. Braungart, *Cradle to Cradle: Remaking the Way We Make Things* North Point Press, New York, 2002.
26. I. S. Jawahir, K. E. Rouch, O. W. Dillon Jr., L. Holloway, A. Hall, and J. Knuf. "Design for Sustainability (DfS): New Challenges in Developing and Implementing a Curriculum for Next Generation Design and Manufacturing Engineers," Proceedings of CIMEC (CIRP) Int. Conf. on Manufacturing Education, 59–71, 2005.
27. "Reduce, Reuse, and Recycle Concept (the "3Rs") and Life-cycle Economy." UNEP/GC. 23/INF/11. Twenty-third Session of the Governing Council/Global Ministerial Environment Forum, Governing Council of the United Nations Environment Programme, 2005.
28. The University of Bolton, Online Postgraduate Courses for the Electronics Industry, Life-cycle Thinking, http://www.ami.ac.uk/.
29. J. Liew, O. W. Dillon Jr., K. E. Rouch, S. Das, and I. S. Jawahir. "Innovative Product Design Concepts and a New Methodology for Sustainability Enhancement in Aluminum Beverage Cans," Proceedings of the 4th International Conference on Design and Manufacture for Sustainable Development. New Castle Upon Tyne, United Kingdom, July 2005.
30. K. Joshi, A. Venkatachalam, I. H. Jaafar, and I. S. Jawahir. "A New Methodology for Transforming 3R Concept into 6R for Improved Sustainability: Analysis and Case Studies in Product Design and Manufacturing." Proceedings of the 4th Global Conf. on Sustainable Product Development and Life-cycle Engineering: Sustainable Manufacturing. Sao Paulo, Brazil October 2006.
31. N. Nasr, and M. Thurston, "Remanufacturing: A Key Enabler to Sustainable Product Systems," Proceedings of 13th CIRP International Conference on Life-cycle Engineering, 15–18, 2006.
32. J. Fiksel, J. McDaniel, and D. Spitzley. "Measuring Product Sustainability." *J. of Sustainable Product Design*, 7–16, 1998.
33. I. S. Jawahir, and P. C. Wanigarathne, "New Challenges in Developing Science-Based Sustainability Principles for Next Generation Product Design and Manufacture." Keynote Paper, Proceedings of the 8th International/Expert Conference Trends in the Development of Machinery and Associated Technology (TMT), 1–10, 2004.
34. I. S. Jawahir, N. De Silva, and J. Liew. "A New, Comprehensive Method for Measurement and Quantification of Product Sustainability." *Int. J. Product Development* (Special Issue on Sustainable Product and Services Design), InderScience Publishers, United Kingdom (Accepted for publication).

35. I. S. Jawahir, P. C. Wanigarathne, and X. Wang. "Chapter 12: Product Design and Manufacturing Processes for Sustainability." In Myer Kutz (ed.), *Mechanical Engineering Handbook*, 3rd ed., vol. 3, *Manufacturing and Management* pp. 414–443. John Wiley, (Hoboken, NJ, 2005).
36. W. Schmidt, and A. Taylor. "Ford of Europe's Product Sustainability Index." Proceedings of the 13th CIRP International Conference on Life-cycle Engineering, 5–10, 2006.
37. International Organization for Standardization (ISO 14040), *Life-cycle Assessment—Principles and Framework* (1997).
38. Society of Environment Toxicity and Chemistry (SETAC). "Environmental Life-cycle Costing" submitted to SETAC publications as the results of the Life-cycle Costing Working Group of SETAC Europe, 2006.
39. T. Engen, "Committed to the Sustainable Management of Water, One of Our Most Precious Resources." A Position Paper in Recognition of the International Year of Freshwater, Alcan Inc., 2003.
40. "Aluminum Recycling: The Road to High Quality Products, European Aluminum Association." Organization of European Aluminum Refiners and Remakers, 2004.
41. F. Sissine, "The Partnership for a New Generation of Vehicles (PNGV)." Congressional Research Service Report (CRS) for Congress, 96–191, Spring 1996.
42. A. K. Gupta, et al., "The Development of Microstructure in 6000 Series Aluminum Sheet for Automotive Outer Body Panel Applications." Fourth International Conference on Aluminum Alloys: Their Physical and Mechanical Properties, Vol. III, 177–186, September 1994.
43. General Aluminum Facts, The Aluminum Association, Inc., http://www.autoaluminum.org/.
44. R. Borns, and D. Whitacre. "Optimizing Designs of Aluminum Suspension Components Using an Integrated Approach." SAE Paper 05M-2, 2005.
45. "Recyclability and Scrap Value," The Aluminum Association, Inc., http://www.autoaluminum.org/.
46. "Aluminum Content for Light Non-Commercial Vehicles assembled in North America, Japan, and the European Union in 2006." The Aluminum Association, Inc. Accessed October 2006 at http://www.autoaluminum.org/downloads/duckerppt.pdf/.
47. C. A. Ungureanu, "Total Life Cycle Analysis Comparison between Aluminum and Steel in Passenger Cars," M.S. thesis, University of Kentucky (2006).
48. De Silva, N., I. S. Jawahir, O. W. Dillon Jr., and M. Russel. "A New Comprehensive Methodology for the Evaluation of Product Sustainability at the Design and Development Stage of Consumer Electronic Products." Proc. of 13th CIRP International Conference on Life-cycle Engineering, 335–340, 2006.

CHAPTER 3

LIFE-CYCLE DESIGN

Abigail Clarke and John K. Gershenson
Life-cycle Engineering Laboratory Department of Mechanical Engineering—Engineering Mechanics Michigan Technological University Houghton, Michigan

Design for the life-cycle (DfLC) is a dynamic and proactive means of improving the environment through product design. Historical events have inspired and influenced the way life-cycle design has been developed and adopted by industry and society. Market drivers and legislation have also motivated design for the life-cycle efforts. This chapter outlines the principles that characterize the goals of design for the life-cycle, as well as the methods that provide the means to achieve these goals and the tools that support life-cycle compatible design decisions. A few illustrative implementations of DfLC are shown to highlight the achievement of these goals. However, it is clear that the future holds new and important challenges to the current state of design for the life-cycle.

> *"Designers actually have more potential to slow environmental degradation than economists, politicians, businesses and even environmentalists"*[1]

1 HISTORICAL INFLUENCES ON THE DEVELOPMENT OF DESIGN FOR THE LIFE-CYCLE

"Green design has a long pedigree, and before the Industrial Revolution it was the norm for many cultures."[1] However, the Industrial Revolution forced a complete rethinking of the field. The rapid pace of industrialization took great toll on the environment.[1,2] As the Industrial Revolution reached full flame in Britain, the British Arts and Crafts movement led one of the first outcries against the new industrial society and its environmental damage. Many critics of industrialization and champions of nature have famously stated their concerns for the environment and society, including William Wordsworth, William Blake, George Perkins Marsh, Henry David Thoreau, John Muir, Aldo Leopold, Buckminster Fuller, Rachel Carson, Victor Papanek, and Fritz Schumacher.[1,3] Organizations have also sprung up to protect nature from industry, including the Sierra Club, Environmental Defense, World Wildlife Federation, Natural Resources Defense Council, and BUND. In addition, environmental science and environmental engineering are direct outgrowths of the Industrial Revolution for the study and remediation of the effects of emissions to the environment.

More recently, the 1970 oil crisis created a public outcry leading to environmental legislation that created design for the environment (DfE) and energy-efficiency appliance labeling.[4,5] Meanwhile, life-cycle engineering and concurrent engineering entered the American business consciousness.[6] The discovery of the ozone hole over Antarctica in 1985 spurred phase-outs and bans of particularly hazardous compounds.[7] Then, the 1986 nuclear reactor meltdown in Chernobyl spread environmental concerns across Europe.[8] At about the same time, companies began to put business and the environment together; 3M started its Pollution Prevention Pays program, where business cost savings were derived from environmental improvements. In 1987, the United Nations sponsored the World Commission on Environment and Development to publish *Our Common*

Future, which called for sustainable development.[3] Nils Peter Flint initiated the Global O2 Network in 1988, which spreads environmentally responsible design ideas to support the growing need.[8,9]

In the 1990s, landfill closings and public resistance to new landfill sites increased attention to solid waste disposal and increased consumer demand for producer responsibility.[1,5] The 1992 United Nations Rio de Janeiro Conference tackled environmental problems and further inspired people to improve industrial impact.[3] On the design side, design for the life-cycle (DfLC) concepts were further developed by the American Electronics Association, yielding a DfE manual for electronics.[10] The field of industrial ecology surged in popularity in the 1990s,[11] and academia began to embrace the concept of design for X (DfX), including design for the environment (DfE). Development of DfE tools surged, in part through collaborations between diverse stakeholders such as Philips Electronics, the Dutch government, and the Technical University of Delft, which simplified life-cycle assessment through IdeMat software, leading to EcoScan, Eco-It, and SimaPro.[1] Events such as the 2002 World Summit on Sustainable Development in Johannesburg have kept the spotlight on environmental concerns,[1] and public concern about global climate change continues to increase.[12]

2 DESIGN FOR THE LIFE-CYCLE DEFINITIONS

The Industrial Revolution was devastating in its environmental impact. As greater environmental concerns arose, designers, writers, scientists, governments, and eventually industrialists initiated efforts to address these problems. These efforts led to the creation, adoption, and practice of life-cycle engineering and design for environment (DfE) programs.

Design for the life-cycle (DfLC) considers the entire system within which the product is created, used, and discarded when making design decisions. "Life-cycle design is a proactive approach for integrating pollution prevention and resource conservation strategies into the development of more ecologically and economically sustainable product systems."[13] Typically, the product life-cycle is considered from inception to retirement, or *cradle to grave*,[13] including product and packaging design, materials extraction and processing, product manufacture and assembly, product distribution, product use and service, and product End-of-Life (see Figure 1).[6] This product-design approach aims to reduce environmental impact and risk through considering the effects of each and every stage of a product's life-cycle[6] and to reduce environmental impact over the entirety of a product's life-cycle, effectively striking a balance between improving individual life-cycle stages.[14] In the last decade, ecological damage, resource flows, liability concerns, and subsequent costs have become increasingly important considerations in each product life-cycle stage.[15] With the significant increase in all environmental concerns, the term *DfLC* has expanded to *DfE* across the product life-cycle. *Life-cycle design, DfE, ecodesign, environmentally conscious design,*

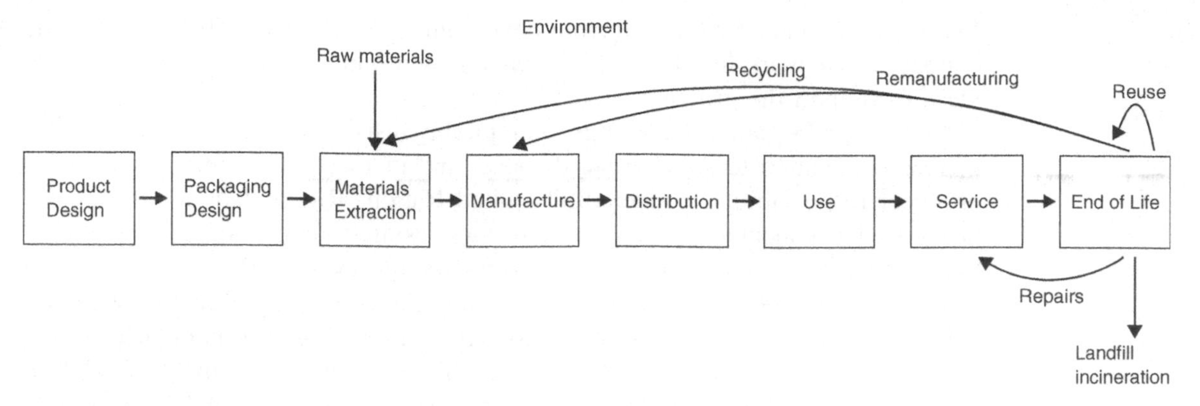

Figure 1 Product life-cycle. (Adapted from Ref. 6.)

and *green design* all consider the life-cycle of a product and its environmental effects holistically. Hence, these terms are for the most part interchangeable.[6]

2.1 Design for Environment

Instead of considering only the product and its environmental effects, DfE looks at the entire system within which the product is made, operates, and is discarded.[16,17] A product is evaluated over its entire life-cycle and at each individual life-cycle phase to decrease environmental harm.[18] DfE is one of several life-cycle *DfX* initiatives that include design for assembly, manufacturability, reliability, serviceability, and cost.[18] DfE stems from concurrent engineering, which seeks to improve design characteristics over and above functionality, including environmental performance, through multiattribute design decision making.[2,4,13,19] DfE entails optimizing a design for its environmental improvement (minimizing environmental impact), instead of just fulfilling regulatory product environmental performance requirements.[4] The common aim of DfE, ecodesign, environmentally conscious design, and green design and engineering is to forecast and remedy environmental effects throughout the systematic, informed choices within a design process.

2.2 Ecodesign

Ecodesign is also part of the lexicon of DfLC. It is more commonly used throughout Europe than DfE, environmentally conscious design, or green design. Ecodesign is the synthesis of ecological concerns with traditional product concerns in the design of a product. The point of ecodesign is to satisfy both economic and ecological principles by considering the entirety of the product life-cycle when designing products. Ecodesign aims to create products with less strain on material and energy resources or physical land mass and to reduce emissions and solid waste production during their life-cycle.[20]

2.3 Environmentally Conscious Design

Environmentally conscious design is an approach that is typically integrated with manufacturing. "Environmentally conscious design and manufacturing is a view of manufacturing which includes the social and technological aspects of the design, synthesis, processing, and use of products."[21] Perhaps more than other terms, environmentally conscious design places more emphasis on manufacturing than the design efforts undertaken.

2.4 Green Engineering and Green Design

Whereas DfE, ecodesign, and environmentally conscious design tend to consider mainly consumer products and industrial processes, green engineering seeks to improve environmental performance of products, processes, and systems encompassing issues ranging in scale from chemistry to land use planning.[22] Green engineering incorporates a holistic systems view that utilizes environmental performance as design targets, so that throughout the life-cycle, environmental objectives are given equal footing to other DfX requirements.[19] Green engineering addresses environmental problems by changing the underlying composition of a product, process, or system or by changing the context in which the system operates. Green product design encompasses methods of realizing design changes or remediation strategies for environmental impact reduction and prevention.[4] The analysis of environmental performance and the creation of more environmentally benign materials, tools, and processes are all a part of the green design methods.[2]

2.5 Industrial Ecology and Cradle-to-Cradle Design

Industrial ecology entails modeling the industrial flows of energy, materials, and capital within ecosystems that can contain many connected processes.[6,23,24] DfE seeks to address the environmental impact of an entire product life-cycle, whereas industrial ecology seeks to address the environmental impact of companies producing many products.[6] This view of an ecosystem in terms of cradle-to-cradle design extends the scope of the previous strategies by also emphasizing that all process outputs are either usable inputs to other processes or inputs to the larger ecosystem, motivating the concept of *waste equals food*.[3] Cradle-to-cradle design aims to develop multiple regenerative product life-cycles where byproducts can be cleanly metabolized by natural systems or fed back into the original processes.[3] Materials unsafe for ecosystems must be eliminated or stored until clean decomposition techniques are created.[3,25,26]

Practitioners use DfLC, DfE, green design, and ecodesign to denote tools, methods, and strategies that minimize the ecological impacts of the product life-cycle.[27] Hence, throughout this chapter, DfLC, life-cycle design, and DfE are used interchangeably.

3 MOTIVATIONS FOR DESIGN FOR THE LIFE-CYCLE

The chief motivation for DfLC is to foresee environmental effects and mitigate them through product design. Many different players affect the environmental impact of products, motivated by a wide range of more specific concerns about emissions and wastes deposited in the air, on the land, and in the water. Resource depletion, rising human populations, climate change, and toxins are several of many important environmental problems that need addressing. Global, regional, and national governments have stated goals and enacted legislation that have had a profound impact on the implementation of DfLC. Corporations have also felt the push from competition, leadership, employees, cost pressures, liability, innovation needs, product performance, and production improvements to start implementing DfLC practices.[4] External pressures applied by consumers, the public, industry and trade organizations, governments, and nongovernmental organizations have resulted in increased regulation and competition, more complex and controlled supply-chain dynamics, and a significant need for improving corporate images with respect to environmental impact. Many standards, certifications, and eco-labels have been put in place to encourage product design improvements and affect consumer buying habits. All of these motivators necessitate the adoption of DfLC approaches.

3.1 Who Is Responsible for Environmental Impact?

Humans are the major cause of environmental damage problems. Among the many issues, because of the ways in which industries produce the goods that society wants, substances are released that alter air, land, and water in ways that can harm people. The overuse of resources and the subsequent releases to ecosystems are what cause environmental damage.[6] People's actions at each stage of a product's life-cycle, from the designers to the trash collectors, add to the total environmental impact of a product. Population increases have, in part, warranted the use of more materials and greater production levels.[23] As the human population continues to grow, many people desire an improved material quality of life. That desire presents many challenges to environmental health.[28] Likewise, poverty is a significant problem globally that leads to ecological damage.[29] However, it is those that enjoy the most benefits from industrial and consumer society who have the greatest ability to remedy the environmental destruction in the past and prevent harm in the present and future.

3.2 Effects to the Environment

Graedel stated the following four widely held goals as preliminary aims that justify the creation of environmentally responsible practices: (1) prevent the extinction of people; (2) preserve the ability to increase quality of life for people; (3) sustain biodiversity; and (4) protect nature's beauty.[30] Graedel also identified several practices that promote these goals: employing renewable resources in ways that allow enough regeneration time; utilizing resources with finite availability in ways that allow enough time to discover renewable alternatives; causing no consequential impacts to global biodiversity; and emitting harmful substances at the rates that global ecosystems can metabolize the compounds.

Table 1 illustrates how different parts of the product life-cycle can lead to different environmental impacts. Certain product life-cycle activities lead to environmental stresses that threaten the lives of organisms. The goal for environmental improvement is to allow industrial activity to benefit society, while similarly allowing the environment to prosper. Since the environment provides the context within which industry and human society flourish, activities that balance environmental and human concerns are the ultimate aim of DfLC improvements.[2]

Environmental impacts can be characterized as primarily acting on a local, regional, or global scale. Several representative environmental impacts to land, air, water, and organism health are collated in Table 2. All effects have ramifications for humans. However, the seriousness differs by time scale and location, as well as ability of particular individuals to handle burdening.[1,14,23,31–35]

3.3 Regulatory Motivators

Regulation has long been a motivation for behavioral changes with respect to the environment. Rather recently, legislative efforts have moved from polluter pays, also called *end-of-pipe efforts,* to producer responsibility legislation. *End-of-pipe legislation* pertains to environmental effects after processes are in place; hence, only small environmental improvements can be achieved.[14] *Producer responsibility legislation* can effectively motivate companies to work towards environmental improvement from the start of their design process.[36] Producer responsibility legislation necessitates setting goals for products during the beginning of the product development process, where environmental problems can be designed out.[14] Regulated product take-back or producer responsibility legislation is one such compelling force to adopt DfLC approaches.[31,36] Legislative efforts at the state, national, regional, and global levels all influence the adoption of DfLC. Table 3 lists several legislative initiatives that have had important ramifications on product design and the product development process.[5,6,7,37–42,84] For example, the German Packaging Ordinance of 1991 addressed lack of landfill space by putting extended producer responsibility into law through obligating producers to consider product end-of-life options involving reuse during the product-development process.[5,6] Importantly, the efforts of one country or region influence other nations to adopt environmental regulations. The Waste Electrical and Electronic Equipment (WEEE) Directive makes electronic product manufacturers and retailers (including foreign producers and Internet retailers) financially responsible for electronic waste, motivating companies outside of the European Union to make product design changes and other countries to shun polluting

Table 1 Ecological Issues in Each Life-cycle Phase (gray boxes are stresses that are always present; white boxes denote issues that are sometimes present)

	Materials Choice	Energy Use	Solid Releases	Liquid Releases	Releases
Materials Extraction		Fossil fuel use Global climate change			
Manufacture		Fossil fuel use Global climate change		Reductions in biodiversity Human health Water availability/quality	Human health Ozone depletion
Distribution		Fossil fuel use Global climate change			
Use		Fossil fuel use Global climate change		Reductions in biodiversity Human health	Ozone depletion
End-of-Life	Fossil fuel use Global climate change	Fossil fuel use Global climate change	Fossil fuel use Global climate change		

Note: Adapted from Ref. 30.

Table 2 Representative Environmental Problems

		Scale of Effects		
		Local	Regional	Global
Effect Location	Land	Sludge Siltation	Landfills	Resource depletion Soil dilapidation
	Air(Climate)	Noise Light Smell Heat Photochemical smog	Visibility concerns	Ozone layer destruction Climate change
	Water	Eutrophication	Acidification Groundwater contaminants	Clean water scarcity
	Organisms	Indoor air quality	Toxins	Endocrine disrupters Habitat loss and biodiversity

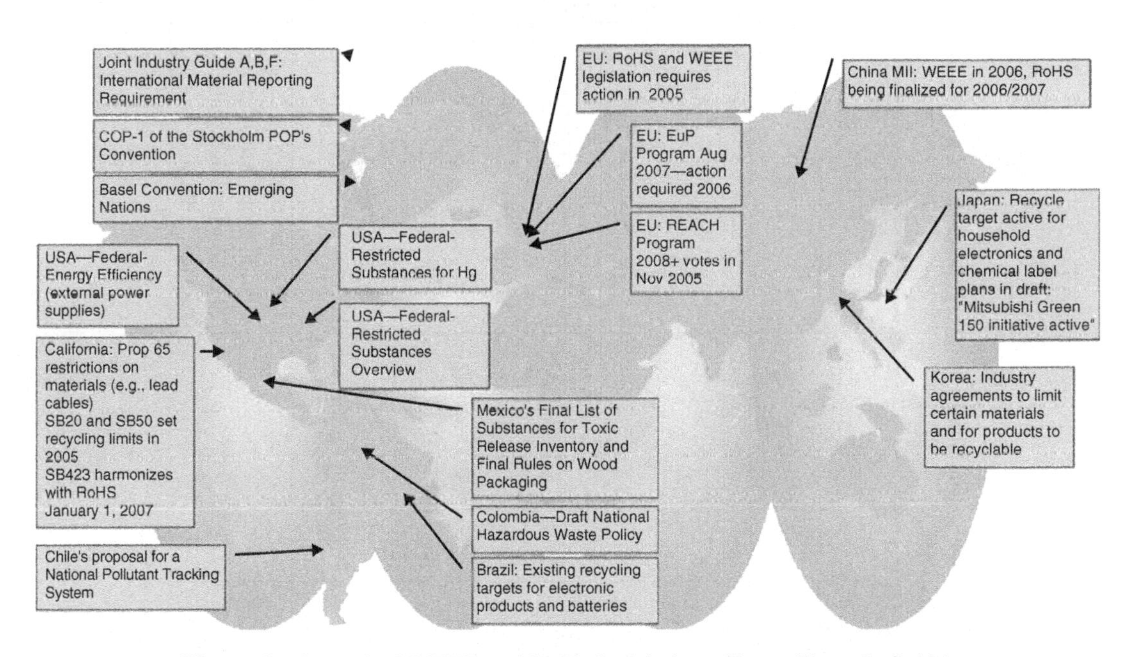

Figure 2 Spread of WEEE and RoHs legislative efforts. (From Ref. 43.)

practices by improving environmental legislation (see Figure 2).[38,39,43] The goal of legislation is not to have companies meet the minimum criteria, but to spur industry to incorporate DfLC initiatives that involve the entire supply chain.[24] For environmental impacts to truly be reduced, designers, governments, producers, product users, and remanufacturers must work in concert.[14]

Table 3 Representative Legislative Initiatives with Ramifications for Product Design (end-of-pipe legislation is in gray, producer responsibility legislation is in white)

	Initiative Level				
	U.S. State	National	Regional	Global	Impacts of Legislation
Examples	1972 Oregon Beverage Container Act				Makes consumers pay deposits on bottles used for food packaging
	1987 Oregon Waste Tire Law				Makes consumers pay taxes on tires to pay for recycling and landfilling
				1987 Montreal Protocol (ratified 1989)	Phases out ozone-depleting compounds
		1990 U.S. Clean Air Act			Reduces emissions and creates phase-outs for particular chemicals
		1990 U.S. Corporate Average Fuel Economy			Reduces fuel use and emissions for vehicles
		1991 German Packaging Ordinance			Makes distributors take back packaging or else pay another organization to do so
			1994 European Commission Directive 94/62 on Packaging and Packaging Waste		Sets packaging take-back standards (recycling percentages including energy recovery) for every member country to meet

Table 3 *(continued)*

	Initiative Level				
	U.S. State	National	Regional	Global	Impacts of Legislation
Examples				1997 Kyoto Protocol (ratified 2005)	Reduces climate-changing gaseous emissions
			2003 European Commission Restriction of Hazardous Substances (RoHS) Directive		Limits and phases out the use of particular hazardous compounds
			2003 European Commission Waste Electrical and Electronic Equipment (WEEE) Directive		Makes electronic product manufacturers and retailers financially responsible for the collection, recycling, and disposal of electronic waste
			2005 European Commission Directive 2005/32/EC on the Ecodesign of Energy-using Products (EuP)		Mandates the environmentally responsible design of products that consume electricity and providing information to users about environmentally responsible use
	2006 Maine enacted electronic product take-back legislation				Makes producers financially responsible for the recycling and disposal of computer monitors and televisions

Table 4 Corporate Motivators

Motivators		Impacts	Results
External	Reduce risk and liability	Eliminate dangers to workers and the environment to avoid lawsuits and regulations	Avoid lawsuits and regulations
	Better public relations	Improve public perception of the company	Maintain market share
	Meet consumer demands	Respond to consumer demand for environmentally preferable products	Retain and attract customers
	Beat competition	Create environmentally preferable products and gain more market share	Maintain market share
	Match supply chain demands	Implement environmental improvements dependent on all members of the supply chain improving environmental performance	Maintain market share
	Attain standards	Set targets for products to reach to attain the credibility of a particular environmental designation	Attract customers
Internal	Enhance product performance	Improve products to increase customer satisfaction	Retain and attract customers
	Reduce costs	Reduce environmental violations and use resources more efficiently	Improve business performance
	Reinvigorate employee commitment	Form a rallying cause that expresses employee values	Improve business performance

3.4 What Motivates Business

Corporations make possible the creation and distribution of products and work hard to compel people to buy these products. These aims directly lead to environmental destruction through resource use and emissions.[30] Corporate leadership that acknowledges and accepts its role in creating ecological damage can steward DfLC aims.[31] Corporations stand to gain tremendously by improvements in business and product performance that come from implementing DfLC.[44] As shown in Table 4, a variety of external and internal motivations propel companies to adopt DfLC.[2,4,6,14,15,21,31,36,44–47] DfLC product standards and certifications are one example set of these motivators (Figure 3).[23]

4 PRINCIPLES OF DESIGN FOR THE LIFE-CYCLE

The purpose of DfLC is to create products that positively affect the environment,[48] thus decreasing ecological damage. This is a broader aim than the elimination, reduction, and prevention of waste.[6,17,18] Each product life-cycle stage

Figure 3 Eco-labels from around the world. (From Ref. 23.)

has its own guidelines or principles that come together to achieve this goal. Trade-offs among these stages, as well as between DfLC principles and other design objectives, must be balanced so that products are optimized for environmental performance over the entire life-cycle. DfLC principles can act in concert to better environmental performance; addressing one aspect of product efficiency can lead to efficiency improvements in other aspects.[9]

Corporations and designers are responsible for the effects of products while in end users' hands and afterward.[6] "Designers actually have more potential to slow environmental degradation than economists, politicians, businesses and even environmentalists."[1] Clearly, designers are responsible for applying DfLC principles, but changes in design alone have limited effects on reducing environmental impact of products and services.[27] Corporations need "to take responsibility both for the environmental consequences of [their] production and for the ultimate disposal of [their] products."[49] The stakeholders who have the greatest ability to enact changes (i.e., manufacturers and designers) have the greatest

Table 5 Product Design Principles

	Explanation	Who Applies Principle	Example
Measure Environmental Performance	Assess resource use and risks	Design team	Use life-cycle assessment to identify and benchmark environmental impacts
Consider All Costs	Determine all product life-cycle costs	Design team	Employ life-cycle costing to capture all costs incurred by a product
Minimize and Eliminate	Choose designs that facilitates recycling	Design team	Upgrade the technology in a product, improve product durability, or employ aesthetics that people will enjoy long-term

responsibility to design out environmental damage and accept liability for the consequences.[5]

4.1 Product Design Principles

The product-design process cements many details of a product and hence determines many of the possibilities for how other life-cycle design principles can be applied (see Table 5).[1,2,4,9,17,23,24,26,31,36,49,50,51] During the product design stage, designers must measure environmental performance iteratively and make design decisions accordingly. The costs of a product throughout its life-cycle must be predicted, including environment-related expenditures. Lastly, by increasing the useful life of a product with appropriate technical and aesthetic life spans in mind, designers can attain many environmental benefits.[51]

4.2 Packaging Design Principles

The design of the product includes the design of packaging. Following environmentally responsible design principles such as those in Table 6[23,31] is an important step toward achieving environmentally benign packaging. Besides the product covering and marketing materials, all transportation packaging must be considered as well. A good way to reduce packaging needs is for design and transportation engineers to communicate about product concerns and design packaging to fit both points of view.[23] Setting up a deposit or refund for packaging (e.g., bottle returns) or some type of return system between supplier, retailer, and user (e.g., pallet returns) encourages packaging reuse.[31]

Table 6 Packaging Design Principles

	Explanation	Who Applies Principle	Example
Minimize and Eliminate	Decrease packaging, decrease impact	Design team	Reduce the size and amount of material needed for packaging
Biodegrade	Create packaging that decomposes safely	Design team	Make packaging edible
Reuse	Design packaging for multiple uses	Design team	Create durable enough packaging for refilling
Recycle	Choose designs that facilitate recycling	Design team	Choose packaging materials that have established recycling markets
Use Industry Standards	Commonalize the packaging to make reuse or recycling more economically feasible	Design team	Select the industry preferred packaging

4.3 Material Design Considerations

Guidelines for choosing the most environmentally responsible materials, such as those in Table 7, depend on product and packaging structure and requirements.[1,2,4,9,14,17,23,24,26,31,36,50,51,52] DfLC materials considerations involve the types of materials chosen and how those materials should best be employed. Material choice can diminish or improve product performance and environmental impacts, both of which must be considered by designers.[9]

4.4 Product Manufacturing Design Principles

After setting goals for environmental improvement through material, product, and packaging design, designers must examine the production processes. Designers are expected to optimize products to eliminate inefficiencies and wastes during production, as well as reduce process energy inputs for manufacturing.[51] This design goal can result in environmentally responsible process selection and energy-efficient production methods.[18] Designers must work with process engineers using the general design principles outlined in Table 8 to create methods that make ecologically benign objects.[4,9,18,31,44,51] Cleaner production can be achieved through optimization of product design, materials processing, and manufacturing using techniques from design for assembly and lean manufacturing.[31]

Table 7 Material Design Considerations

		Explanation	Who Applies Principle	Example
Material Selection	Choose Abundant Renewable Resources	Avoid dependence on diminishing finite material and energy capital, instead use feedstocks that regenerate	Design team	Avoid dependence on diminishing finite material and energy resources like fossil fuels
	Choose Sustainably Harvested Materials	Ensure that renewable resources remain available and viable	Design team	Pick materials that meet sustainable certification requirements, like wood with the Forest Stewardship Council label
	Choose Recyclable Materials	Extend the life of materials through several cycles	Design team	Avoid using composites; instead choose materials with economically viable recycling markets
	Choose Recycled Materials	Ensure that recyclable materials have a market	Design team	Keep recycled material quality high for multiple uses, use recycled materials in their original colors and textures
	Avoid Hazardous Substances	Ensure that products are safe for human and environmental health	Design team	Choose materials that cause no health or legal concerns
	Reduce Material Process Energy	Account for material production effects in the environmental impact of a product	Material producers and design team	Consider the energy and impact differences for producing materials at a facility instead of outsourcing the finished substances before making manufacturing changes
Material Employment	Eliminate Material Waste	Decrease the amount of material that becomes waste during production	Design team	Design products to make manufacturing offcuts as small as possible
	Dematerialize	Remove material from a product	Design team	Reduce the weight and volume of materials in a product
	Simplify Products	Eliminate the material waste of overdesign	Design team	Eliminate features that are not essential or necessary for a product to function or combine features

Table 8 Product Manufacturing Design Principles

	Explanation	Who Applies Principle	Example
Choose Cleaner Production Processes	Select the production processes with least environmental impact	Manufacturers	Employ lean manufacturing techniques to remove inefficiencies and waste from production and choose suppliers that use the most benign methods
Improve Quality	Ensure that production techniques and methods produce quality products	Manufacturers	Improve production to have fewer rejects and therefore less waste
Choose Clean Power Sources	Utilize power sources that create the least pollution	Manufacturers	Use renewable energy like wind power to generate needed electricity

4.5 Product Distribution Design Principles

After production, the product is distributed. Again, there are design principles—principles for the design of the distribution system and principles for product design—that impact the distribution system (see Table 9)[1,23,31] and reduce a product's environmental impact. Management must consider the following factors to determine the best mode of transportation for each product: number of products, expense, time until a product is needed, length of travel and dependability, and ecological damage incurred. The National Research Council of Canada recommends having designers, shipper/receivers, and sales personnel compare the various modes of transportation with these factors to select the most appropriate method for transporting products.[51] Transportation routes for products can also provide opportunities for packaging and product take-back.[23]

4.6 Product Use Design Principles

The use phase of a product can also contribute significantly to its impact. Designers are responsible for improving the energy efficiency of products (Ashley, 1993; Mont, 2000; NRC, 2003). However, ensuring that products are safe for users and their environment is also important for meeting DfLC principles. The principles in Table 10[1,2,9,23,24,26,31,51] highlight some of the general dos and don'ts that a product designer can control with respect to the environmental impact of product use. Including the hardware for users to reuse consumables with a product,

Table 9 Product Distribution Design Principles

	Explanation	Who Applies Principle	Example
Choose Cleaner Transportation Methods	Select modes of transportation that create the least environmental impact	Design team and manage- ment	Distribute goods throughout a city by bicycle instead of truck, like Peace Coffee in Minneapolis, Minnesota
Reduce Transportation of Products	Decrease transportation of goods to reduce environmental impact	Design team and manage- ment	Optimize distribution routes to deliver larger quantities of goods together, such as several product types, or use local suppliers

Table 10 Product Use Design Principles

	Explanation	Who Applies Principle	Example
Reduce Product Energy Use	Improve product energy efficiency to reduce waste and emissions	Design team	Fix leaks or energy losses and inform consumers how to best use products
Reduce What a Product Consumes	Eliminate or decrease the inputs a product needs over its lifetime to decrease waste and material and energy use	Design team	Use environmentally benign or reusable consumables and inform users how to best utilize consumables
Keep Products Clean	Create products that do not emit pollutants	Design team	Substitute materials used in adhesives to stop product off-gassing

such as rechargeable batteries and battery chargers, can reduce what a product consumes.[31]

4.7 Product Service Design Principles

Serviceable products likely have longer lifetimes than nonserviceable products. Increasing the product lifetime reduces the burden on material and energy resources and, hence, the environment. For this reason, companies should follow

Table 11 Product Service Design Principles

	Explanation	Who Applies Principle	Example
Make Service Easy	Design for ease of servicing to improve chances that maintenance will be performed and product life will be extended	Design team	Provide easy access to parts and clearly label parts that need different maintenance
Use Benign Servicing Consumables	Ensure that all inputs used in servicing are safe for people and the environment	Product servicers	Use environmentally benign or reusable consumables and inform consumers how to best utilize consumables

the principles outlined in Table 11 to provide facilities for servicing products and to design products with service in mind.[23,31]

4.8 Product End-of-life Design Principles

At the end of a product's useful life, end-of-life systems for the collection of broken or unwanted products must be initiated or in place. Designing for product take-back and establishing a unique product take-back system increases the chances of a product being reused, remanufactured, or recycled.[26,31] Sometimes users discard the whole product when only one component fails, so designing all product components to fail at the same time can create less waste. Many end-of-life options exist for products; each has its own advantages and disadvantages. However, following the design principles in Table 12 will lead to reduced life-cycle environmental impact.[1,2,4,9,14,17,23,24,26,31,51]

4.9 Beyond the Principles

There are additional principles that lead to radically new ways in which DfLC can be realized. These principles fit in several broad categories, as shown in Table 13.[1–3,31,44,46,50–57] Using nature as inspiration for product design can lead to reduced environmental impact. Users can also enjoy the function of a product without possessing an object, reducing the necessary production volume while increasing utilization. Designing in multiple life-cycles or industrial ecosystems is another worthy goal. Designing for sustainability incorporates DfLC principles as well as social and economic concerns, a challenge that corporations and designers recognize as *designing for the triple bottom line*.[3,6,58,59] All of these

Table 12 Product End-of-life Design Principles

	Explanation	Who Applies Principle	Example
Design for Disassembly	Create a product to easily come apart into different materials and components to facilitate recycling, reuse, and remanufacturing	Design team and management	Design components in detachable modules that have similar characteristics, such as time to expected failure
Design for Recycling	Design products to facilitate the recycling of undesirables during the entire product life-cycle	Design team and management	Label different components and materials with different colors for easy separation
Design for Reuse	Enable the reuse of products, components, and packaging to reduce waste and resource consumption	Design team and management	Create products for easy cleaning, fixing, or adapting to new improvements, uses, or aesthetics
Design for Remanufacturing	Recover, test, and use unwanted components in the same or different products to reduce resource use and waste	Design team and management	Consider packaging, transportation, and component design that facilitates shipping, tooling requirements, and processing for remanufacturing
Design for Biodegradation	Compost as a viable option for disposing of waste in an environmentally benign manner	Design team and management	Select benign, biodegradable materials for products

product design principles stretch the generally held body of thinking behind DfLC.

5 LIFE-CYCLE DESIGN METHODS

Life-cycle design—that is, product life-cycle design—is at the very heart of the development of the product life-cycle. Many design methods are employed to incorporate DfLC concerns into product development. Including life-cycle design objectives in the design process increases the resources needed for product design and the number of stakeholders.[6] There are three facets to adding life-cycle considerations to the design requirements:

Table 13 Beyond Product Design for the Life-cycle Principles

		Explanation	Who Applies Principle	Example
Nature-Inspired Design Principles	Design Inspiration	Incorporating environmental concerns can lead to product innovation	Design team	Reconsider the underlying suppositions for material and energy use in products
	Sun Fuels All	A set amount of matter cycles in our world, but net gains in energy come from the sun	Design team	Use the benefits of sun power through photosynthesis and plant metabolization or wind turbines and photovoltaic cells
	Use and Render Wisely	Industry impacts the environment when compounds are altered in form and dispersed more quickly than regeneration occurs	Material producers, design team and manufacturers	Replace scarce materials with resources that abound and use renewable resources within ecosystem capabilities for rehabilitation
	Consider the Consequences	The creation of products has many ramifications for ecosystems	Marketers, design team, and management	Trace the beginnings and future of all resource uses and actions that go into a product
Functionality without Possession	Location Is Important	Acknowledge the particular character of each location that is affected by product life-cycle stages	Design team	Design to reflect the uniqueness of place while recognizing and preventing negative effects of design choices
	Products into Services	Customers want the function provided, not an object	Marketers, design team, management, servicers, and product reclaimers	Create services, product systems, and life-cycles of products that meet user expectations instead of designing products
	Immaterialize	Replacing the utilization of products with actions that do not involve resource use	Marketers, design team, and management	Create information and activities to occur electronically, eliminating infrastructure such as working or shopping over the Internet
	Design for Sharing	Sharing reduces consumption and environmental damage	Marketers, design team, and management	Design organizational systems for sharing technical support and encouraging groups that facilitate sharing like libraries

(*continued overleaf*)

Table 13 (*continued*)

		Explanation	Who Applies Principle	Example
Sustainable Design Principles	Use Local Resources	Using local resources supports the economies that lose when local ecosystems suffer	Design team and management	Substitute locally available materials or work with nearby suppliers
	Utilize Natural Advantages	Nature provides particular output energies that can be utilized easily	Design team, manufacturing, and distribution	Employ gravity-fed delivery or temporal temperature and climatic moisture differences for cooling
	Promote Wellness of All People	Create products in such a way that all people benefit	Marketing, design team, and management	Setup fair trade systems for products, replacing toxins with benign materials

1. Managing and measuring material and energy streams within production processes and throughout the product life-cycle
2. Integrating costs to the environment into the financial analysis
3. Considering the entire context in which a product design operates.[4]

It is generally accepted that around 80 percent of a product's life-cycle costs and environmental impacts are decided during product design.[9,19,20] When DfLC is implemented early in product development, as part of a concurrent engineering process, more substantial and viable design impacts and therefore increased environmental benefits are possible.[4]

Various design methods are best applied (or only applicable) during different stages of the product design process. Typically, the product design process is described by four stages, each with unique opportunities for environmental improvements. These stages are described in Table 14 as product definition, conceptual design, embodiment, and detail design. Many design methods work to improve the environmental impact of a product during a single or multiple life-cycle phases, over the entire product life-cycle, or with respect to concerns beyond a product's life-cycle.[6] The design methods shown in Table 15 are general, and hence can be used on a wide variety of products. Each method has different environmental impacts that are realized during life-cycle stages, and each method can be applied at different stages of the product life-cycle. Source reduction can be achieved by analyzing a product design and removing excess material unnecessary for strength requirements.

Table 14 Design and Environmental Considerations during the Product Design Process

	Product Design Application Stage	Product Life-cycle Stages Affected	Environmental Benefits
Customization	Product definition	End-of-Life	Extends product life, thereby reducing waste and conserving resources
Expert Systems	Product definition	Materials extraction, manufacturing, distribution, use, service, and End-of-Life	Sets particular environmental goals to apply throughout the product life-cycle
Sustainable Product and Service Development (SPSD) Method	Product definition	Materials extraction, manufacturing, distribution, use, service, and End-of-Life	Conserves energy and materials by extending product life, eliminates waste, reduces toxicity
Modular Design	Product definition or conceptual design	Manufacturing, service, and End-of-Life	Decreases manufacturing and servicing energy use, conserves resources by updating components of products, allows for the safe disposal of toxins
Material Substitution	Embodiment or detail design	Materials extraction, manufacturing, and distribution	Reduces toxicity and energy intensity and encourages reuse or recycling
Source Reduction	Embodiment or detail design	Manufacturing, distribution, and End-of-Life	Eliminates waste
Environmentally Conscious Product Design: A Collaborative Internet-based Modeling Approach	Embodiment or detail design	Materials extraction, manufacturing, distribution, use, service, and End-of-Life	Variety of environmental benefits

Note: From Refs. 13, 51, 112.

Table 15 Comparison of Representative Design for the Life-cycle Methods

		Product Definition	Conceptual Design	Embodiment Design	Detail Design
Design Process	Design Process Considerations	Define customer requirements and product functioning	Assess competing product concepts for customer satisfaction. Select single product for further development	Design basic product architecture	Fully specify all aspects of the product and components
	Environmental Considerations	Set environmental performance objectives. Challenge basic assumptions regarding product functioning and end user needs	Determine end-of-life options	Engineer how to reduce impact during use. Create efficient transport Enact cleaner production goals and strategies	Consider how to reduce the impact from materials

5.1 Methods Applied during Product Definition and Conceptual Design

Methods that can be applied early in the design process have a greater effect on a product's overall environmental impact.[4] Customization involves designing products to be personal so that people develop the care and interest in maintaining and prolonging the life of a product.[60] Designing for modularity yields a product with clusters of components or modules by similar physical characteristics, such as the same recycling or disposal treatment.[26,61–64] Using expert systems, a company solicits information from product development experts about environmentally responsible products to set goals or requirements and develop a methodology toward producing successful products.[65] Sustainable product and service development examines the functional expectations, life-cycle stages, and supply chain of a product to convert the product into a service.[66]

5.2 Methods Applied during Embodiment or Detail Design

Methods applied during embodiment and detail design can still abate a product's environmental impact. Material substitution entails the exchange of ecologically damaging materials with alternatives that reduce environmental impact. Source reduction involves reducing the amount of waste caused by a product before it leaves the factory.[26] The environmentally conscious product design, collaborative Internet-based modeling approach involves the exchange of design and environmental assessment information between designers and environmental knowledge experts, providing contemporaneous feedback to both parties.[67] All of these methods present different approaches to improving product environmental performance.

6 DESIGN FOR LIFE-CYCLE TOOLS

Design methods provide an overall approach to implementing DfLC in a design or throughout an organization. Design tools perform the specific tasks needed to implement a DfLC method. Tools can provide new information, organize existing information, or present information in a new light. Tools generally help provide logical, accurate, and repeatable predictions that can influence design decisions within multiple subject areas covered.[6,68] Tools perform trade-off analyses between design objectives and assess life-cycle impacts.[6] The tool must discern those design aspects in need of improvement and provide guidance for how to transcend the problems.[68] Tools should allow designers to gain a sense of what leads to better outcomes and how to redesign products in the future to meet more criteria and make better products.[6] The ability to describe and handle complex processes with many linked processes is essential for a good tool.[69] Tools work in different ways to aid designers in making better environmental design choices. Tools can aid designers' decision making by performing the following tasks: environmental assessment, priority setting, supporting idea generation, and integrating additional criteria besides environmental concerns. DfLC tools and indicators and environmental accounting tools all differ in the complexity and time required for use. Some tools help with a single task while others aid several tasks (Figure 4).[70]

Design tools enable designers to make the engineering decisions essential to implementing DfLC. Indicators are a specific group of tools that assess and represent predicted environmental damage of product life-cycle actions. They are typically part of more comprehensive life-cycle design tools. Environmental accounting tools assess the total financial obligations of the product life-cycle, including environment-related costs.

6.1 Design Tools

Design tools aid design engineers in many aspects of product development decision making. Tools are useful in different stages of the product development process, apply to different types of products, and consider distinct stages of the product life-cycle. Tools that are applied in the early stages of product design have greater ability to improve the environmental performance of a product (see

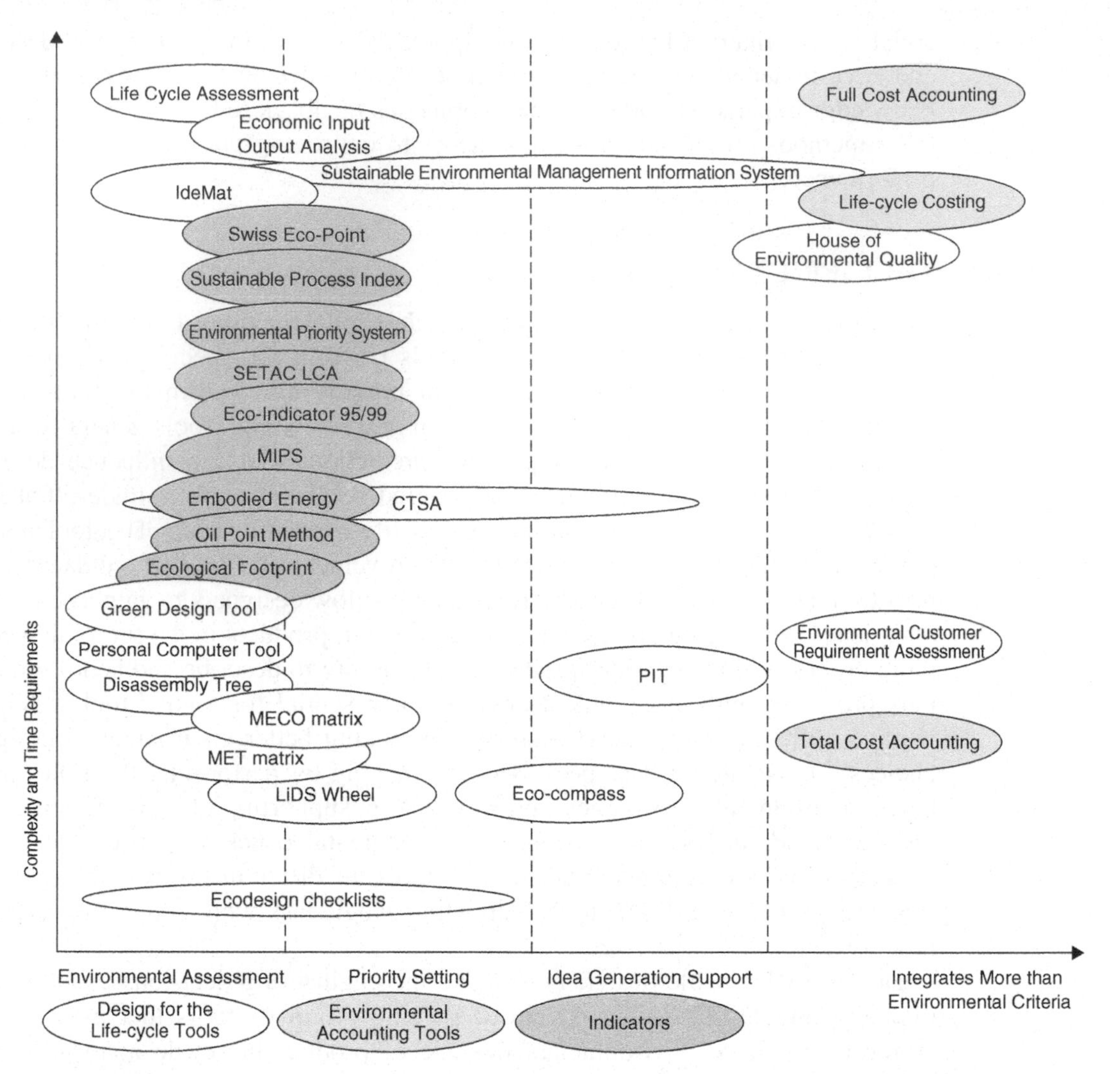

Figure 4 Specific purpose and level of complexity of DfLC tools. (Adapted from Ref. 70.)

Figure 5). Not all of the tools shown here are specific to DfLC. They are, however, tools that can greatly improve the efficiency and efficacy of the DfLC process.

6.2 Product Definition

At the product definition stage of product development, tools are needed to help gather, organize, and apply information leading to product design requirements for improved environmental impact. The tools in Table 16 aid in uncovering customer requirements, outlining the design process, and organizing innovation concepts.

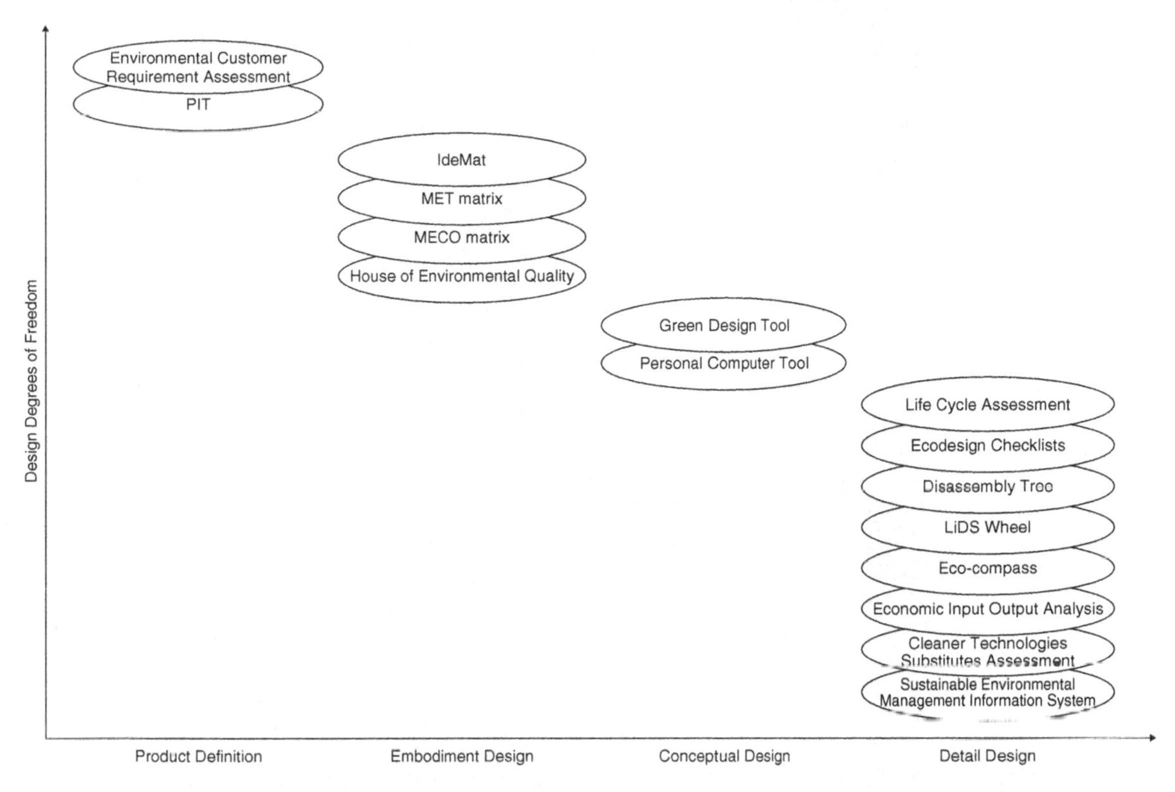

Figure 5 Application of different tools during the product design process stages. (Adapted from Ref. 14.)

Environmental customer requirements can be uncovered with *conjoint analysis*, where users rank products with different attributes, and *contingent valuation*, where willingness to pay for environmental attributes is assessed.[71] The *Kano technique* is a way to interpret the environmental "voice of the customer" into customer requirements through surveys of users. Similar to conjoint analysis and contingent valuation, the designer using the Kano technique must select weightings of each design feature.[71,72] The *Product Ideas Tree* (PIT) diagram (inspired by Mind Maps, the Life-cycle Design Strategy or LiDs Wheel, and the Eco-compass) provide a record and organization strategy for ideas to incorporate into a product concept generated during design brainstorming sessions where ideas are recorded by the most relevant design process stage and environmental impact category affected.[73]

Conceptual Design

Tools that can be applied during conceptual design often require more detailed information about material and energy flows with respect to a product than tools applied during product definition. In turn, these tools also result in greater specificity, as they help to narrow the design solution space. The tools in Table 17 help

Table 16 Product Definition: Design for the Life-cycle Tools

	Picture	Product Design Application Stage	Product Life-cycle Stage Affected	Impacts	Advantages or Disadvantages
Tools for Assessing Environmental Customer Requirements	Differences of Kano categories for a single product attribute *Current perception of an environmental attribute* *Types of attribute* / *When the attribute is present in the product* / *When the attribute is absent from the product* One-dimensional / Satisfied / Dissatisfied Must-be / No feeling / Dissatisfied Attractive / Satisfied, delighted / No feeling Indifferent / No feeling / No feeling Reverse / Dissatisfied when attribute is present or satisfied when attribute is absent	Product definition	Product design, packing design, materials extraction, manufacturing, distribution, use, service, end of life	Customer environmental concerns can be incorporated into product design	Difficult to interpret customer responses into requirements
Product Ideas Tree (PIT)	TOXICITY, RES-USE, MASS, ENERGY, IDEA BRANCH, IDEA CLUSTER, IDEA STARTING POINTS FOR ECO-INNOVATION THE DESIGN PROCESS TRIGGER PRODUCT PLANNING CONCEPT DESIGN EMBODIMENT DESIGN DETAIL DESIGN MANUFACTURE AND LAUNCH	Product definition	Product design, packing design, materials extraction, manufacturing, distribution, use, service, end of life	Innovated ideas can be recorded and placed by what environmental benefit is espoused and when in the design process the idea can be implemented	Complicated

Table 17 Conceptual Design: Design for the Life-cycle Tools

	Picture	Product Design Application Stage	Product Life-cycle Stage Affected	Impacts	Advantages or Disadvantages
IdeMat		Conceptual, embodiment, or detail design	Product design, packing design, materials extraction, manufacturing, distribution, use, service, end of life	Promotes the selection of more environmentally benign materials, processes and components	Only provide a database
MET-matrix	Materials Cycle Input/Output; Energy Use Input/Output; Toxic Emissions Output Production of materials and components Manufacturing Distribution Use: Operation, Service End-of-life: Recovery, Disposal	Conceptual, embodiment, or detail design	Product design, packing design, materials extraction, manufacturing, distribution, use, service, end of life	Displays and organizes resource flows throughout the life cycle	Simple
MECO-matrix		Conceptual, embodiment, or detail design	Product design, packing design, materials extraction, manufacturing, distribution, use, service, end of life	Records and designates relative impact of resource flows throughout the life cycle	Simple, but designers create relative impact weighting
House of Environmental Quality	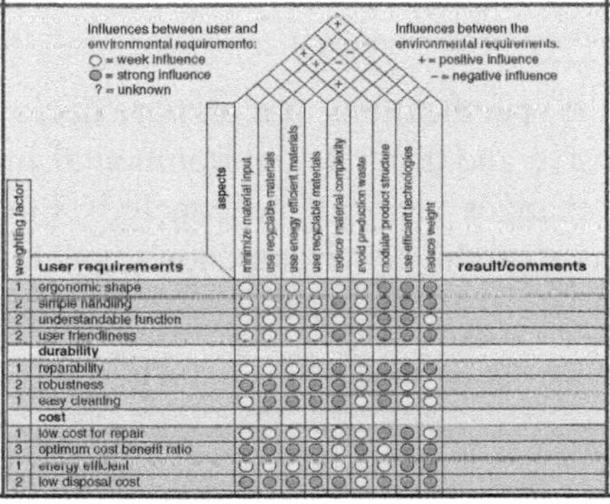 Influences between user and environmental requiremonts: ○ = week influence; ● = strong influence; ? = unknown Influences between the environmental requirements: + = positive influence; – = negative influence aspects: minimize material input; use recyclable materials; use energy efficient materials; use recyclable materials; reduce material complexity; avoid production waste; modular product structure; use efficient technologies; reduce weight weighting factor; user requirements; result/comments 1 ergonomic shape; 2 simple handling; 2 understandable function; 2 user friendliness durability: 1 reparability; 2 robustness; 1 easy cleaning cost: 1 low cost for repair; 3 optimum cost benefit ratio; 1 energy efficient; 2 low disposal cost	Conceptual, embodiment, or detail design	Product design, packing design, materials extraction, manufacturing, distribution, use, service, end of life	Trade-offs between environmental customer requirements and design parameter choices are uncovered	Time consuming

compare product concepts for environmental impact, display potential resource flows throughout the product life-cycle, and note potential conflicts between design parameters because more product design details are known.

IdeMat is a database (which can be used with life-cycle assessment software SimaPro's database) that aids selection of materials, processes, and components based on environmental indicators such as Eco-indicator 95 and the Environmental Priority System.[74,75] The *Material cycle, Energy use, and Toxic emissions (MET) matrix* can track and display inputs and outputs of materials, energy, and toxics that flow through the product life-cycle stages.[76,77] The *Materials-Energy-Chemicals-Other (MECO) matrix* creates an inventory of its namesake flows and ranks the impacts and feasibility of change of the flows throughout the product life-cycle.[9] The *House of Environmental Quality* (derived from the House of Quality)[78] illuminates potential conflicts between design and environmental criteria throughout the product life-cycle using weightings of the importance of particular life-cycle criteria.[79]

Embodiment Design

Tools applied during embodiment design (see Table 18) can yield a more accurate assessment because the basic layout, components, and materials have already been chosen. At this point, designers can address deeper concerns about manufacturing, distribution, use, and End-of-Life. Again, the fact that more product details are already set bars sweeping product design changes.

The *Green Design Tool* allows designers to see the effects of product design attribute choices on process waste creation by comparing environmental impact scores of designs based on end-of-life options, ease of disassembly, labeling, materials, and toxicity.[80] The *Personal Computer DfE* tool rates the ability of a design to be recycled by scoring products for environmental impact of materials, disassembly ease, ease of recycling, and hazardous material content.[68]

Detail Design

Detail design is where all of the specifications of a part are decided. In this design process stage, the most extensive and accurate environmental impact predictions are made and some design changes can still be enacted. Table 19 showcases design tools that can characterize product environmental impact, offer suggestions

Table 18 Embodiment Design: Design for the Life-cycle Tools

	Picture	Product Design Application Stage	Product Life-cycle Stage Affected	Impacts	Advantages or Disadvantages
Green Design Tool	Green Design → Product / Manufacturing Process → Greenness Attributes → Subassembly (Physical, Electrical, Packaging, User' Manual) → Process → Design Sound? No → Alternative Design; Yes → Stop	Embodiment or detail design	Material extraction, manufacturing, end-of-life	A score is given for both the environmental soundness of product and process attributes, which designers can use to choose between design alternatives	Some life-cycle stages are ignored
Personal Computer DfE Tool		Embodiment or detail design	End of life	Recycling is improved	Focus is on single product and life-cycle stage

Table 19 Detail Design: Design for the Life-cycle Tools

	Picture	Product Design Application Stage	Product Life-cycle Stage Affected	Impacts	Advantages or Disadvantages
Life-cycle Assessment		Detail design	Product design, packaging design, materials extraction, manufacturing, distribution, use, service, end of life	Uncovers the more environmentally responsible choice of two or more options.	Detail and time intensive; it and has relative results.
Ecodesign Checklists	**Reuse/Recycling (closing technical material and energy cycles)** • recycling strategy in place? • guarantee for take back in place? • re-use of the complete product (e.g. second-hand, recycling cascade) • recycling of components (e.g. upgrading, reuse of components) • recycling of materials • dismantling of products • separability of different materials • low diversity of materials • low and materials energy input for reuse/recycling **Final Disposal** • compostable, fermentable products (closing biological cycles) • combustion characteristics • environmental aspects at deposition	Detail design	Product design, packaging design, materials extraction, manufacturing, distribution, use, service, end of life	Shows which design for environment strategies have not been employed.	Some questions may not be relevant to product and neglects trade-offs between employing different strategies.
Disassembly Trees		Detail design	End of life	Shows product disassembly process allows for optimization.	Time-consuming.
LiDS Wheel	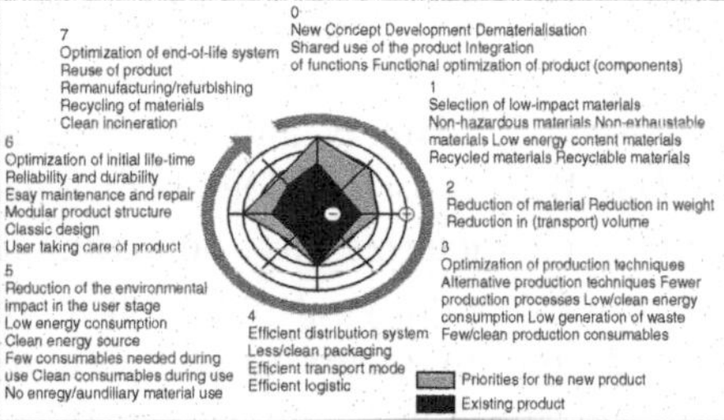 0 New Concept Development Dematerialisation Shared use of the product Integration of functions Functional optimization of product (components) 1 Selection of low-impact materials Non-hazardous materials Non-exhaustable materials Low energy content materials Recycled materials Recyclable materials 2 Reduction of material Reduction in weight Reduction in (transport) volume 3 Optimization of production techniques Alternative production techniques Fewer production processes Low/clean energy consumption Low generation of waste Few/clean production consumables 4 Efficient distribution system Less/clean packaging Efficient transport mode Efficient logistic 5 Reduction of the environmental impact in the user stage Low energy consumption Clean energy source Few consumables needed during use Clean consumables during use No enregy/aundiliary material use 6 Optimization of initial life-time Reliability and durability Esay maintenance and repair Modular product structure Classic design User taking care of product 7 Optimization of end-of-life system Reuse of product Remanufacturing/refurbishing Recycling of materials Clean incineration Priorities for the new product Existing product	Detail design	Product design, packaging design, materials extraction, manufacturing, distribution, use, service, end of life	Maps effort extended at each life-cycle stage and allows comparison to see which stages could use more effort.	Simple, but relationship between effort extended and environmental impact reduction is not clear.
Eco-Compass	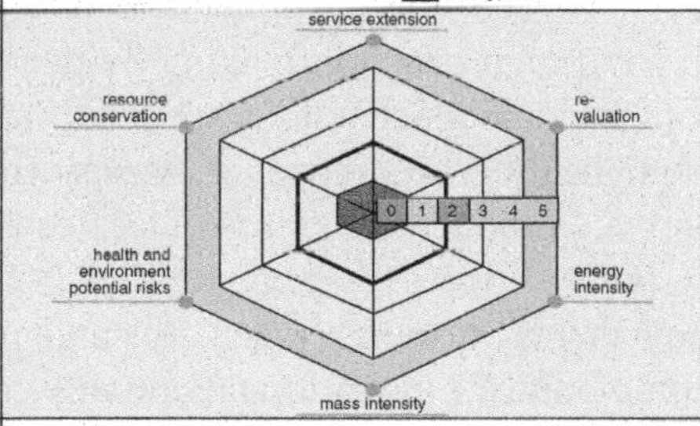 service extension re-valuation energy intensity mass intensity health and environment potential risks resource conservation 0 1 2 3 4 5	Detail design	Product design, packaging design, materials extraction, manufacturing, distribution, use, service, end of life	Displays improvement in different categories of environmental impact relative to each other for a product or between design choices.	Simple, but all scores are relative and based on design team knowledge.
Economic Input Output Analysis		Detail design	Product design, packaging design, materials extraction, manufacturing, distribution, use, service, end of life	Show the environmental and financial significance of aggregate production within the U.S. economy of a particular commodity.	Provides big-picture view of industrial environmental impact, but data carry some uncertainty and represent U.S. only.
Cleaner Technologies Substitutes Assessment	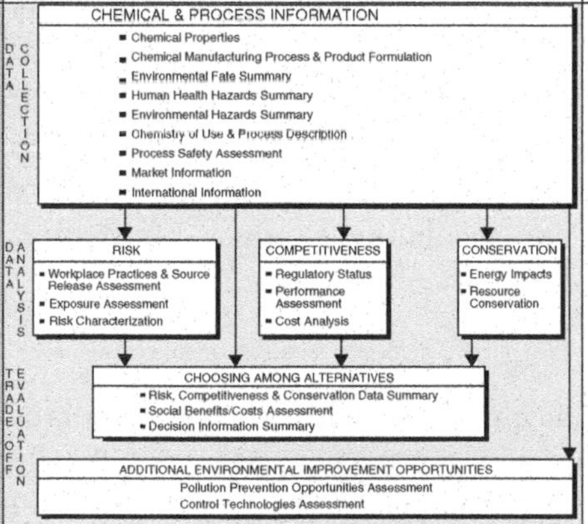 DATA COLLECTION CHEMICAL & PROCESS INFORMATION • Chemical Properties • Chemical Manufacturing Process & Product Formulation • Environmental Fate Summary • Human Health Hazards Summary • Environmental Hazards Summary • Chemistry of Use & Process Description • Process Safety Assessment • Market Information • International Information DATA ANALYSIS RISK • Workplace Practices & Source Release Assessment • Exposure Assessment • Risk Characterization COMPETITIVENESS • Regulatory Status • Performance Assessment • Cost Analysis CONSERVATION • Energy Impacts • Resource Conservation TRADE-OFF EVALUATION CHOOSING AMONG ALTERNATIVES • Risk, Competitiveness & Conservation Data Summary • Social Benefits/Costs Assessment • Decision Information Summary ADDITIONAL ENVIRONMENTAL IMPROVEMENT OPPORTUNITIES Pollution Prevention Opportunities Assessment Control Technologies Assessment	Detail design	Product design, packaging design, materials extraction, manufacturing	Presents alternatives for meeting product or process need with environmentally responsible technologies.	Detail and time intensive.
Sustainable Environmental Management Information System		Detail design	Product design, packaging design, materials extraction, manufacturing, distribution, use, service, end of life	This software aids material and component selection, calculates environmental impact, suggests product end-of-life options and records environmentally related costs.	Helps organize many different types of environmental information for easy access.

for environmental improvement, or display product disassembly because product specifications are known.

Life-Cycle assessment (LCA) sets a goal, scope, and product function (functional unit) based on an inventory of all inputs and outputs, impact assessment, and an interpretation of results. LCA is a relative measure for comparing environmental performance of two or more products useful for benchmarking environmental performance, setting environmental goals, aiding material/component design decisions, and uncovering environmental impacts that are not obvious.[9]

Ecodesign checklists, of which there is an abundance, ask a series of questions related to DfLC strategies for each product life-cycle stage (e.g., for product End-of-Life, is a product take-back strategy in place?) to compare different designs or highlight areas needing improvement.[31,81] *Disassembly trees* are diagrams that show the chronology of part removal to facilitate product disassembly for a variety of product end-of-life options.[65,69] Brezet and van Hemel[31] created the *Life-cycle Design Strategies (LiDS) Wheel* for the Dutch Promise Manual and UNEP Ecodesign Manual to quickly characterize environmental concerns and to note the level of effort extended toward employing the environmental strategies at each life-cycle stage, allowing design engineers to compare which life-cycle stages need more design effort.[6,73,82] The *Eco-Compass* was developed by Dow Chemical Company in Europe for design decision making by denoting a numerical environmental impact score for a product with respect to mass and energy intensity, human and ecosystem health risk, reuse of wastes (revalorization), resource conservation, and improved product functioning (service extension).[73,82] *Economic Input Output Analysis Life-Cycle Assessment (EIO-LCA)*, a software tool that follows financial movements, resource intensity, and releases to the environment due to specific commodities in a national economy, was developed by the Carnegie Mellon Green Design Institute.[83] The *Cleaner Technologies Substitutes Assessment (CTSA)* was created and refined by the U.S. Environmental Protection Agency in conjunction with industry, nonprofits, and academia to present companies with alternatives for meeting their product or process needs with different environmentally responsible technologies.[84] The Ricoh Group created the *Sustainable Environmental Management Information System* software for aiding the management of materials, purchasing, the supply chain, resource flows, environmental impact, and accounting product information.[85]

6.3 Indicators

Indicators assess environmental impact, which is then incorporated into comprehensive life-cycle design tools and approaches. Many of the life-cycle design indicators are used to establish the impact and weighting for life-cycle assessment (LCA) or similar tools. Indicators quantify particular categories of impact, resulting in a predicted environmental damage.[67] In general, indicators provide a single impact score by evaluating a single parameter or assessing many impact parameters and combining their values. The science used to measure environmental impacts is incomplete and complex. Hence, indicators can be inaccurate impact assessors.[9] Each indicator is a balance between the effort extended to achieve the indicator score and the knowledge provided from that result. Different indicators take more time and thought to complete while presenting more or less aid to the designer for choosing between product designs and attributes. When choosing to use an indicator, designers must balance the level of indicator complexity with the helpfulness of results gained (see Figure 6).

Indicators can be characterized by focusing on a single or multiple categories of impact. Single-parameter indicators may be simpler to execute, but their results may miss crucial areas of environmental impact. Multiple-parameter indicators involve complex assessment and may present scores that are difficult to interpret. Both types of parameters can portray environmental impacts in a way that furthers some understanding of product design choices.

Single-Parameter Indicators

Single-parameter indicators have the potential to provide results that are more meaningful, since all impacts are rated by the effect to one unit of measure such as mass flow in kilograms for material input per service unit (MIPS). However, these indicators also have the disadvantage of overlooking environmental impacts not well characterized by that unit of measure. For MIPS, the differences in materials such as toxicity are ignored.[9] Table 20 shows how single parameter

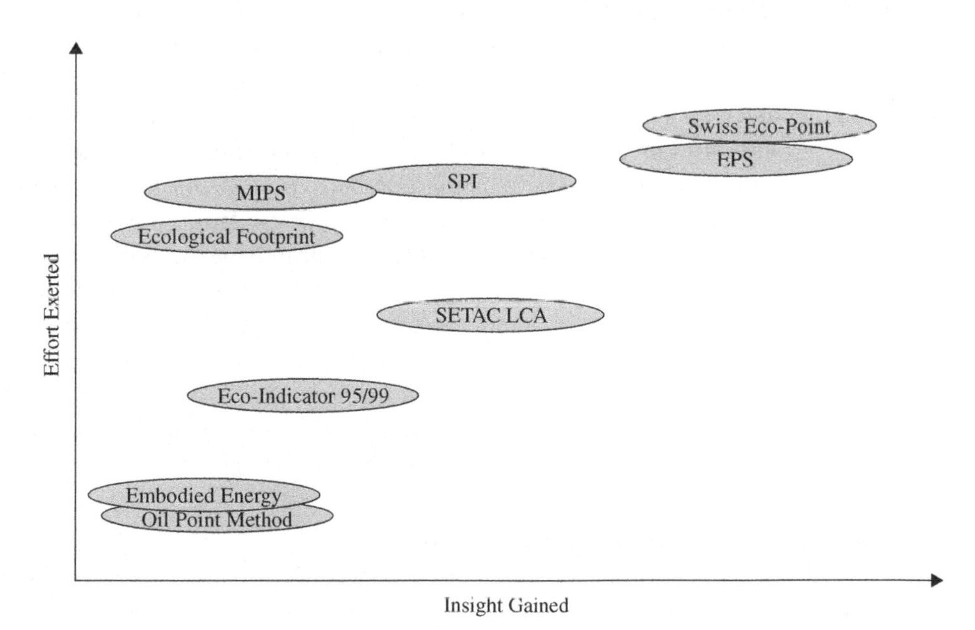

Figure 6 Comparison of indicators for insight gained from effort exerted. (Adapted from Ref. 9.)

Table 20 Single-parameter Indicators

	Unit of Comparison	Impact Categories Characterized	Data Requirements	Advantages or Disadvantages
Ecological Footprint	Hectares of land	Land use	Product life-cycle inventory	Lacks data needed for analysis and only focuses on effects to land
Embodied Energy	Energy	Energy use	Product life-cycle inventory	Simple measure allows easy comparison between products
Material Input Per Service-Unit (MIPS)	Mass flow in kilograms	Resource use	Extensive data with high accuracy	Neglects differences between materials
Oil Point Method	Energy content of one kilogram of crude oil called an oil point (OP)	Resource use	Product life-cycle inventory	Gives qualitative results that needs careful holistic interpretation

indicators differ by units, environmental impact categories considered, and data requirements.[9,45,86,87]

The *ecological footprint*[88] assesses environmental impact by calculating the total area of land *bioproductivity*, or productive capacity used to support an activity. Factors relate different types of resource extraction (such as fossil fuel use) into areas of land productive capacity required for that activity. All effects can be represented by a single number with the unit hectares of land required.[9,86] *Embodied energy* is a concept that comes from input/output analysis, which assesses the total amount of energy required for the product life-cycle.[9] *Material input per service unit* (MIPS)[89] accounts for specific material and energy flows throughout a product life-cycle to reduce material throughput.[9,87] *The Oil Point Method* (OPM) quantifies environmental impact by the energy content of one kilogram of crude oil, called an *oil point.* LCA methodology is used to uncover the energy used in each part of the product life-cycle, and conversion information from energy into oil point indicators is provided.[45]

Multiple-Parameter Indicators

Multiple-parameter indicators combine designated values of environment impact in several areas into a single score. The implication of the individual indicator can be buried in the combined score because of the trade-offs among the different effects of each environmental impact category. For that reason, multiple-parameter indicators can also be more difficult to assess than single-indicator values. However, multiple-parameter assessments have the potential to account for a larger variety of environmental impacts. Table 21 shows how multiple-parameter indicators differ by units, environmental impact categories considered, and data requirements.[4,9,87,88]

The *Swiss eco-point* (SEP) measures location-specific impacts by a relative comparison measure, *eco-point*, derived from ecosystem health-quality levels. Fourteen categories are evaluated for relative distance from a target for the impact category, such that values that lie further from the target value for a category are given higher weightings.[9,87] The *Environmental Priority System* (EPS) was created by Volvo, the Federation of Swedish Industries, and the Swedish Environmental Research Institute to combine factors from several impact categories. Willingness to pay measures are used to quantify the importance of particular impacts.[4,9,87] Very little input information is required to perform this analysis, but errors in input information can affect the analysis strongly.[87] *Eco-indicator 95* was created by Pré Consultants and the Dutch government in 1995 and improved in 1999 as *Eco-indicator 99*.[90] The weightings for the indicator are measured by the distance from an impact category target similar to the Swiss eco-point. Only a few impact categories are used; hence, certain impacts like acidification are overemphasized by the Eco-indicator.[9,88] The *Sustainable Process Index* sets out to quantify pollution taxation on the environment using land or area as the measure, similar to the ecological footprint except looking at several impact areas. Data must be fairly accurate to beget a pertinent outcome. The Society of Environmental Toxicology and Chemistry's Life-Cycle Impact Assessment (*SETAC LCA*) characterizes impacts by assigning each a relative score in a particular impact categories. Not much information is needed to perform this analysis.[87]

6.4 Environmental Cost Accounting Tools

Often, environmental costs are neglected by companies during accounting. Part of this omission comes from not including costs incurred by products over the entire product life-cycle.[90] Environmental cost accounting tools differ from indicators because of the focus on the financial obligations of each part of the product life-cycle. Table 22 shows where several environmental cost accounting tools are applied during the life-cycle and what life-cycle stages these tools affect, as well as the impacts using such tools can have.

Total cost accounting (TCA) incorporates financial obligations associated with liability, and was created alongside the idea of cleaner production in the late 1980s. Some dynamic decision making is incorporated into the tool by trying to reflect how costs might differ when relationships with users and suppliers change. Most costs considered in TCA come from manufacturing.[90] *Life-cycle*

Table 21 Multiple-parameter Indicators

	Unit of Comparison	Impact Categories Characterized	Data Requirements	Advantages or Disadvantages
Swiss Eco-point (SEP)	Eco-point	Resource conservation, toxicity, global climate change, ozone generation, resource depletion, etc (14 categories total)	Location-specific data with a fair amount of accuracy	Emissions outside of Switzerland are ignored
Environmental Priority System (EPS)	Environmental load unit per kilogram	Human health, biological diversity, manufacturing, resource conservation, and aesthetics	Little data needed, but high accuracy required	Errors in input information strongly affect analysis
Sustainable Process Index (SPI)	Meter squared	Resource conservation, toxicity, global climate change, ozone depletion, *etc.*	Location-specific data with a fair amount of accuracy	Robust to some errors in input information

Table 21 (*continued*)

	Unit of Comparison	Impact Categories Characterized	Data Requirements	Advantages or Disadvantages
Society of Environmental Toxicology and Chemistry's Life-cycle Impact Assessment (SETAC LCA)	Relative scale	Global climate change, ozone creation and depletion, human and environmental toxicity, acidification, eutrophication, and effects to land, living creatures, and natural resources	Little data input needed	Gives consistent results
Eco-indicator 95 and 99	Numeric value	Human health, ecosystem health, and resources	Product life-cycle inventory	Incorporates fate of emissions and degree of effects on receiving ecosystems, but it overemphasizes acidification while deemphasizing land use and biodiversity concerns

Table 22 Environmental Cost-Accounting Tools

	Product Design Application Stage	Product Life-cycle Stages Affected	Impacts	Advantages or Disadvantages
Total Cost Accounting (TCA)	Detail design	Product design, packaging design, materials extraction, and manufacturing	Incorporates financial obligations from liability and stakeholder dynamics into conventional accounting	Most costs included come from manufacturing life-cycle stage
Life-cycle Costing (LCC)	Detail design	Product design, packaging design, materials extraction, manufacturing, distribution, use, service, and End-of-Life	Uncovers how costs affect different life-cycle stages	Cost advantages in the market due to life-cycle design measures can be revealed
Full Cost Accounting (FCA)	Detail design	Product design, packaging design, materials extraction, manufacturing, distribution, use, service, and End-of-Life	Shows how particular parties are affected by product life-cycle costs	Societal costs are difficult to assess, so willingness-to-pay measures are often used

costing (LCC) takes into account the dynamic effects of cost at all product life-cycle stages. Additionally, influences to costs are considered, such as the price of capital, labor, materials, energy, and disposal. The overall aim of LCC is to reveal how influences on costs may create advantages for different parties because of how costs affect different life-cycle stages.[90] *Full cost accounting* (FCA) expands LCC by considering how particular parties are affected by the costs incurred during the product life-cycle. Hence, FCA considers what costs are paid by society because of environment degradation. The cost of this detriment is hard to assess, so contingent valuation or willingness to pay measures are often used for assessment.[90]

7 IMPLEMENTATION OF DESIGN FOR THE LIFE-CYCLE

Implementing DfLC is complicated. There is no single path to incorporating DfLC concerns in a product, but approaches exist to help companies incorporate life-cycle design into their structure. "The idea of green design seems simple, but there is no rigid formula or decision hierarchy for implementing it."[16] There are many different approaches to DfLC because there are many ways to view life-cycle design, and there are many different situations in which to apply it.

7.1 How to implement DfLC within a Company

Most businesses start with a finished design and try to improve its environmental performance.[66] This strategy does not beget the most benefit from life-cycle design approaches for a product. Several approaches for implementing DfLC in a company are described in this section.

The *ISO 14000 standards* provide a standard for environmental management systems and allow companies to achieve and receive recognition.[91] This system helps companies identify and put in place their own environmental policy, including planning, actions, and reviews. All parts of the management system work together toward continuous environmental performance improvement.[65]

Integrated environmental management systems (IEMS)[92] are management approaches that address the following actions:

- Evaluate changes with full cost accounting.
- Instate environmentally preferable processes.
- Improve handling and risk assessment of toxic compounds.
- Assess process and material inputs and outputs to improve operation performance.
- Decrease multi media environmental harm created.
- Implement extended producer responsibility.
- Combine environment and employee health and safety requisites.

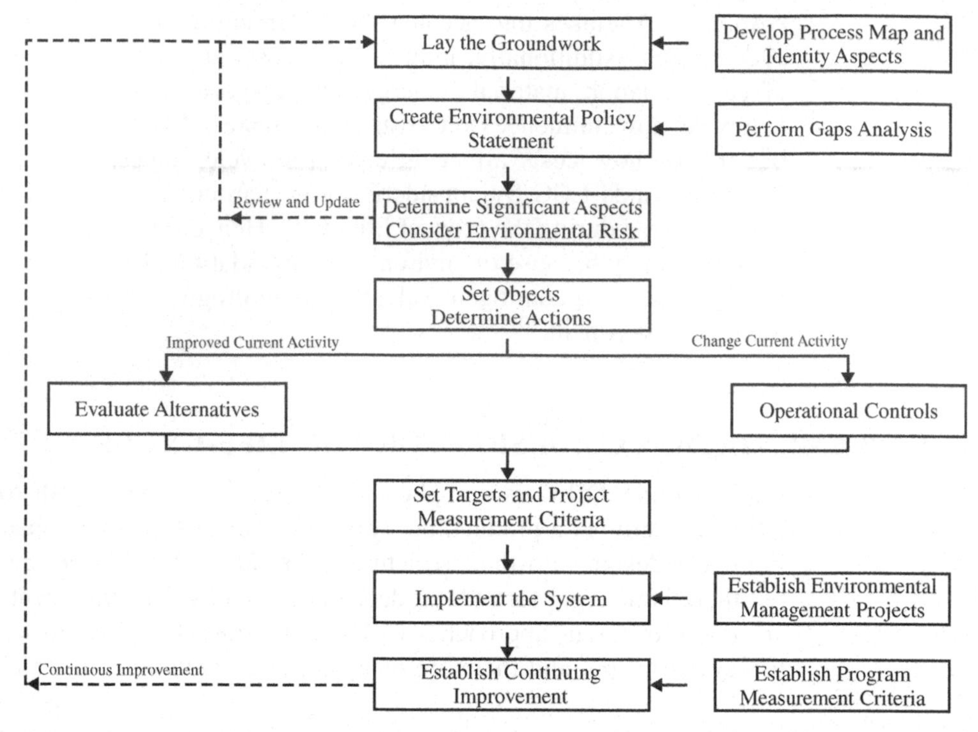

Figure 7 Integrated environmental management implementation process. (Adapted from Ref. 111.)

The EPA's *Integrated Environmental Management Systems Guide* divides the task of creating these systems into 10 steps (Figure 7).[92] The first step is to consider the fundamentals behind environmental management, including DfLC principles, quality management, and characterizing environmental impact at different life-cycle stages. Using this information, the organization then create a corporate environmental policy. It sets goals and actions that take into account the risk of environmental performance currently. Substitute technologies or compounds are next considered for implementation. From the substitutions assessment, the organization creates specific aims and indicators of achievement. Operating procedures for instituting these aims are then decided upon. The organization chooses timelines for implementing the procedures, and responsibility for specific actions is designated throughout an organization.[92]

Each company must adapt methodologies for implementing DfLC to its own corporate and employee specifications. Some companies have created their own DfLC management systems. Electrolux and Philips include ecodesign in their product-orientated environmental management system.[66]

7.2 Industrial Life-cycle Design Methods

Table 23 highlights two particular DfLC methods developed by corporations. Hewlett-Packard incorporates a product steward on each design team who considers potential environmental impacts at several life-cycle stages. The product steward is responsible for providing the environmental customer voice for the product and ensuring that environmental goals are included in product design.[93] Philips Sound and Vision developed the Selection of Strategic Environmental Challenges (STRETCH) methodology to push environmental goals into marketing approaches, the product design process, and the corporate organization to improve product environmental performance. STRETCH involves a diverse group of participants from within the company to do market data acquisition, set environmental goals, plan changes to the product-development process, create and rank potential designs, and assess the environmental impact for final implementation of new products and management plans.[73]

8 IMPLEMENTATION EXAMPLES

Despite the difficulties in applying DfLC principles, many products exhibit environmental excellence. Finally, we present several DfLC examples, a wide variety of product designs with differing methods and impacts. The chosen product examples affect either single or multiple life-cycle stages. Some implementation examples go far beyond meeting DfLC principles over the life-cycle. Each example in Table 24 is described in the sections that follow.

8.1 Single Life-cycle Stage Concerns

McDonough and Braungart design consultants helped Röhner Textil create Climatex® Life-cycle™ fabric so that any waste fabric could be composted (instead of designated hazardous waste), and effluent water quality during manufacturing was improved by using natural fibers and replacing harmful dyes with help from Ciba Specialty Chemicals.[3,46] To improve the fabric, Braungart worked with the Clariant chemical corporation to create a flame-retardant safe for people and the environment. Austrian fiber producer Lenzing helped to develop a way to apply the flame retardant to avoid off-gassing by incorporating the compound directly into a beechwood fiber. The new flame retardant and fiber together created a new fabric, Climatex® LifeguardFR™. Fabric offcuts are still compostable, but now have a second life as the collapsible plyFOLD container (see Figure 8).[94,95]

The Re-Define furniture project by the National Centre for Design at RMIT University developed a three-seat sofa (see Figure 9) and lounge chair by

Table 23 Industrial Life-cycle Design Methods

	Product Design Application Stage	Product Life-cycle Stages Affected	Quantified Impact
Hewlett-Packard Product Steward	Detail design	Materials extraction, manufacturing, distribution, use, service, and End-of-Life	Recycled material incorporation, resource conservation, waste and emissions reduction, energy efficiency, disassembly ease, and ability to recycle
STRETCH—Selection of Strategic Environ-mental Challenges	Product definition, conceptual design, embodiment design, and detail design	Materials extraction, manufacturing, distribution, use, service, and End-of-Life	Setting of environmental goals, assessment of goals are set by a diverse management team

following environmental guidelines. MID Commercial Furniture designed the furniture and Wharington International made the sofa mold from Recopol™ instead of wood. Recopol™ is recycled ABS (Acrylonitrile Butadiene Styrene) from large appliances, automobiles, and electronics that are typically landfilled.[96,97] Wharington takes back the sofa shells for reuse, and the steel and recycled-PET fabric are also recyclable.[9,97] No toxic materials are used in the furniture. Production costs are on par with similar furniture, though the development costs were greater.[97]

8.2 Multiple Life-cycle Stage Concern

Recycline has several products that are made with 100 percent recycled handles, including the Preserve® toothbrushes, tongue cleaners, and razors (Figure 10).[98] Preserve® toothbrushes are produced with polypropylene from 65 percent or more recycled Stonyfield Farm® yogurt cups, and the toothbrush bristles are made of virgin nylon.[98] The price of the Recycline products includes the envelope and postage for returning to the company. Once the products are returned to Recycline, the handles become deck furniture.[46] Hence, the products have a guaranteed second use and two disposable product life-cycles (yogurt cups and personal hygiene products) are successfully extended.

Table 24 Design for the Life-cycle Implementation

	Key Players	Product Life-cycle Stages Affected	Impacts
Climatex® LifeguardFR™ Fabric	Röhner Textil, Design Tex (Steelcase), CIBA Specialty Chemicals, Clariant, Lenzing, Ply Designs, William McDonough, and Michael Braungart	Manufacturing	Benefits include reducing toxicity of product and processing effluents and wastes and secondary use of processing waste
Re-Define Furniture	Wharrington International, MID Commercial Furniture, National Centre for Design at RMIT University, EcoRecycle Victoria	End-of-Life	Benefits include incorporating recycled materials into product, creating completely recyclable products, and developing product take-back programs
Preserve® Toothbrush	Recycline, Stonyfield Farm®	Materials extraction, product design, and End-of-Life	Benefits include using recycled materials, extending materials life for a secondary product, taking products back at End-of-Life
Patagonia Clothing	Patagonia, Teijin	Materials extraction, manufacturing, service, and End-of-Life	Benefits include incorporating recycled and benignly grown materials, utilizing renewable energy during manufacturing, servicing products, performing product take-back to recycle products at End-of-Life
Timberland Shoes	Timberland	Manufacturing	Customers are informed about social and environmental product attributes
Lifeline Aid/relief Radio	Free Play Energy	Use	Impoverished people receive radios, products can operate with human power or renewable energy

Figure 8 PlyFOLD container. (From Ref. 94.)

Figure 9 Re-Define sofa. (From Ref. 112.)

Figure 10 Preserve™ products. (From Ref. 98.)

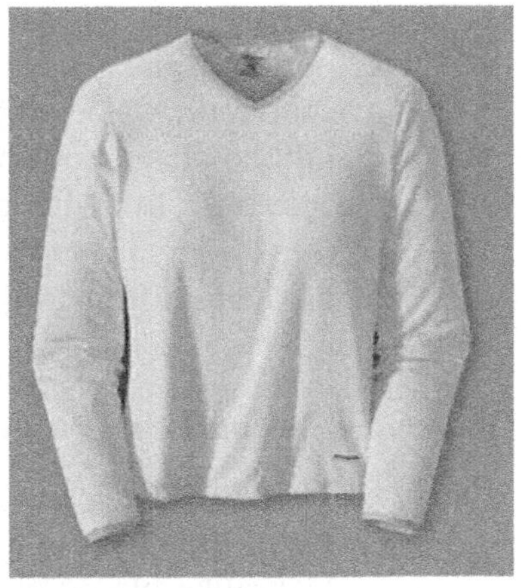

Figure 11 Patagonia garments before and after recycling. (From Refs. 99, 101.)

Our Footprint **Notre Empreinte**	
Environmental Impact **Impact sur l'environnement**	
Energy to Produce: (per pair)*	2kWh
Energie utilisée (per pair)*	2kWh
Renewable energy (Timberland-owned facilities):	5%
L'énergie renouvelable (sites Appartenant á Timberland) :	5%
Community Impact **Impact sur la communauté**	
Hours served in our communities:	119,776
Nombre total d'heures données :	119,776
% of factories assessed against code of conduct:*	100%
% d'usines évaluées pour leur conformité au code de conduite :*	100%
Child labor:*	0%
Main-d'oeuvre enfantine :*	0%
Manufactured **Fabriqué à**	
Shingtak, China Shingtak, Chine	
* metrics based on global footwear production for 2005	
* informations fondées sur production totale de chaussures en 2005	
FOR MORE INFORMATION VISIT WWW.TIMBERLAND.COM/CSRREPORT	
POUR PLUS D'INFORMATIONS : WWW.TIMBERLAND.COM/CSRREPORT	

Figure 12 Timberland's shoe eco-nutrition facts. (From Ref. 103.)

Patagonia, an outdoor clothing manufacturer, creates fleece with 90 percent post-consumer recycled polyester from soda bottles at a factory using 100 percent renewable wind energy. "Since 1993, Patagonia has diverted over 100 million plastic soda bottles from landfills."[46] The company also uses only organic cotton.[100] In addition, customers can now return worn out Capilene® garments (see Figure 11) to Patagonia for recycling into new clothing by a Japanese textile production firm, Teijin, through the Common Threads Recycling Program.[102]

8.3 Concerns beyond the Product Life-Cycle

Timberland is going above and beyond life-cycle design principles by communicating its environmental efforts to consumers through a nutritional label look-alike placed on all shoeboxes (Figure 12). The label consists of a table of environmental and social impacts and states where the shoe was manufactured. Environmental information provided considers production power requirements, including the renewable energy acquired. Social information on the label includes the time employees volunteer and the percentage of facilities that are evaluated for meeting ethical and child labor standards. Timberland added the label to display corporate values and change consumer opinions.[103]

Free Play Energy produces radios (Figure 13), lights, cell phone chargers, and car battery chargers that are powered by human power and AC/DC suited for impoverished regions. The company encourages prospective buyers to consider

Figure 13 Lifeline aid/relief radio. (From Ref. 104.)

using solar or wind energy if using electricity instead of muscle power.[108] In addition, the company distributes radios to impoverished areas throughout the world to facilitate education, health, emergency relief, peacemaking, and agriculture.[104]

9 THE FUTURE OF DESIGN FOR THE LIFE-CYCLE

There are some trends in product design that we believe will significantly impact life-cycle product design in the coming years. Deciding the end-of-life scenario of a product at the beginning of product design development will become standard.[14] More economically and technically feasible solutions to environmental problems will come about through *eco-innovation*.[9] To achieve eco-innovation, current products must be reconsidered from their very core, instead of sprucing up products already in existence.[24] A true concurrent, life-cycle engineering environment is necessary for such considerations during product development. Even with a properly structured design process, the information gap—the lack of appropriate information for making informed engineering decisions at each stage of the design—must be overcome in terms of databases and engineering information systems to efficiently access these databases. In addition to these general needs, we anticipate the rise of new design parameters and an increased prominence of new life-cycle design influences.

9.1 Design Paradigm Changes

The design process in the future will be influenced more greatly by life-cycle design. Change can occur through expanding the scope of DfLC to look beyond just the product life-cycle, incorporating robust design and sharing into product design. Industrial ecology perspectives are likely to be further incorporated in many aspects of the design process. In addition, tools that can handle the uncertainty of environmental information and can incorporate DfLC information within computer-aided design tools will greatly change life-cycle design.

Expanded Robust Design

Robust products continue to perform by evolving when the systems and environments the products interact with change in two ways—over the life of the product as an artifact and through the product itself as a design. Fiksel uses the term *resilience,* "the ability to resist disorder," to mean stability in the face of perturbations.[44] "Although industrial systems are nonlinear and dynamic, most design and management methods are based on a linear, static worldview. As a result, our systems are brittle—vulnerable to small, unforeseen perturbations—and isolated from their environments."[105] The uncertainties in potential future changes to the environment and the need to minimize environmental impact in all environments are significant drivers for robust design. Thermodynamic efficiency, safety, and a supply chain's ability to cope with unpredicted changes must be addressed.[105] Designing for resilience seeks to address this deficiency by promoting diversity (many available options), efficiency (high output for low input), adaptability (ability to adjust to stimuli), and cohesion (unity between distinct elements).[44]

Life-cycle design tools must also become more robust. Early on in the design process, materials, product architecture, and other attributes are still undecided, and environmental impacts are hard to characterize. For the conceptual design stage, efficient and accurate product life-cycle characterization must be possible with little input data and over a variety of potential designs. In embodiment design, fewer designs are considered and more details are known, but still uncertainty in design exists. Life-cycle design tools must be capable of assessment with a low volume of information and many different design options, and must increase in accuracy as product attributes become finalized and fewer design options are being considered.[108]

Uncertainty in Design for the Life-Cycle

The lack of reliable data with regards to environmental impact and emissions for components, materials, and substances makes design decision making difficult. Gathering this data is expensive and can be subjective. Uncertainties in data must be cleared for more robust designs and better design decisions early in the design process. The vagueness of final product details during the very early stages of product development also makes design analysis for environmental concerns difficult. More robust life-cycle design tools would help address these uncertainties. An increase in intelligent computer-based life-cycle design tools and their gradual commercialization will increase design engineering capabilities. These tools will most likely be categorized by their application (*e.g.*, automobile tools, computer hardware tools, and household electronics tools).[65]

Implementing Industrial Ecology

Designing product and industry ecologies is still a new challenge for design engineers.[23] Creating industrial systems patterned after ecosystems can yield unique goals for design, including "buildings that, like trees, produce more energy than they consume and purify their own wastewater[,] factories that produce effluents that are drinking water", and "products that, when their useful life is over, can be tossed onto the ground to decompose and become food for plants and animals and nutrients for soil".[3]

Moreover, it is essential for organizations to design industrial processes to be a part of the solution by not creating environmental problems in the first place, and to participate in environmental clean-up. Creating systems that encourage and make possible the trading and use of byproducts between industries is one way of many that needs more work for the implementation of industrial ecology to become real.[24]

Designing Products for Multiple Lives

Designing products for multiple lives through reuse is part of the industrial ecology future. "Designing a product so that it has subsequent lives is a very difficult thing to do. It cannot be achieved generically."[107] Retrofitting old products and materials to meet human needs will become increasingly important. When designers are compelled to use available resources instead of creating new ones, they create results that incorporate component reclamation and reuse. Achieving products that are new in the marketplace that can reuse existing manufacturing technology also supports environmental and industrial ecology goals.[108] Currently, very few life-cycle design tools can evaluate environmental effects that span greater than one design, one producer, or one life-cycle. Tools that have a broader perspective could support the kind of energy and materials sharing prescribed by industrial ecology or support sustainable development goals.[6]

Sharing

Many DfLC and consumer culture critics extol the value of sharing products. Examples of people sharing products already take place in society such as libraries, which can be set up to lend everything from books to tools. Setting up community-based product lending/sharing or reuse centers is one way to encourage product sharing.[60] Sharing products engenders certain types of social attitudes and requires social changes. The cohousing movement in Denmark and Sweden sets up living situations where people live close to each other and in ways that facilitate the sharing of goods.[60] There is also a significant impact on product design, including the need for more robust products, longer life-cycles, and increased end user customization.

Designing for End User Repair of Products

Products that require frequent maintenance and user interaction would lead to "owner-builder[s] [who] would develop an understanding of how the thing works, making trouble-shooting, repair, and replacement of parts easier."[60] Involving the end user in repairs will prolong the life of products and increase the products' value to the owner. However, designing products for end user repair runs counter to most current product trends and requires significant design changes for ease of diagnostics, disassembly, repair, and assembly.

9.2 Future Design for the Life-cycle Influences

Changes to companies, the supply chain, and industry all affect life-cycle design implementation in the future. The drive to incorporate life-cycle design will increase, for reasons beyond the clear need to reduce environmental impact for the good of society. Companies and designers will be increasingly liable for product environmental effects. Industrializing nations and small manufacturers will press for help in adopting DfLC principles. In addition, the market risk of uncertainty in design will play a larger role in organizational decisions.

Increased Liability Concerns for Designers and Corporations

The future brings liability concerns for both corporations and designers, who will have to take responsibility for the entire product life-cycle and its consequences because of changes in legislation and public expectations. Greater reconciliation between environmental and economic interests must be put into practice. Organizations must designate a responsible party to preserve designed intentions with respect to environmental impact, product take-back, and so on, and must protect designers' intentions, thereby reducing liability concerns and environmental damage.[109]

Product Information Sharing along Supply Chain

Future legal compliance will become more than merely meeting set environmental targets. Legislation will emphasize improvement through the sharing of information throughout the product life-cycle among supply chain members. Information about how and where a product is manufactured and used can help companies optimize environmental impact over the life-cycle. This will require new tools to facilitate the collection and sharing of product creation, use, and service information.[24]

Product Service Systems

The nature of business and transactions involving products will significantly change. A move exists to create and convert products as product service systems. Likely, fewer end users will own products. More user needs will be met by the service economy.[9] Changing products into services "can have significant environmental benefits in terms of reducing the volume of products manufactured while maintaining or increasing profits for the company through service provision."[66] In addition to rethinking their marketing, companies will have to change their product design, manufacturing, service, and end-of-life strategy to convert products into services. The conversion will involve new approaches, different life-cycle design tools, and new technologies.

DfLC Incorporation by Industrializing Nations

"The globalisation of markets is also extending the industrialised model of development (lifestyles, behavior and consumption patterns) to developing countries" while population increases globally.[9] A 90 to 95 percent reduction in consumption and resource use must occur in the next 50 years to maintain resource availability for the future. This reduction in resource use is possible if industrializing countries design a higher quality of life with fewer resources than industrialized countries have done so far. Industrialized countries must cut resource use to 5 percent of current levels. Clearly social, cultural, and technical changes must be made to achieve these goals.[9] These will have a significant impact on life-cycle product design.

Life-cycle Design Help Needed for Small and Medium Producers

Large companies with many resources are the most able to put DfLC changes into place. Small and medium-sized companies lack the financial flexibility and resources to make sweeping changes in organizational structure and design processes. Small and medium-sized producers need help in incorporating life-cycle design initiatives.[6] Governments and trade organizations may have to help the smaller companies adapt to meet expectations of improved product environmental performance. In addition, commercial tools that are generic in their application but accurate in their assessment must be made available to a wider community.

Design for the Life-Cycle becomes Sustainable

DfLC will begin to consider more than just a particular product's environmental performance.[24] DfLC must become sustainable design, with a systems view.[9,24] Industry cannot solve product environmental impact problems alone. "It is important for the practice of DfLC to go beyond the confines of industry into the large society, since a fully successful implementation of DfLC is a cooperative effort of both industry and society."[65] The key to creating sustainable design is to involve perspectives from many different areas of study with a focus on natural science.[60]

Localization of DfLC Information

Characterizing how industrial systems affect specific ecological systems needs to be improved upon to really reduce industrial impacts on the environment. Considering localized toxicity instead of general large-scale environmental effects must occur. The localization of LCA and the connection of ecosystem models and LCA will support these needs. Similarly, it is important to recognize the connection between the product and production and their impacts on specific physical spaces and ecosystems.[24]

Localization of Production and Markets

Papanek calls for localization of manufacturing and product markets in the United States for societal and environmental gain, another need that is somewhat counter to current trends.[60] The localization of manufacturing and product markets could provide a variety of environmental benefits from reduced transportation and smaller geographical areas over which to characterize environmental impacts. The flexibility of a dispersed and diversified manufacturing infrastructure will also have significant economic benefits through the potential to increase product variety and regional modification of products. Localization will also require life-cycle design tools that can make design and manufacturing decisions based on locally available resources and environmental capital.

REFERENCES

1. A. Fuad-Luke, *The Eco-design Handbook: A Complete Sourcebook for the Home and Office*, Thames and Hudson, London, 2002.
2. D. Navinchandra, "Design for Environmentability," *Design Theory and Methodology (DTM '91)*, **31**, 119–125 (September 1991).
3. W. McDonough and M. Braungart, *Cradle to Cradle: Remaking the Way We Make Things*, North Point Press, New York, 2002.
4. S. Ashley, "Designing for the Environment," *Mechanical Engineering*, **115**(3), 52–55 (1993).
5. G. A. Davis, C. A. Wilt, P. S. Dillon, and B. K. Fishbein, "Extended Product Responsibility: A New Principle for Product-oriented Pollution Prevention—Introduction and Chapters 1 through 4" (1997). Retrieved March 18, 2006, from University of Tennessee, Center for Clean Products and Clean Technologies, Publications Web site: http://eerc.ra.utk.edu/clean/pdfs/eprn1-4.pdf.
6. B. Bras, "Incorporating Environmental Issues in Product Design and Realization," *Industry and Environment*, **20**(1–2), 7–13 (1997).
7. Ozone Secretariat, "Montreal Protocol" (2004). Retrieved April, 20, 2006, from United Nations Environment Programme Web site: http://ozone.unep.org/Treaties_and_Ratification/2B_montreal_protocol.asp.
8. B. Kiser, "A Blast of Fresh Air: The History of O_2" (2000). Retrieved April 26, 2006 from O2 Global Network Web site: http://www.o2.org/media/document/Kiser.pdf.
9. H. Lewis and J. Gertsakis, *Design + Environment: A Global Guide to Designing Greener Goods*, Greenleaf, Sheffield, 2001.
10. American Electronics Association, T*he Hows and Whys of Design for the Environment—A Primer for Members of the American Electronics Association*, American Electronics Association, Washington, D.C., 1993.
11. S. Erkman, "The Recent History of Industrial Ecology," in R.U. Ayres and L.W. Ayres (eds.), *A Handbook of Industrial Ecology*, pp. 27–35, Edward Elgar, Cheltenham, U.K., 2002.
12. J. Kluger, "Global Warming," *Time,* **167**(14), 28–42 (2006).
13. G. A. Keoleian and D. Menery, *Life cycle Design Guidance Manual,* EPA Publication No. EPA 600/R-92/226, U.S. Government Publishing Office, Washington, DC, 1993.
14. C. M. Rose, "Design for environment: a method for formulating product end of life strategies," Doctoral dissertation, Stanford University, 2000.
15. C. L. Henn, (1996). "Design for Environment in Perspective," in *Design for Environment: Creating Eco-efficient Products and Processes* J. Fiksel (ed.), McGraw-Hill, New York, 1996, pp. 473–490.
16. G. Eyring, "Policy Implications of Green Product Design," Proceedings of the 1993 IEEE International Symposium on Electronics and the Environment, Arlington, Virginia, pp. 160–163, May 10–12, 1993.
17. J. Fiksel, "Design for Environment: An Integrated Systems Approach," Proceedings of the 1993 IEEE International Symposium on Electronics and the Environment, Arlington, Virginia, pp. 126–131, May 10–12, 1993.

18. T. E. Graedel, and B. R. Allenby, *Design for Environment*, Prentice Hall, Upper Saddle River, NJ, 1996.
19. S. B. Billatos, and N. A. Basaly, *Green Technology and Design for the Environment*, Taylor and Francis, Washington, DC, 1997.
20. U. Tischner, "Introduction: Ecodesign in Practice," in *How to Do EcoDesign?* U. Tischner, E. Schmincke, F. Rubik, M. Prösler, B. Dietz, S. Maßelter, and B. Hirschl (eds.), Verlag form GmbH, Frankfurt, 2000, pp. 9–14.
21. H. C. Zhang and T. C. Kuo, "Environmentally Conscious Design and Manufacturing: Concepts, Applications, and Perspectives," *Proceedings of the 1997 ASME International Mechanical Engineering Congress and Exposition*, Dallas, Texas, pp. 179–190, November 16–21, 1997.
22. P.T. Anastas and J. B. Zimmerman, "Design through the 12 Principles of Green engineering," *Environmental Science and Technology*, 37(5), pp. 94A–101A, 2003.
23. T. E. Graedel, and B. R. Allenby, *Industrial Ecology*, Prentice Hall, Upper Saddle River, NJ, 1995.
24. T. E. Graedel, and B. R. Allenby, *Industrial Ecology*, Prentice Hall, Upper Saddle River, NJ, 2003.
25. K. N. Blue, N. E. Davidson, and E. Kobayashi, "The 'Intelligent Product' System," *Business and Economic Review*, **45** (2), 15–20 (1999).
26. O. Mont, *Product-Service Systems*. Swedish Environmental Protection Agency, AFR-REPORT 288, Stockholm, 2000.
27. W. J. Glantschnig, "Green Design: A Review of Issues and Challenges," Proceedings of the 1993 IEEE International Symposium on Electronics and the Environment, Arlington, Virginia, pp. 74–78, May 10–12, 1993.
28. L. Alting and J. B. Legarth, Life cycle engineering and design. *Annals of the CIRP-Manufacturing Technology*, **44**(2), 569–580, 1995.
29. World Commission on Environment and Development (WCED), *Our Common Future*, Oxford University Press, Oxford, U.K., 1987.
30. T. E. Graedel, "The Grand Objectives: A Framework for Prioritized Grouping of Environmental Concerns in Life cycle Assessment," *Journal of Industrial Ecology,* **1**(2), 51–64 (1997).
31. H. Brezet and C. van Hemel, *Ecodesign: A Promising Approach to Sustainable Production and Consumption*, United Nations Environment Programme, Paris, 1997.
32. N. Bruce R. Perez-Padilla, and R. Albalak, "Indoor Air Pollution in Developing Countries: A Major Environmental and Public Health Challenge," *Bulletin of the World Health Organization*, **78**(9), 1078–1092 (2000).
33. T. Colborn, D. Dumanoski, and J. P. Myers, *Our stolen future*. Penguin Books, New York, 1997.
34. Committee on Health Risks of Exposure to Radon (BEIR VI), Board on Radiation Effect Research, Commission on Life Sciences, National Research Council, *Health effects of exposure to radon*, Washington, DC, National Academy Press, 1999.
35. B. Watson, et al., "Climate Change 2001: Synthesis Report" (2001). Retrieved March 22, 2006, from Intergovernmental Panel on Climate Change Web site: http://www.ipcc.ch/pub/un/syreng/spm.pdf.
36. D. Mackenzie, *Design for the Environment*, Rizzoli, New York, 1991.

37. Container Recycling Institute, "Beverage Container Deposit Systems in the United States: Key Features," in *Bottle Bill Resource Guide*, 2005. Retrieved May, 19, 2006, from Web site: http://www.bottlebill.org/legislation/usa_deposit.htm.
38. European Parliament, "Directive 2002/95/EC of the European Parliament and of the Council of 27 January 2003 on the restriction of the use of certain hazardous substances in electrical and electronic equipment," *Official Journal of European Communities, L 037* (13/02/2003), pp. 0019–0023 (2003a). Retrieved April 20, 2006, from EUROPA Web site: http://europa.eu.int/smartapi/cgi/sga_doc?smartapi!celexapi!prod!CELEXnumdocandlg=ENandnumdoc=32002L0096andmodel=guichett.
39. European Parliament, "Directive 2002/96/EC of the European Parliament and of the Council of 27 January 2003 on waste electrical and electronic equipment (WEEE)," *Official Journal of European Communities, L 037* (13/02/2003), pp. 0024–0039 (2003b). Retrieved April 20, 2006, from EUROPA Web site: http://europa.eu.int/smartapi/cgi/sga_doc?smartapi!celexapi!prod!CELEXnumdocandlg=ENandnumdoc=32002L0095andmodel=guichett.
40. European Parliament, "Directive 2005/32/EC of the European Parliament and of the Council of 6 July 2005 establishing a framework for the setting of ecodesign requirements for energy-using products and amending Council Directive 92/42/EEC and Directives 96/57/EC and 2000/55/EC of the European Parliament and of the Council." *Official Journal of European Communities, L 191* (22/07/2005), pp. 29–58 (2005). Retrieved April 20, 2006, from EUROPA Web site: http://europa.eu.int/comm/enterprise/eco_design/directive_2005_32.pdf.
41. United Nations Framework Convention on Climate Change (UNFCCC), "Kyoto Protocol" (2006). Retrieved December 7, 2006 from http://unfccc.int/kyoto_protocol/items/2830.php.
42. E. Royte, "E-Waste@Large," *New York Times* (January 27, 2006), p. 23.
43. G. Roos, "Environmental Compliance Poses Several Business Risks," *GreenSupplyLine* (December 12, 2005). Retrieved March 26, 2006, from http://www.greensupplyline.com/howto/showArticle.jhtml;jsessionid=5CQPRS4HZHTUGQSNDBECKHSCJUMEKJVN?articleID=174918050.
44. J. Fiksel, "Designing Resilient, Sustainable Systems," *Environmental Science and Technology,* **37**(23), 5330–5339 (2003).
45. C. Berner, B. Bauer, and J. Dahl, *New Tools for the Design of Green Products* (2005). Retrieved March 26, 2006, from Danish Environmental Protection Agency site: http://www.mst.dk/publica/projects/2003/87-7972-585-6.htm.
46. E. Datschefski, *The Total Beauty of Sustainable Products*, Rotovision, Crans-Près-Céligny, Switzerland, 2001.
47. C. Madu, *Handbook of Environmentally Conscious Manufacturing*, Kluwer Academic Publishers, Boston, 2000.
48. C. Beard and R. Hartmann, "Naturally Enterprising—Eco-design, Creative Thinking, and the Greening of Business Products," *European Business Review*, **97**(5), 237–243 (1997).
49. D. J. Richards, "Environmentally Conscious Manufacturing," *World Class Design to Manufacture,* **1**(3), 15–22 (1994).
50. D. Wann, *Biologic, Environmental Protection by Design*, Johnson Books, Boulder, 1990.

51. National Research Council of Canada (NRC), "*Design for Environment Guide*" (2003). Retrieved May 17, 2006, from National Research Council of Canada Web site: http://dfe-sce.nrc-cnrc.gc.ca/.

52. J. Todd, *From Eco Cities to Living Machines: Ecology as the Basis of Design*, North Atlantic Press, Berkeley, CA, 1994.

53. R. Bedrossian, "Green Design," *Communication Arts* (May/June 2005). Retrieved March 19, 2006, from http://www.commarts.com/CA/feadesign/green/.

54. J. Benyus, *Biomimicry: Innovation Inspired by Nature*. Morrow, New York, 1997.

55. Change Design, "What is D|MAT design?" (2004c). Retrieved March 20, 2006, from EcoDesign Foundation, now Change Design, Web site: http://www.changedesign.org/DMat/DMatWhatMain.htm.

56. P. Hawken, A. Lovins, and L. H. Lovins, "The Next Industrial Revolution, in *Natural Capitalism,* Little, Brown and Co., Boston, 1999, pp. 1–21. Retrieved March 20, 2006, from Web site: http://www.natcap.org/images/other/NCchapter1.pdf.

57. World Congress of Architects Chicago, "Chicago Declaration of Interdependence for a Sustainable Future," in *The Sustainable Design Resources Guide* (1993). Retrieved March 27, 2006, from American Institute of Architects Web Site: http://www.aiasdrg.org/sdrg.aspx?Page=5.

58. J. R. Mihelcic, J. C. Crittenden, M. J. Small, D. R. Shonnard, D. R. Hokanson, Q. Zhang, H. Chen, S. A. Sorby, V. U. James, J. W. Sutherland, and J. L. Schnoor, "Sustainability Science and Engineering: The Emergence of a New Metadiscipline," *Environmental Science and Technology,* **37**, 5314–5324 (2003).

59. M. Z. Hauschild, J. Jeswiet, and L. Alting, "Design for Environment—Do We Get the Focus Right?" *Annals of the CIRP,* **53**(1), 1–4 (2004).

60. V. J. Papanek, *The Green Imperative*. Thames and Hudson, New York, 1995.

61. F. Guo, and J. K. Gershenson, Comparison of Modular Measurement Methods based on Consistency Analysis and Sensitivity Analysis, Proceedings of the 2003 ASME Design Engineering Technical Conferences, Chicago, Illinois, 393–401, September 2–6, 2003.

62. F. Guo and J. K. Gershenson, "A Comparison of Modular Product Design Methods Based on Improvement and Iteration," Proceedings of the 2004 ASME Design Engineering Technical Conferences, Salt Lake City, Utah, September 28–Ocotber 2, 2004.

63. P. J. Newcomb, B. Bras, and D. W. Rosen, "Implications of Modularity on Product Design for the Life cycle," *Journal of Mechanical Design,* **120**(3), 483–490 (1998).

64. X. Qian, Y. Yu, and H. Zhang, "A Semi-quantitative Methodology of Environmentally Conscious Design for Electromechanical Products," Proceedings of the 2001 IEEE International Symposium on Electronics and the Environment, Denver, Colorado, pp. 156–160, May 7–9, 2001.

65. J. Sun, B. Han, S. Ekwaro-Osire, and H. Zhang, "Design for Environment: Methodologies, Tools, and Implementation, *Journal of Integrated Design and Process Science,* **7**(1), 59–75 (2003).

66. D. Maxwell and R. van der Vorst, "Developing sustainable products and services," *Journal of Cleaner Production,* **11**(8), 883–895 (2003).

67. N. Borland and D. Wallace, "Environmentally Conscious Product Design: A Collaborative Internet-based Modeling Approach," *Journal of Industrial Ecology* 3(2 and 3), 33–46 (1999).

68. H. C. Zhang, and S. Y. Yu, "Environmentally Conscious Evaluation/Design Support Tool for Personal Computers," Proceedings of the 1997 IEEE International Symposium on Electronics and the Environment, San Francisco, California, pp. 131–136, May 5–7, 1997.

69. C. Mizuki, P. A. Sandborn, and G. Pitts, "Design for Environment—A Survey of Current Practices and Tools," Proceedings of IEEE International Symposium on Electronics and the Environment, Dallas, Texas, 1–6, May 6–8, 1996.

70. U. Tischner, and B. Dietz, "Checklists," in *How to Do EcoDesign?* U. Tischner, E. Schmincke, F. Rubik, M. Prösler, B. Dietz, S. Maβelter, and B. Hirschl (eds.), Verlag form GmbH, Frankfurt, 2000d, pp. 102–118.

71. D. L. Thurston and W. F. Hoffman III, "Integrating Customer Preferences into Green Design and Manufacturing," Proceedings of the 1999 IEEE International Symposium on Electronics and the Environment, Danvers, Massachusetts, pp. 209–214, May 11–13, 1999.

72. M. Finster, P. Eagan, and D. Hussey, "Linking Industrial Ecology with Business Strategy: Creating Value for Green Product Design," *Journal of Industrial Ecology,* **3**(1), 107–125 (2001).

73. E. Jones, D. Harrison, and J. McLaren, "Managing Creative Eco-innovation—Structuring Outputs from Eco-innovation Projects," *The Journal of Sustainable Product Design,* **1**(1), 27–39 (2001).

74. Design for Sustainability Program, Delft University of Technology, "Product info." (2005). Retrieved April 3, 2006, from IdeMat Online, Design for Sustainability Program, Delft University of Technology from Web site: http://www.io.tudelft.nl/research/dfs/idemat/Product/pi_frame.htm.

75. S. Maßelter and U. Tischner, "Software tools for ecodesign," in *How to Do EcoDesign?* U. Tischner, E. Schmincke, F. Rubik, M. Prösler, B. Dietz, S. Maßelter, and B. Hirschl (eds.), Verlag form GmbH, Frankfort, 2000, pp. 147–149.

76. U. Tischner and B. Dietz, "Spider or Polar Diagrams," in *How to Do EcoDesign?* U. Tischner, E. Schmincke, F. Rubik, M. Prösler, B. Dietz, S. Maßelter, and B. Hirschl (eds.), Verlag form GmbH, Frankfurt, 2000b, pp. 91–97.

77. Environment Canada, "The Netherlands' Promise," *EcoCycle*, **5** (2003). Retrieved March 26, 2006, from http://www.ec.gc.ca/ecocycle/issue5/en/p15.cfm.

78. Y. Akao, *Quality Function Deployment: Integrating Customer Requirements into Product Design*, G. Mazur, Trans., Productivity Press, Cambridge, MA, 1990.

79. U. Tischner and B. Dietz, "Tools for Cost Estimation/Environmental Cost Accounting," in *How to Do EcoDesign?* U. Tischner, E. Schmincke, F. Rubik, M. Prösler, B. Dietz, S. Maßelter, and B. Hirschl (eds.), Verlag form GmbH, Frankfurt, 2000f, pp. 142–146.

80. B. Kassahun, M. Saminathan, and J. C. Sekutowski, "Green Design Tool," Proceedings of the 1993 IEEE International Symposium on Electronics and the Environment, Orlando, Florida, pp. 118–125, May 10–12, 1995.

81. U. Tischner, and B. Dietz, "MET Matrix and Ecodesign Checklist," in *How to Do EcoDesign?* U. Tischner, E. Schmincke, F. Rubik, M. Prösler, B. Dietz, S. Maßelter, and B. Hirschl (eds.), Verlag form GmbH, Frankfurt, 2000a, pp. 86–90.

82. U. Tischner and B. Dietz, "The Toolbox: Useful Tools for Ecodesign," in *How to Do EcoDesign?* U. Tischner, E. Schmincke, F. Rubik, M. Prösler, B. Dietz, S. Maßelter, and B. Hirschl (eds.), Verlag form GmbH, Frankfort, 2000c, pp. 65–70.

83. H. S. Matthews, J. Garrett, A. Horvath, C. Hendrickson, M. Legowski, M. Sin, J. Mayes, H. H. Ng, K. McCloskey, R. Ready, J. Knupp, V. Hodge, and S. Griffin, "eiolca.net" (2005). Retrieved March 26, 2006, from Carnegie Mellon University, Green Design Institute Web site: http://www.eiolca.net/index.html.
84. U.S. Environmental Protection Agency (EPA) Design for the Environment (DfE) Program and the University of Tennessee Center for Clean Products and Clean Technologies, "Cleaner Technologies Substitutes Assessment—Executive Summary" (2006). Retrieved March 26, 2006, U.S. Environmental Protection Agency Web site: http://www.epa.gov/opptintr/dfe/pubs/tools/ctsa/exsum/exsum.htm., U.S. Environmental Protection Agency (EPA), "A Quick Reference Guide to State Scrap Tire Programs: 1999 Update" (1999). Retrieved May, 19, 2006, from Web site: http://www.epa.gov/epaoswer/non-hw/muncpl/tires/scrapti.pdf.
85. Ricoh, "*Ricoh Group Sustainability Report (Environment)*" (2005). Retrieved April 4, 2006, from Ricoh Group Web site: http://www.ricoh.com/environment/report/pdf2005/49-50.pdf.
86. N. Chambers, C. Simmons, and M. Wackernagel, *Sharing in Nature's Interest: Ecological Footprints as an Indicator of Sustainability*, London, 2000.
87. E. G. Hertwich, W. S. Pease, and C. P. Koshland, "Evaluating the Environmental Impact of Products and Production Processes: A Comparison of Six Methods," *The Science of the Total Environment,* **196** (1), 13–29 (1997).
88. B. Jansen and A. Vercalsteren, "Eco-KIT: Web-based Ecodesign Toolbox for SMEs," Proceedings of the 2nd International Symposium on Environmentally Conscious Design and Inverse Manufacturing (EcoDesign'01), Tokyo, Japan, pp. 234–239, December 11–15, 2001.
89. M. Wackernagel, "Ecological footprint and appropriated carrying capacity: a tool for planning toward sustainability," Doctoral dissertation, University of British Columbia, 1994.
90. Pré Consultants, "Eco-Indicator 99" (2006). Retrieved May 18, 2006, from Web site: http://www.pre.nl/eco-indicator99/eco-indicator_99_introduction.htm.
91. F. Schmidt-Bleek and R. Klüting, *viel Umwelt braucht der Mensch?: MIPS, das Mass für ökologisches Wirtschaften*, Birkhäuser Verlag, Berlin, 1994.
92. U. Tischner and B. Dietz, "Tools for Setting Priorities, Making Decisions and Selecting," in *How to Do EcoDesign?* U. Tischner, E. Schmincke, F. Rubik, M. Prösler, B. Dietz, S. Maßelter, and B. Hirschl (eds.), Verlag form GmbH, Frankfurt, 2000e, pp. 130–141.
93. International Organization for Standardization, "Environmental Management: The ISO 14000 Family of International Standards" (2002). Retrieved May 18, 2006, from http://www.iso.org/iso/en/prods-services/otherpubs/iso14000/index.html.
94. U.S. Environmental Protection Agency (EPA), "Integrated Environmental Management Systems Partnership" (2006). Retrieved March 26, 2006, from U.S. Environmental Protection Agency, Office of Pollution Prevention and Toxics (OPPT), Design for Environment (DfE) Web site: http://www.epa.gov/opptintr/dfe/pubs/projects/iems/index.htm.
95. T. Korpalski, "Role of the 'Product Steward' in Advancing Design for Environment in Hewlett-Packard's Computer Products Organization," Proceedings of IEEE International Symposium on Electronics and the Environment, Dallas, Texas, pp. 37–41, May 6–8 1996.

96. Röhner Textil AG, "*Product Climatex® Life cycle*." Retrieved April 21, 2006, from Röhner Textil AG Web site: http://www.climatex.com/en/products/felt_climatex_life-cycle.html.
97. P. Storey, "Exploring New Horizons in Product Design" (2002). Retrieved April 21, 2006, from McDonough Braungart Design Chemistry, LLC. Web site: http://www.mbdc.com/features/feature_june2002.htm.
98. schamburg+alvisse, "Eco-evolution." Retrieved March 26, 2006, from Schamburgalvisse Web site: http://www.schamburgalvisse.com.au/about/eco-design.htm.
99. Environment Australia, "Product Innovation—The Green Advantage: An Introduction to Design for Environment for Australian Business" (2001). Retrieved March 26, 2006, from Australian Government, Department of Environment and Heritage Human settlements Web site: http://www.deh.gov.au/settlements/industry/finance/publications/producer.html.
100. Recycline, "Preserve." Retrieved March 29, 2006, from Recycline Inc. Web site: http://www.recycline.com.
101. Patagonia, "It Lives on" (2006a). Retrieved April 21, 2006, from Patagonia, Inc. Web site: http://www.patagonia.com/recycle/itlives_on.shtml.
102. Patagonia, "Organic Cotton" (2006b). Retrieved April 21, 2006, from Patagonia, Inc. Web site: http://www.patagonia.com/enviro/organic_cotton.shtml.
103. Patagonia, "Wear It Out" (2006c). Retrieved April 21, 2006, from Patagonia, Inc. Web site: http://www.patagonia.com/recycle/wearit_out.shtml.
104. J. McLain, "Patagonia seeks to recycle used capilene products into new clothing," *Ventura County Star (California)* (August 23, 2005), p. 1. Retrieved March 29, 2006, from http://www.enn.com/today.html?id=8591.
105. J. Makower, "Timberland Reveals its "Nutritional" Footprint" (2006). Retrieved March 29, 2006, from Web site: http://makower.typepad.com/joel_makower/2006/01/timberland_reve.html
106. Freeplay Foundation, "Lifeline Self-powered Radios Provide Sustainable Access to Our Five Areas of Focus." Retrieved March 29, 2006, from Freeplay Energy Plc. Web site: http://www.freeplayfoundation.org/.
107. Center for Resilience, "Concepts". Retrieved March 19, 2006, from The Ohio State University, Center for Resilience Web site: http://www.resilience.osu.edu/concepts.html.
108. P. Fitch and J. Cooper, "Life cycle Modeling for Adaptive and Variant Design. Part 1: Methodology," *Research in Engineering Design,* **15**(4), 216–228 (2005).
109. Freeplay Energy, "Lifeline: Aid/Relief radio." Retrieved March 29, 2006, from Freeplay Energy Plc. Web site: http://www.freeplayenergy.com/index.php?section=productsand-subsection=lifeline.
110. Change Design, "Multiple Life Products" (2004b). Retrieved March 29, 2006, from EcoDesign Foundation, now Change Design, Pathfinding Projects Web site: http://www.edf.edu.au/Sustainments/What_are/WhatAreNearlyMain.htm.
111. Change Design, "How We Came to Realize that There Was a Need for This Notion" (2004a). Retrieved March 29, 2006, from Ecodesign Foundation, now Change Design, Sustainments Web site: http://www.changedesign/Sustainments/What_are/WhatAreMain.htm.

112. G. Pahl, and W. Beitz (1996). *Engineering design: A systematic approach.* London: Springer.

113. U.S. Environmental Protection Agency (EPA), "Integrated Environmental Management Systems Implementation Guide" (2000). (EPA 744-R-00-011). Retrieved April 21, 2005, from U.S. Environmental Protection Agency Web site: http://www.epa.gov/dfe/pubs/iems/iems_guide/.

114. Environment Australia, "Design for Environment—Wharington 'Re-Define' Furniture" (2000). Retrieved May 18, 2006, from Australian Government, Department of Environment and Heritage, Human settlements Web site: http://www.deh.gov.au/settlements/industry/corporate/eecp/case-studies/wharington-dfe.html.

CHAPTER 4

FUNDAMENTALS AND APPLICATIONS OF REVERSE ENGINEERING

Kemper E. Lewis
University at Buffalo–SUNY, Buffalo, New York

Michael Castellani
University at Buffalo–SUNY, Buffalo, New York

Timothy W. Simpson
Penn State University, University Park, Pennsylvania

Robert B. Stone
University of Missouri–Rolla

William H. Wood
United States Naval Academy, Annapolis, Maryland

William Regli
Drexel University, Philadelphia, Pennsylvania

1 INTRODUCTION

Reverse engineering is finding more and more uses in modern engineering design research and application. It has moved well beyond its origin as largely a process of taking stuff apart for benchmarking purposes and is now a field marked by challenging research areas, strategic product-development life-cycle issues, innovative university-level pedagogical initiatives, and integration with state-of-the-art digital technologies and tools. Reverse engineering is in fact the exact opposite of design: Design maps a desired set of functions into new forms (products), while reverse engineering examines the existing form of an object to infer its function. In this chapter, we discuss the motivation behind reverse engineering and *product teardown* and describe the basic reverse engineering process using a one-time-use camera as an illustrative example. We conclude with our thoughts on future trends in reverse engineering research, implementation, and technologies.

1.1 Motivation for Reverse Engineering

Defined as the prediction of what a product should do, followed by modeling, analysis, dissection, and experimentation of its actual performance,[1] reverse engineering of competitors' products is used by many companies for both benchmarking and as a rich source of ideas for both product and process design.[2] For instance, General Motors' Vehicle Assessment and Benchmarking Activity Center dissects and analyzes nearly 40 of its competitors' vehicles each year using a tear-down process that takes about six weeks to complete.[3] Hoffman quotes auto industry analyst Lindsay Brooke, who emphasizes the importance of competitive teardowns: "As much as you think you know, nothing beats picking up the parts, feeling them, weighing them, and knowing the processes that made them."[3] General Motors uses information gathered from dissecting products to figure out how much each component costs to produce and how to cut costs on parts, shed weight, and improve manufacturing. In addition, important market trends and breakthroughs can be identified, allowing companies to make strategic decisions. Other U.S. automakers join General Motors in using competitive intelligence and teardowns: DaimlerChrysler disassembles competitors' products within its Competitive Teardown Operations department, and Ford has competitive intelligence teams in its Automotive Strategy and Corporate Strategy offices.[4,5] Automobile manufactures are joined by suppliers such as Lear, Johnson Controls, TRW, and Motorola, which conduct competitive intelligence activities with teardown rooms, competitor product databases, and part performance analyses.[6]

As operating costs and market pressures increase, automobile manufacturers may no longer be able to dedicate resources to perform teardowns in-house. Conover and Day describe the trend toward the use of independent product assessment companies that perform contract teardown and measurement work for automakers and suppliers.[5] This trend extends well beyond the automobile industry. For instance, Chipworks Inc. specializes in reverse engineering, including product teardowns of electronics.[7]

Some companies are turning to product dissection or product teardown to help improve their own product lines. Reverse engineering is an important step in the product redesign process,[1,8] which has been reported to account for as high as 90 percent of all product design in some industries.[9] Whirlpool Corporation holds an annual Supplier Innovation Challenge, during which suppliers have the opportunity to disassemble Whirlpool's products.[10] The goal of the competition is to identify ways to reduce costs, improve quality, and identify innovative ideas. In

1999, Ticona Corporation, one of Whirlpool's plastic suppliers, spent two days evaluating a high-volume washer and electric dryer. Selecting 12 of Ticona's 30 recommendations to investigate further, Whirlpool identified an aggregate potential savings of $7 million.[10]

Dofasco Inc., a fully integrated steel producer, has spent many years perfecting its appliance teardown process aimed at cost reduction for its customers.[11] Dofasco became more involved in the product-development processes of the appliance original equipment manufacturers (OEMs) that it services in the late 1980s, because customers were demanding cost-savings measures be taken by suppliers. Likewise, Henkel Corporation, maker of Loctite® and other commercial adhesives, offers appliance manufacturers a teardown service that identifies opportunities to substitute adhesives for costly mechanical fastening methods. The process thereby reduces component inventory and decreases total manufacturing costs.[12] Companies like Dofasco Inc. and Henkel Corporation realize that providing appliance OEMs with cost-reduction ideas and innovative solutions is a valuable customer service that sets them apart from other suppliers.

The emphasis on reverse engineering activities in industry and the recognition of the significant learning opportunities that occur when reverse engineering is properly conducted led to a resurgence of engineering dissection activities in U.S. universities that can be traced to Professor Sherri Sheppard's ME 99 Mechanical Dissection at Stanford in 1991.[13,14] Funded by the National Science Foundation–sponsored Synthesis Coalition, the course objective was to give mechanical engineering students an understanding of mechanical artifacts by answering the question, "How did others solve a particular problem?" Several courses followed, drawing heavily on the materials and activities in ME 99.[15–23] These course developments were in response to a general agreement by U.S. industry, engineering societies, and the federal government that there had been a decline in the quality of undergraduate engineering education over the previous two decades.[24,25] As a result, there was a push toward providing both intellectual and physical activities (such as dissection) to anchor the knowledge and practice of engineering in the minds of students.[16,26] Product dissection enables hands-on activities to couple engineering principles with significant visual feedback.[23] Dissection can also be used to increase awareness of the design process,[1] and such *learning-by-doing activities* encourage the development of curiosity, proficiency, and manual dexterity, three desirable traits of an engineer.[18] Dissection also gives students early exposure to functional products and processes, and introducing such experiences early in the students' academic careers has been shown to increase motivation and retention.[21]

Beyond the obvious pedagogical benefits and industry-specific learning that occurs, reverse engineering provides a wealth of benefits to establishing and maintaining marketplace competitiveness while also addressing sound environmental practices. These benefits are discussed in the following section.

1.2 Needs and Benefits

Initially, most reverse engineering was directed toward extending the useful life of existing products by providing the knowledge necessary to produce replacement parts. However, there are a number of needs for sound reverse engineering principles and processes in the context of environmentally friendly design.

First, many significant product and process performance issues such as manufacturability, ease of assembly, maintainability, serviceability, reusability, and level of recycling are difficult to simulate and require the collection and application of empirical experience to predict current and future product performance. Reverse engineering provides this platform to integrate empirical insights with design theory principles in order to better support a wide range of product decisions.

Second, in a knowledge economy, there is a clear need to create, capture, store, and reuse diverse forms of product and process knowledge. Reverse engineering provides an effective framework to extract and capture valuable knowledge in a wide range of design domains. An important challenge both to those doing reverse engineering and to those attempting to apply its lessons is how to capture the embedded knowledge in such a way that it can be both readily identified and applied to new design contexts.

Third, the manufacturability of a product and its environmental impact is not only influenced by design function and performance, but is primarily influenced by decisions about product configuration, design materials, and production processes that are made early in the design process. Insights gained from holistic reverse engineering practices can make a significant impact in reducing the environmental impact of a product and the manufacturing processes required to produce it. Viewing product design with a life-cycle perspective, reverse engineering should not merely reside at the CAD level of design representation but must recognize the relationships between product function, form, and performance. In addition, as *green* legislation spreads throughout the world economy, lessons drawn from industries that have already adapted to new regulations become quite valuable.

In the following section, the reverse engineering process is illustrated in the context of a product design application.

2 THE REVERSE ENGINEERING PROCESS WITH EXAMPLE

At the most basic level, the reverse engineering process consists of the following three steps:

1. Dissect the product.
2. Gather relevant product data.
3. Digitize the product.

To help demonstrate these steps, we use a Kodak one-time-use camera as an illustrative example. Once considered to be ecologically offensive by many environmental groups due to the initial disposable Kodak "Fling" perception, these cameras have since become the cornerstone of Kodak's recycling, remanufacturing, and reuse efforts, providing the best example of a closed-loop recycling program in the world, shown in Figure 1(a). According to Kodak's *2004 Annual*

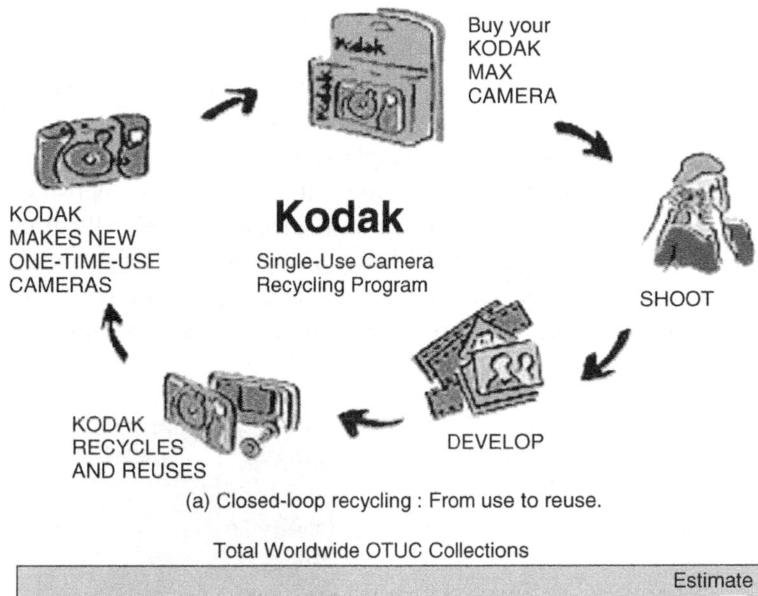

(a) Closed-loop recycling : From use to reuse.

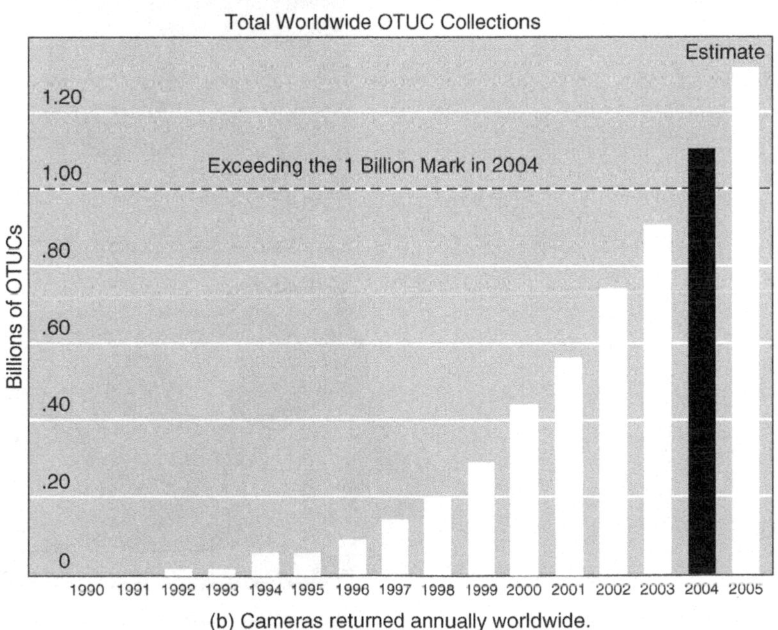

(b) Cameras returned annually worldwide.

Figure 1 Kodak one-time-use cameras. (From Kodak.com).

Report, one-time-use cameras are recycled at a rate of 74 percent in the United States (60 percent worldwide), surpassing that of corrugated containers (73 percent), aluminum cans (63 percent), and glass bottles (33 percent), with total worldwide returns exceeding one billion in 2004 (see Figure 1(b)). By weight, 77 to 90 percent of the material in a one-time-use camera is recycled, reused, or remanufactured, and the time it takes to recycle a one-time-use camera once it has been collected is about half that of an aluminum can (approximately 30 days versus 60 days). The extent to which these cameras can be reused and recycled continues to be improved, thanks to ongoing redesign efforts that were launched more than a decade ago by an integrated product development team composed of design, business, manufacturing, and environmental personnel. Consequently, dissection and analysis of different models of Kodak's one-time-use cameras provides interesting insight into these products as they have evolved to facilitate disassembly and part reuse while continuing to ensure high-quality photographs.

The one-time-use camera selected for this illustrative example is the Kodak Water & Sport model, shown in Figure 2. This model has a scratchproof lens, a large rubber band that serves as a wrist-strap, and a rugged, durable, shockproof shell made of rubber that seals the camera, making it waterproof up to 50 feet.

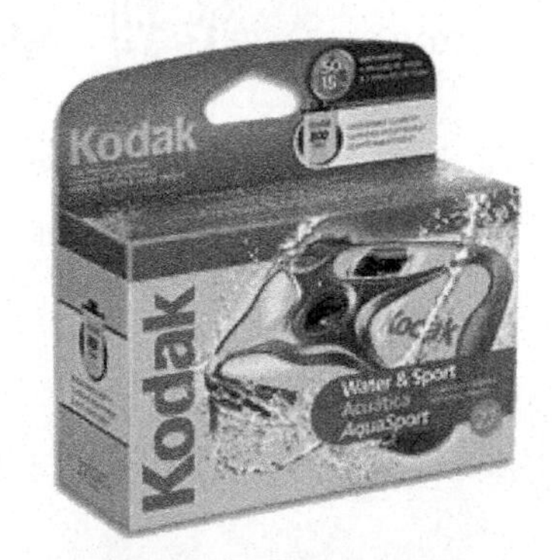

(a) Water & Sport packaging.

(b) Water & Sport one-time-use camera.

Figure 2 Kodak Water & Sport one-time-use camera.

It has 27 exposures, uses new 800-speed film, and does not have a flash, as it is intended for outdoor use. It is currently available in the market, selling for about $14.99 at retailers like Wal-Mart and Target.

2.1 Step 1: Dissect the Product

After the product has been identified for analysis and a team of experts assembled for this analysis, the first step is to dissect (disassemble) the product. Table 1 provides a comparison of the steps that are prescribed by four different product

Table 1 A Comparison of Steps in Product Dissection Process

	Steps in Product Dissection Process	Otto & Wood [1]	Ingle [27]	Whitney [29]	Ulrich & Pearson [30]
1	Predict design issues: customer needs, product function.	X			
2	Examine distribution and packaging.	X			
3	Prepare and document order of disassembly.	X	X		
4	Label removed components.	X		X	X
5	List components/subassemblies in hierarchical form; note product architecture.	X		X	
6	Record tools used.	X		X	
7	Take photographs of disassembly.	X		X	
8	List access direction.	X			
9	List orientation of product.	X			
10	Acknowledge/document expected permanent deformation.	X			
11	Create bill-of-materials (BOM).	X		X	X
12	Draw exploded view.	X		X	
13	Identify main function of components and subassemblies.	X		X	
14	Identify degrees-of-freedom (DOF).	X		X	
15	Understand inclusion of parts/features.	X		X	
16	Understand assemblability of product.	X	X	X	X
17	Record physical connections (liaison or assembly diagram).			X	
18	Record difficult steps or special considerations.	X		X	
19	Experiment with overall product.	X			

dissection processes that are published in the literature. The differences between them stem primarily from the intended level of detail that one wishes to capture, be it understanding simply how the components assemble together[27] or creating a detailed CAD model and functional representation of the product.[1] Regardless of the process selected, a plan for disassembly should be constructed first, listing the expected order of disassembly, tools for use, access direction, orientation of product (to prevent components from falling out), and any expected permanent deformation that may be caused by disassembly.[28]

Factors such as distribution and packaging should be included in the analysis, since these also contribute to the overall product cost. If possible, the means used to acquire parts, contain them, ship, distribute, and market the product are inspected and documented.[1] Photographs should be taken of the product before dissection begins and frequently during dissection to aid in product reassembly. As the product is disassembled, the function of every component can be determined by using, for instance, the Subtract-and-Operate Procedure (SOP) developed by Otto and Wood:[1,28]

1. Disassemble (subtract) one component from the assembly.
2. Operate the system through its full range.
3. Analyze the effect of removing the component through visual inspection or measurements.
4. Deduce the component's subfunction; compare to bill-of-material.
5. Replace the component; repeat for all parts in the assembly and all assemblies in the product.

An SOP table is created while performing the procedure, recording the part number, part description, and effect of removing the component.

For the Kodak Water & Sport one-time-use camera, the disassembly plan listed in Figure 3 is created to guide the product dissection process. The internal components of Kodak's one-time-use cameras are very similar; the unique aspect of this camera is its outer shell, which can be difficult to pry apart, as a tight, waterproof enclosure is needed. As noted in step 4 of Figure 3, the internal components for the shutter and film-advance mechanisms should be removed slowly and carefully, noting the correct assembly order so that the camera can be easily reassembled. These small parts are easily lost and can be difficult to reassemble due to their size.

2.2 Step 2: Capture Relevant Product Data

As the product is being dissected, relevant product data should be captured and stored. Table 2 summarizes in detail the information that is collected for each component or subassembly during the dissection processes for the same four methods listed in Table 1. The level of detail captured should be tailored to accommodate the objective(s) for which dissection is being performed.

(a) Step 1
Pop film advance wheel up and out from body; separate the outer shell—use a screw driver to pry apart the four hinges circled in the photo—and then separate the front and back covers from the body.

(b) Step 2
Remove film canister and film spindle.

(c) Step 3
Remove lens holder, outer lens shutter cap, shutter spring, shutter, eyehole, and inner lens. Be careful when removing the shutter spring as it is stretched very tightly, causing it to "fly away" easily.

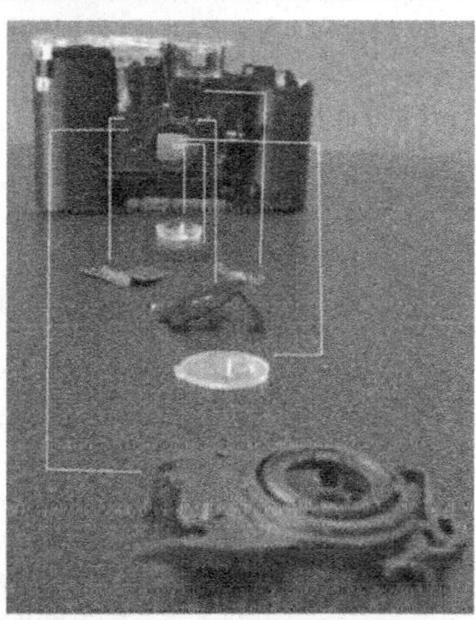

(d) Step 4
Remove view finder, frame counter, internal film advance wheel shutter lever, advance lock, camshaft, lever spring, and sprocket. Make sure to note the correct assembly order as well as the location of the ends of the helical spring for the film advance mechanism to work.

Figure 3 Disassembly plan for the Kodak Water & Sport one-time-use camera.

Returning to our camera example, the bill-of-materials (BOM) for the Kodak Water & Sport one-time-use camera is given in Table 3. Each row in the BOM lists a component found in the camera along with a brief description of it, its color, mass, material, manufacturing process, and up to three defining dimensions in this example (e.g., length, width, and height; diameter and thickness). The purpose of dissection and the desired level of detail dictate the specificity of this data (e.g., a component is plastic versus a component made from polycarbonate, which may be very important when considering a material's ability to be recycled). Physical connections between each of these components are captured within the

Table 2 Comparison of Data Recorded during Dissection

	Information Collected for Each Component/Subassembly	Otto & Wood [1]	Ingle [27]	Whitney [29]	Ulrich & Pearson [30]
1	Part number	X		X	X
2	Material	X	X	X	X
3	Description	X		X	
4	Effect of removing component	X			
5	Key dimensions	X	X		X
6	CAD model		X		
7	Quantity	X		X	X
8	Mass/weight	X	X		X
9	Color	X			X
10	Surface finish	X	X		X
11	Manufacturing method	X		X	X
12	Cost				X
13	Geometric complexity (rated from 1 to 5)				X
14	Geometric, spatial, and parametric tolerances		X	X	
15	Classification (type of component)			X	X
16	Function	X		X	
17	Name and location of manufacturer			X	X
18	Standard or unique part			X	X
19	Interfaces and connections			X	
20	Multiple states (e.g., on/off)			X	
21	Photograph	X			X

Design Structure Matrix given in Table 4 (*Note*: This matrix is symmetric about the diagonal; hence, only the lower half is shown). A Design Structure Matrix provides a matrix representation of an assembly diagram for a product.[31,32] In practice, a detailed assembly diagram would accompany this matrix to illustrate the types of connections between components, which is especially important for more complex products that employ a variety of connection types (e.g., screws, rivets, welding, soldering, brazing, snap fits, press fits). In this camera, the outer shell and housing use snap fits, as seen in Figure 3, while the remainder of the components simply slide into position or snap on and off to facilitate disassembly when the camera is returned to Kodak for recycling.

Table 3 Bill of Materials for the Kodak Water & Sport One-Time-Use Camera

Artifact Name	Part Number	Sub Artifact of	Quantity	Description	Artifact Color	Mass (kg)	Material	Manufacturing Process	Physical Parameters Label 1	(m)	Label 2	(m)	Label 3	(m)
Camera	1		1											
Front panel	2	Camera	1	Housing, contains other parts	Black with multi label	0.0093	Plastic	Injection molding	Length	0.1016	Width	0.0270	Height	0.0009
Back panel	3	Camera	1	Housing, contains other parts	Black with multi label	0.0118	Plastic	Injection molding	Length	0.1032	Width	0.0270	Height	0.0572
Film	4	Camera	1	Kodak 800 film, captures and stores image	Black, white, brown	0.0190	Various		Outer diameter	0.0250	Height	0.0450		
Film holder	5	Camera	1	Holds unexposed film	Black	0.0025	Plastic	Injection molding	Outer diameter	0.0191	Height	0.0365		
Lens cover	6	Camera	1	Holds outer lens in place	Black	0.0008	Plastic	Injection molding	Length	0.0380	Width	0.0100	Height	0.0290
Lens 1	7	Camera	1	Focus light	Clear	0.0003	Plastic	Injection molding	Outer diameter	0.0160	Height	0.0030		
Shutter cover	8	Camera	1	Holds lens in place	Black	0.0002	Plastic	Injection molding	Length	0.0190	Width	0.0080	Height	0.0160
Spring 1	9	Camera	1	Provides force to return shutter to original position	Copper	0.0000	Metal		Outer diameter	0.0020	Length	0.0070		
Shutter	10	Camera	1	Used to control exposure of the film	Dark grey	0.0003	Metal	Stamping	Length	0.0180	Width	0.0020	Height	0.0140
Lens 2 cover	11	Camera	1	Restricts light between inner and outer lens, covers inner lens	Black	0.0000	Plastic	Injection molding	Outer diameter	0.0140	Thickness	0.0001		
Lens 2	12	Camera	1	Focus light	Clear	0.0003	Plastic	Injection molding	Outer diameter	0.0135	Thickness	0.0048		
Film base	13	Camera	1	Main structural base of camera, holds other parts in alignment	Black	0.0147	Plastic	Injection molding	Length	0.1190	Width	0.0300	Height	0.0570

Table 3 *(continued)*

Artifact Name	Part Number	Sub Artifact of	Quantity	Description	Artifact Color	Mass (kg)	Material	Manufacturing Process	Physical Parameters					
									Label 1	(m)	Label 2	(m)	Label 3	(m)
Film advance wheel	14	Camera	1	Advance film	Blue	0.0046	Plastic	Injection molding	Length	0.0365	Width	0.0365	Height	0.0222
Washer	15	Camera	1	Makes film advance wheel watertight and allows it to rotate easier	Orange	0.0003	Rubber	Injection molding	Length	0.0143	Width	0.0143	Height	0.0024
Viewfinder	16	Camera	1	Allows user to frame picture and magnifies the counter wheel	Clear	0.0028	Plastic	Injection molding	Length	0.0619	Width	0.0238	Height	0.0175
Exposure counter	17	Camera	1	Counts the remaining exposures left on the film	Black and white	0.0003	Plastic	Injection molding	Length	0.0238	Width	0.0238	Height	0.0079
Film advance wheel 2	18	Camera	1	Internal wheel that turns to advance film when users turn outside wheel	Black	0.0014	Plastic	Injection molding	Length	0.0270	Width	0.0064	Height	0.0270
Advance arm	19	Camera	1	Stopped by the cam to prevent the film from over advancing	Black	0.0002	Plastic	Injection molding	Length	0.0175	Width	0.0175	Height	0.0048

Arm retainer	20	Camera	1	Mechanism's overall execution	Silver	0.0002	Metal	Stamping	Length	0.0803	Width	0.0064	Height	0.0021
Shutter arm	21	Camera	1	Cocks the shutter, preparing it for next picture	Black	0.0003	Plastic	Injection molding	Length	0.0206	Width	0.0127	Height	0.0159
Spring 2	22	Camera	1	Provides force to the shutter arm, helping with the mechanism's execution	Silver	0.0001	Metal	Wire drawn	Length	0.0095	Width	0.0143	Height	0.0024
Cam	23	Camera	1	Activates the shutter arm and locking	Black	0.0002	Plastic	Injection molding	Length	0.0095	Width	0.0079	Height	0.0206
Film advance gear	24	Camera	1	Uses the film motion to turn the cam	Black	0.0002	Plastic	Injection molding	Length	0.0143	Width	0.0143	Height	0.0016
Rubber band	25	Camera	1	Way for user to hold onto camera	Blue	0.0036	Rubber	Banded	Length	0.1524	Width	0.0016	Height	0.0064
Identific-ation label	26	Camera	1	Displays instructions and product info	Yellow, blue, red		Plastic	Stamping						
Waterproof back cover	27	Camera	1	Keeps water out of camera	Clear	0.0291	Plastic	Injection molding	Length	0.1191	Width	0.0222	Height	0.0746
Waterproof front cover	28	Camera	1	Keeps water out of camera	Clear	0.0354	Plastic	Injection molding	Length	0.1111	Width	0.0349	Height	0.0794

Table 4 Design Structure Matrix for the Kodak Water & Sport One-Time-Use Camera

Kodak Water & Sport	Advance arm	Shutter arm	Arm retainer	Back panel	Cam	Exposure counter	Film	Film advance gear	Film advance wheel	Film advance wheel 2	Film base	Film holder	Front panel	Identification label	Lens 1	Lens 2	Lens 2 cover	Lens cover	Rubber band	Shutter	Shutter cover	Spring 1	Spring 2	Viewfinder	Washer	Waterproof back cover	Waterproof front cover
Advance arm																											
Shutter arm																											
Arm retainer	1																										
Back panel																											
Cam	1	1																									
Exposure counter																											
Film																											
Film advance gear					1		1																				
Film advance wheel	1						1																				
Film advance wheel 2									1																		
Film base	1	1	1		1	1	1																				
Film holder							1				1																
Front panel				1																							
Identification label				1									1														
Lens 1																											
Lens 2											1																
Lens 2 cover																1											
Lens cover															1												
Rubber band																											
Shutter																	1										
Shutter cover															1					1							
Spring 1											1									1							
Spring 2		1									1																
Viewfinder											1																
Washer										1																	
Waterproof back cover				1						1									1								
Waterproof front cover										1			1												1	1	

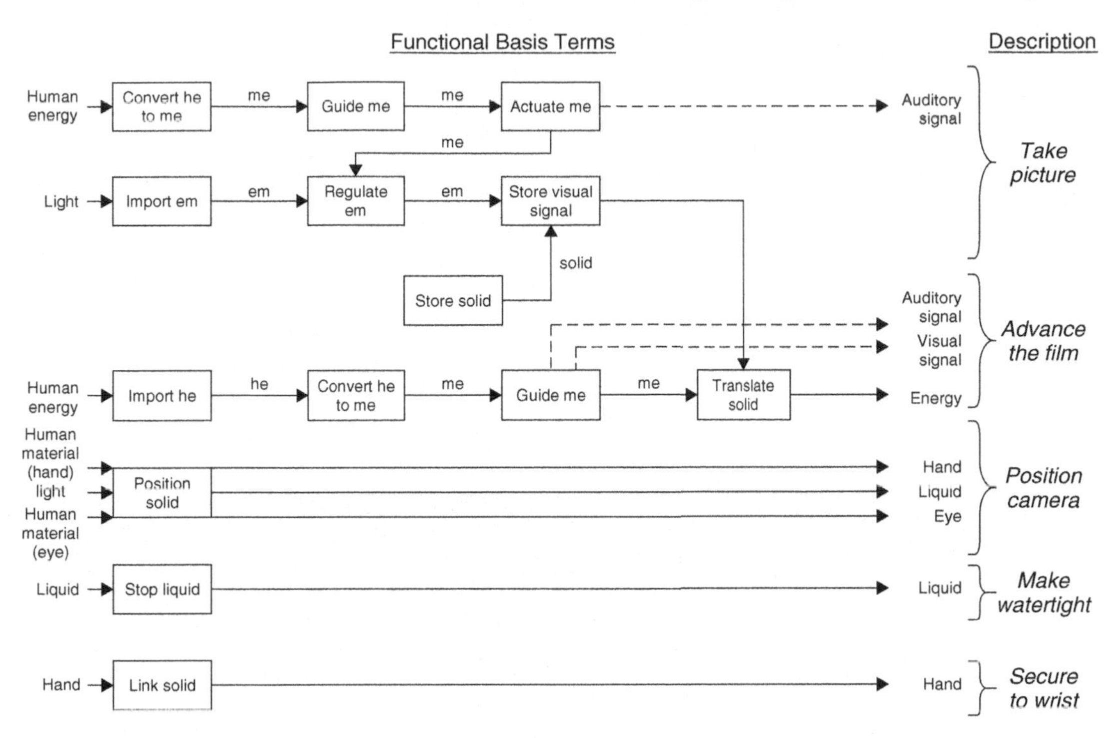

Figure 4 Function structure for the Kodak Water & Sport one-time-use camera (he = human energy, me = mechanical energy, em = electromagnetic energy).

The function structure for the camera is shown in Figure 4. The function structure seeks to provide a form-independent representation of the product that describes how the product functions at a sufficient level of abstraction.[33] The function structure in Figure 4 uses terms from the *functional basis* to describe the functionality shown on the far right-hand side of the figure.[34] Combining this information with the components listed in the BOM creates the function-component matrix shown in Table 5. This matrix provides a mapping between a product's components and its subfunctions[35] and represents the product architecture for the camera as it indicates how the different functions of the camera are mapped to its physical components.

2.3 Step 3: Digitize the Product

The final step in the reverse engineering process typically involves creating a 3D digital representation of the product. Traditionally, this has been a rather painstaking task that involves measuring all of the component's dimensions by hand and then manually recreating each component using CAD software (e.g., AutoCAD, SolidWorks, Pro/Engineer, CATIA) to generate a 3D solid model or a set of 2D drawings. Figure 5 gives examples of several CAD models that have been created directly from actual components in the Kodak Water & Sport

Table 5 Function-Component Matrix for the Kodak Water & Sport One-Time-Use Camera

KODAK Water & Sport	Arm retainer	Advance arm	Shutter arm	Waterproof front cover	Cam	Exposure counter	Film advance gear	Spring 2	Battery	Back panel	Film	Film advance wheel	Film base	Film holder	Front panel	Viewfinder	Lens 1	Lens 2	Lens 2 cover	Shutter	Shutter cover	Spring 1	Film advance wheel 2	Rubber band	Waterproof back cover	Washer	Identification label
Actuate me		1	1					1												1	1	1					
Convert he to me				1																			1				
Guide me	1				1	1						1		1													
Import em																1	1	1									
Import he				1																			1				
Position solid				1											1										1		
Regulate em																	1	1	1	1							
Store solid											1		1														
Store visual										1	1				1												1
Translate solid							1																1				
Link solid																								1			
Stop liquid				1																					1	1	

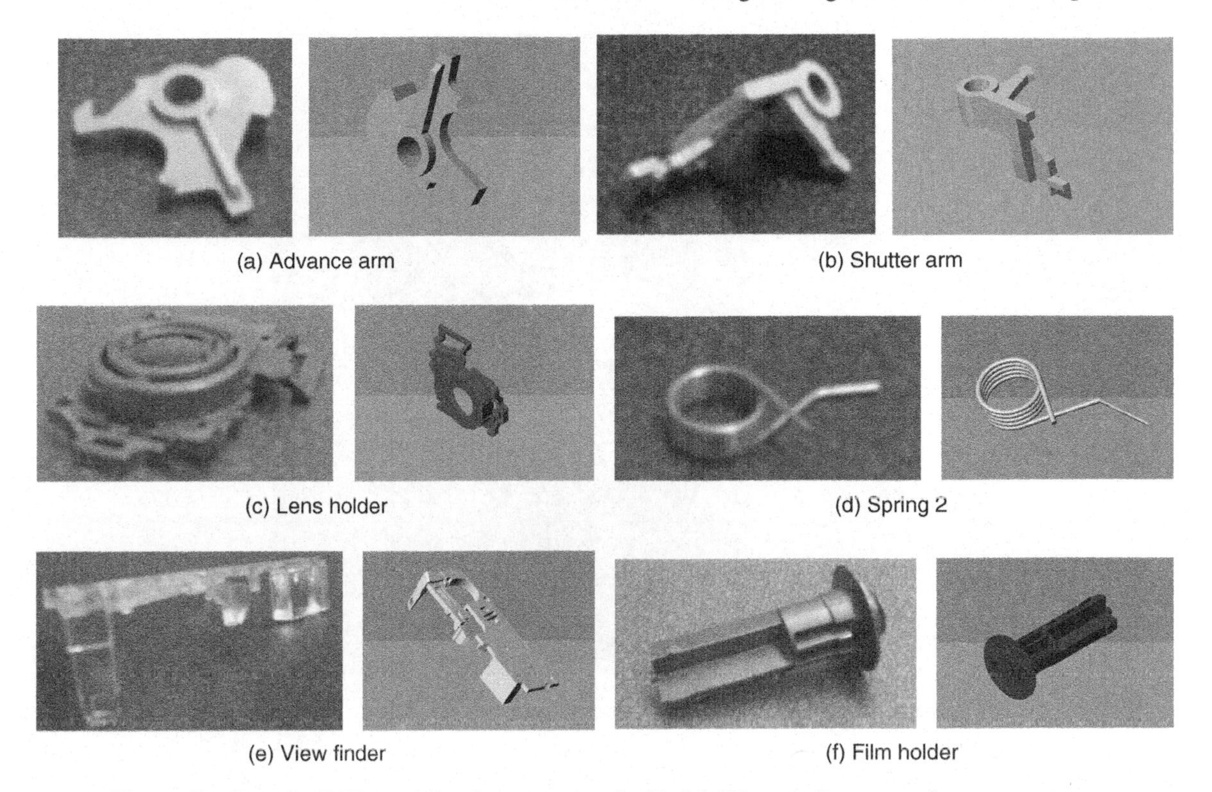

Figure 5 Sample CAD models of components in Kodak Water & Sport one-time-use camera.

one-time-use camera. These models can often shed light on the designer's intent, but it is impossible to recreate the parts exactly, as there is no way to measure or know the tolerances or exact specifications for each part. The models can be used, however, to create assembly drawings of the camera, which might provide a starting point for new design ideas. As an example, Figure 6 shows an assembly drawing of the internal housing that contains many of the parts in Figure 5.

The availability of 3D digitization technologies has made it considerably easier to acquire accurate 3D models of existing artifacts automatically. In the recent past, models could be acquired by inspection with a point-probe inspection machine, but these machines are very costly, depending on the accuracy requirements. Furthermore, these machines must be programmed based on a well-defined inspection plan designed to probe each model surface adequately enough to get reliable estimates of curvature, as well as other geometric and topological parameters.

Currently, advances in 3D laser scanning and stereovision have created huge reductions in the cost of 3D data acquisition. 3D laser scanners operate by tracing a laser beam across an artifact and sampling distances to points that the laser strikes on the object's surface. The distance is measured by calibrating the laser's

Figure 6 Assembly drawing of internal housing of Kodak Water & Sport One-time-use camera.

optics and calculating the angle between the laser emitter and a separate camera that finds where it strikes the surface. This information is used to produce a set of point locations in 3D space, where each point is an estimate of where the laser struck the surface of the object, as shown in Figure 7(a). The resulting data structure is referred to as a *point cloud.* Unlike traditional CAD/CAM data structures, the point cloud is just a set of points in 3D with no reference to topology (other than a vector that approximates the normal vector on the model surface at that location). To build a full 3D model, the scanning process may involve a turntable that rotates the object in regular increments, shown in Figure 7(b). In this way, the scanner can acquire a complete 3D model of the artifact by getting different point clouds for each angle (e.g., 5-degree increments would mean 72 point clouds) and performing *object registration* to connect and align the boundaries of these point clouds to create a single 3D model. Laser scanners are typically configured for acquiring objects that fit in $1 m^3$ areas, usually smaller. Acquiring larger models requires custom set-ups, hardware, and calibration.

Rarely, however, is this process completely automatic. The laser scanner may miss internal geometry, features that arise due to the genus of the artifact, and geometry that is *occluded* (i.e., obstructed from the line of sight). Additional set-ups or human editing may be needed to fit points to the top and bottom of the model (i.e., the portions of the model that would be resting on the turntable

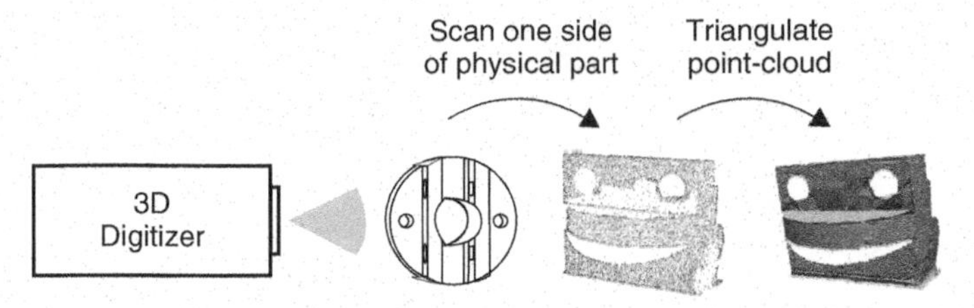

(a) Single scan: Take one scan from a single view.

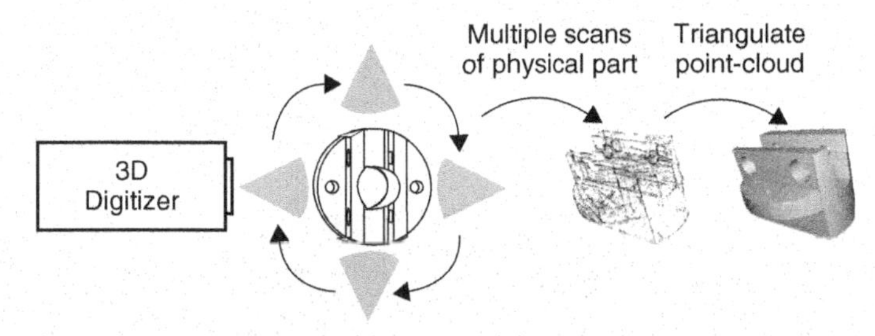

(b) Full scan: Take multiple scans from multiple views and register these scans together.

Figure 7 The 3D scanning process: Single vs. multiple scans.

and thus not scanned by the laser). The point clouds resulting from 3D scanning typically require human editing in order for them to be useful for downstream applications. The most common downstream application is to fit a surface to the points in order to impose a mesh or other regular topology to the otherwise unstructured point data.

As an example, Figure 8 and Figure 9 illustrate the 3D digitization process for the front outer shell and front housing inside the shell, respectively, of the Kodak Water & Sport one-time-use camera. Both figures include two views of raw data from an isometric scan (a & b), the merged and partially cleaned-up data (c), and the actual component that was scanned (d). These scans were produced using a Minolta Vivid 910 3D Scanner and GeoMagic Studio 6 software to merge and edit the data.

Before discussing some of the future directions of reverse engineering, we offer some observations about Kodak's one-time-use cameras, based on our experiences with reverse engineering these products over the past decade.

2.4 Some Observations about Kodak One-Time-Use Cameras

One of the first things that you notice upon opening the camera is that the film is unwound from the film canister and prewound into a roll in the camera; opening the camera without advancing through all the pictures will expose the film and ruin it. However, this allows the film to rewind into the canister as pictures are

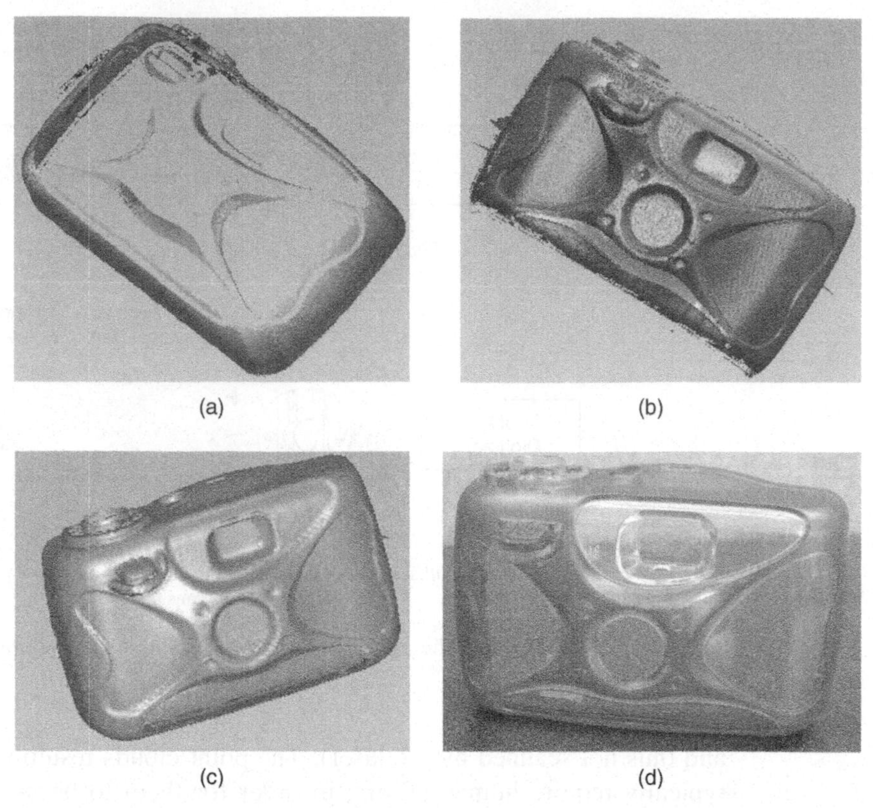

Figure 8 Digitization of front outer shell (a) Raw scan—Top, (b) Raw scan—side, (c) Merged images, and (d) Actual component.

taken, eliminating the need for a separate film rewind mechanism in the camera. Once all of the pictures have been taken, the camera is dropped off at a photo finisher, who removes the film canister to develop your pictures and returns the camera body to Kodak for reuse and recycling (manufacturers are paid a small fee as incentive to recycle). If the camera has a flash, the batteries can be removed by the photo finisher and sold to another industry, since they typically have more than half of their useful life left; if returned to Kodak, they donate them to various charities for use. Meanwhile, the processing center disassembles each camera and proceeds through the process similar to that shown in Figure 10. Lenses, viewfinders, and external housings are ground up and combined with raw materials to make new external covers. The chassis and camera mechanism (and electronic flash system, if present) are tested, inspected, and reused, if possible; otherwise, they are scrapped. New lenses—to ensure optical purity for high-quality photographs—and new film (and new batteries if the camera has a flash) are added to make a "new" one-time-use camera, which is packaged and shipped to a retailer to begin the cycle all over again.

Figure 9 Digitization of front housing inside outer shell: (a) Raw scan—top, (b) Raw scan—side, (c) Merged images, and (d) Actual component.

It is interesting to note that initially some photo finishers were recycling the cameras directly, without returning them to Kodak. As one can image, without proper inspection and testing, the quality of the photographs could degrade very quickly if the cameras are simply reused time and again. Hence, Kodak designed a special film canister that only it produces for its one-time-use cameras to prevent

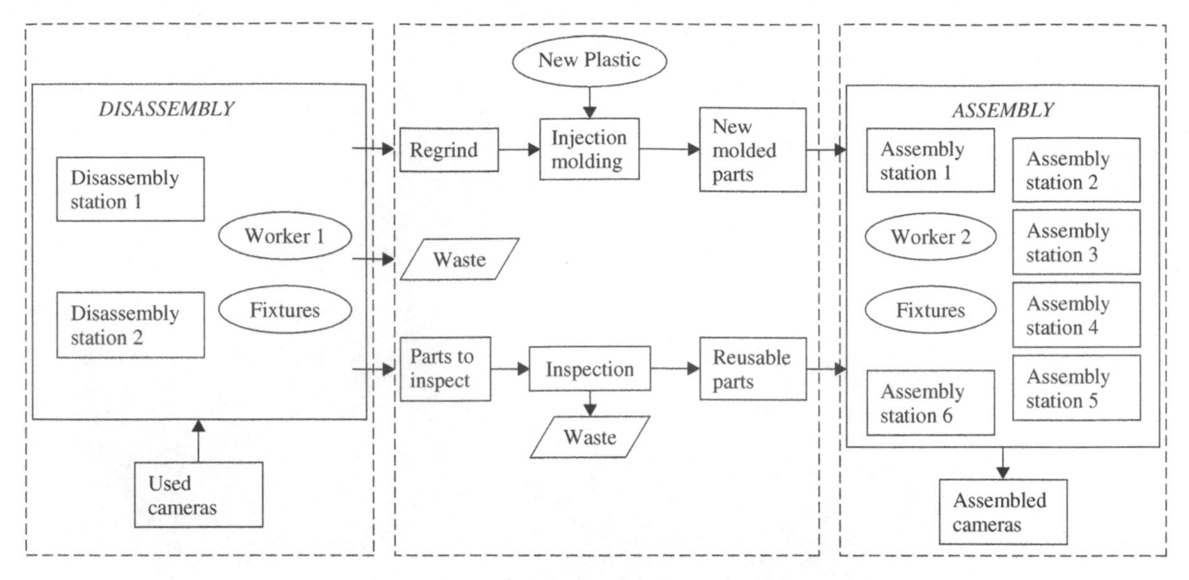

Figure 10 Camera disassembly and recycling process.

Figure 11 Film canister for one-time-use cameras.

this from happening (i.e., this film canister cannot be purchased separately from the camera itself). Figure 11 shows the unique saw-tooth pattern on top of the film canisters that are used in all Kodak one-time-use cameras, which mates perfectly to the underside of the film advance wheel, that is also shown in Figure 11.

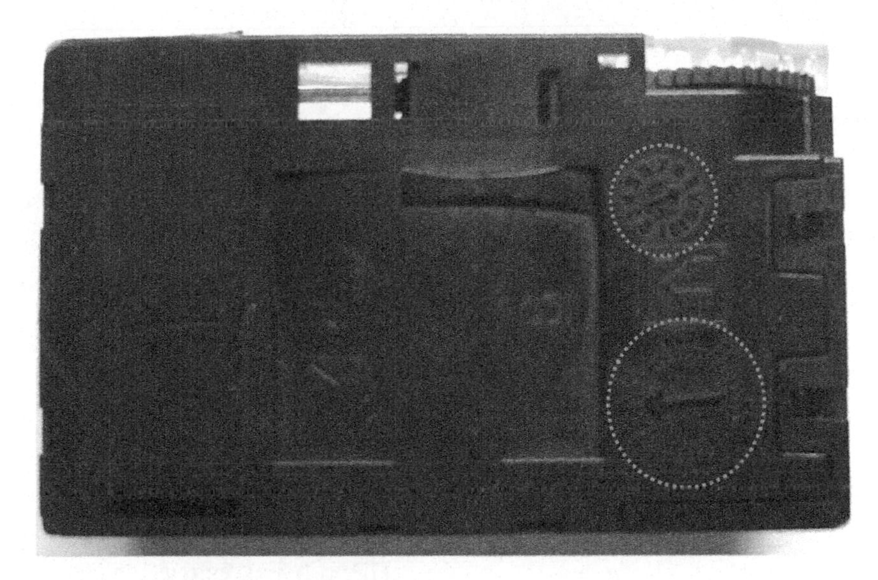

Figure 12 Date stamp on rear housing.

Another interesting feature that can be found on many of the large, plastic camera components is a date stamp, indicating when the part was made. As circled in Figure 12, this date stamp often takes the form of a clock with an arrow pointing to the month in one (top circle in figure) and day of the month in the other (bottom circle in figure). This form of a date stamp is easy to change in the mold itself before parts are made and helps Kodak keep track of when particular parts are made (e.g., what percentage of new and recycled material was used in a given batch of plastic on a particular day), providing useful information for the recycling process since the time between when a camera is made and when it is returned to the photo finisher is highly variable.

Finally, it has been interesting to observe the changes that the one-time-use cameras have undergone since they were first manufactured in 1988 to improve their recyclability and facilitate their disassembly. An exploded view of the internal chassis of the original FunSaver camera is shown in Figure 13. The components of the camera mechanism itself have remained largely unchanged since first introduced, as can be seen in Figure 14, which shows the camera mechanism for the Kodak Water & Sport camera that is available in the market over 15 years later.

One part that has noticeably evolved over time is the top cover and view finder, which was originally a single piece, as seen in Figure 13. This monolithic piece has become smaller and has since evolved into two separate pieces, as can be seen in Figure 15. As this part has evolved, the percentage of clear transparent plastic used in the view finder has decreased as well, enabling Kodak to recycle a higher percentage of plastic. The clear view finder uses raw (virgin) material to

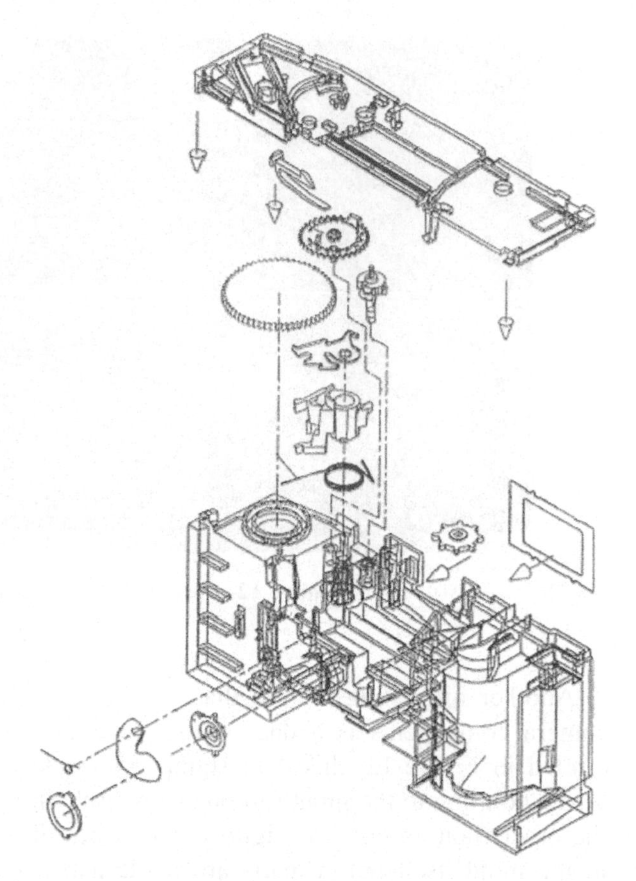

Figure 13 Exploded view of chassis of Kodak's FunSaver one-time-use camera.

Figure 14 Camera mechanism in Kodak Water & Sport one-time-use camera.

(a) Top view of finder on one-time-use cameras.

(b) Close-up of view finder in one-time-use cameras.

Figure 15 View finder in four of Kodak's one-time-use cameras.

ensure the highest possible quality, while the shutter actuation button—circled for each camera in Figure 15(b)—can now be made from recycled plastic. The dates listed beside the model name in Figure 15 are the copyright date listed on the packaging, and while it may appear that the evolution took a step backward in the Water & Sport camera that we analyzed, we have found that this model tends to lag behind the other models by about one generation. In fact, there is usually a strong resemblance between the camera enclosed within the outer waterproof shell and the outdoor (nonflash) model of the previous generation's one-time-use cameras.

3 FUTURE OF REVERSE ENGINEERING

A critical issue in the successful and continued implementation of reverse engineering is the ability to capture and represent product information in a consistent, standardized way. Digital repositories are emerging as the preferred medium for archiving product data. These repositories are digital storehouses or libraries where information is cataloged for effective storage, efficient retrieval, and secure use. The main purpose of product repositories is to capture and store design information in a format that (1) facilitates analyses or empirical studies of historical product data, (2) promotes consistency of design information, and (3) allows past design data to be reused for design synthesis activities. These repositories offer an improvement over current methods in which both the resolution and allocation of product information may fluctuate during reverse engineering activities due to the perceived issues of importance from the person performing the task. By capturing product data systematically at each individual level, engineers can achieve a more comprehensive view of the product.

In hopes of providing guidance to product representation research, the National Institute of Standards and Technology (NIST) has developed a set of information models for modeling product knowledge at varying levels of detail. There are several data entities that allow for a variety of aspects of a product description to be represented. The classes specified in the NIST Core Product Model include: artifact, function, transfer function, flow, form, geometry, material, behavior, specification, configuration, relationship, requirement, reference, and constraint.[36] Along with these classes, there is a set of specific information needed with each item and a specified type of value that can be entered; however, all of NIST's identified information models exceed the representational capabilities of current commercial function-based computational design tools.

Two distinct, yet complementary, repositories that have excellent capabilities in the context of reverse engineering have evolved from NIST's Design Repository Project: the University of Missouri–Rolla's Design Repository and the National Design Repository hosted at Drexel University.[37–39] Initial efforts for the UMR Design Repository were based on the NIST Core Product Model,[36] and recent efforts have established an enhanced bill of materials (EBOM) that handles entry, management, and export of repository knowledge. The UMR Design Repository (sample screenshot shown in Figure 16) focuses primarily on design artifacts, which may be composed of additional subartifacts, serving as an archive of expert knowledge to support novice designers and concept generation techniques.

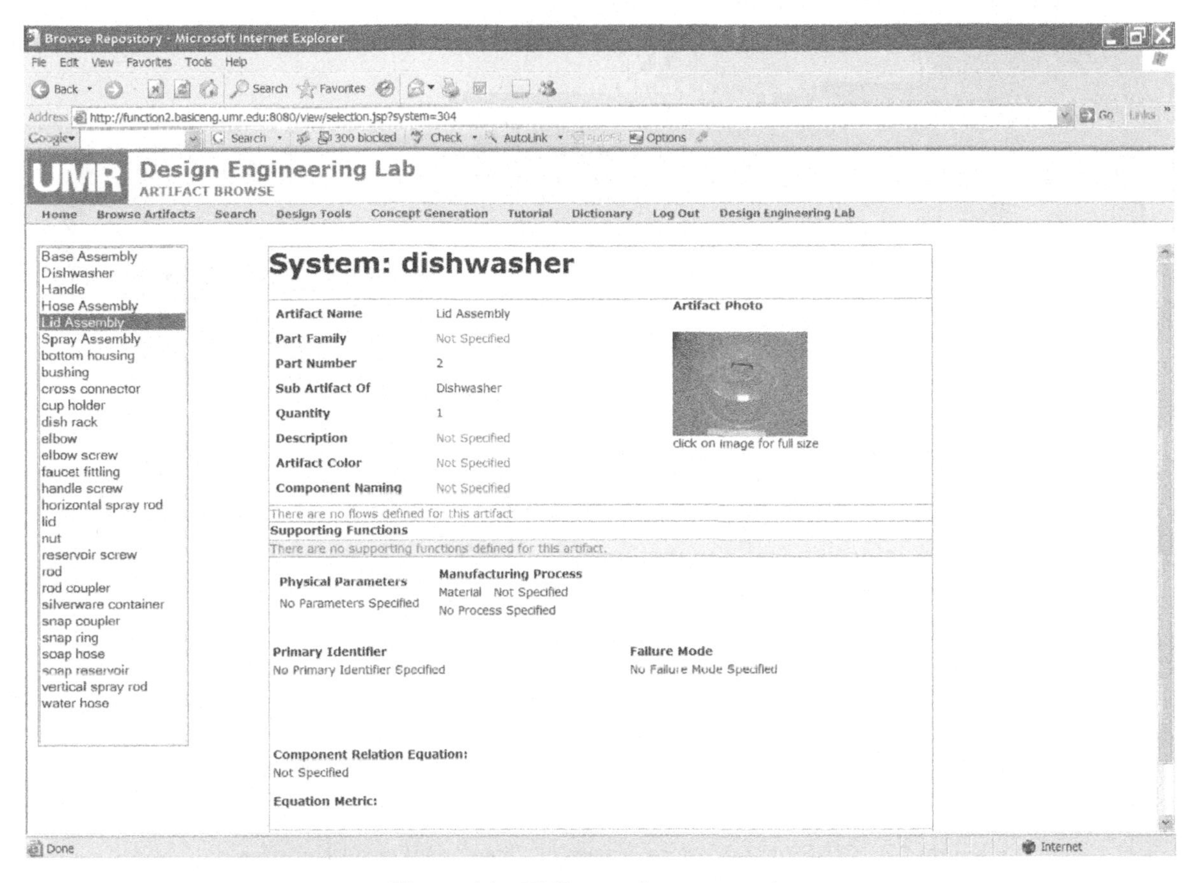

Figure 16 UMR repository screenshot.

The National Design Repository at Drexel University is a digital library of Computer-Aided Design (CAD) models and engineering data from a variety of domains, consisting of more than 55,000 CAD models and assemblies.[40] This repository currently serves about 1,000 users a month. Parts from the Repository have become standard benchmarks in reverse engineering, graphics, computer-aided design, engineering design, and process planning research. A screenshot of the repository from the Cyber-Infrastructure-Based Engineering Repositories for Undergraduates (CIBER-U) national research project between Penn State University, the University of Missouri–Rolla, Drexel University, and the University at Buffalo is shown in Figure 17.

Further capitalizing on digital technologies, researchers at Iowa State are developing virtual assembly technologies in an interactive and haptic feedback environment, as shown in Figure 18.[41,42] This kind of technology would allow design engineers to engage in reverse engineering processes without the cost of acquiring both actual products to dissect and product assembly procedures in order to rapidly check assembly rules, geometric constraints, and collision detection.

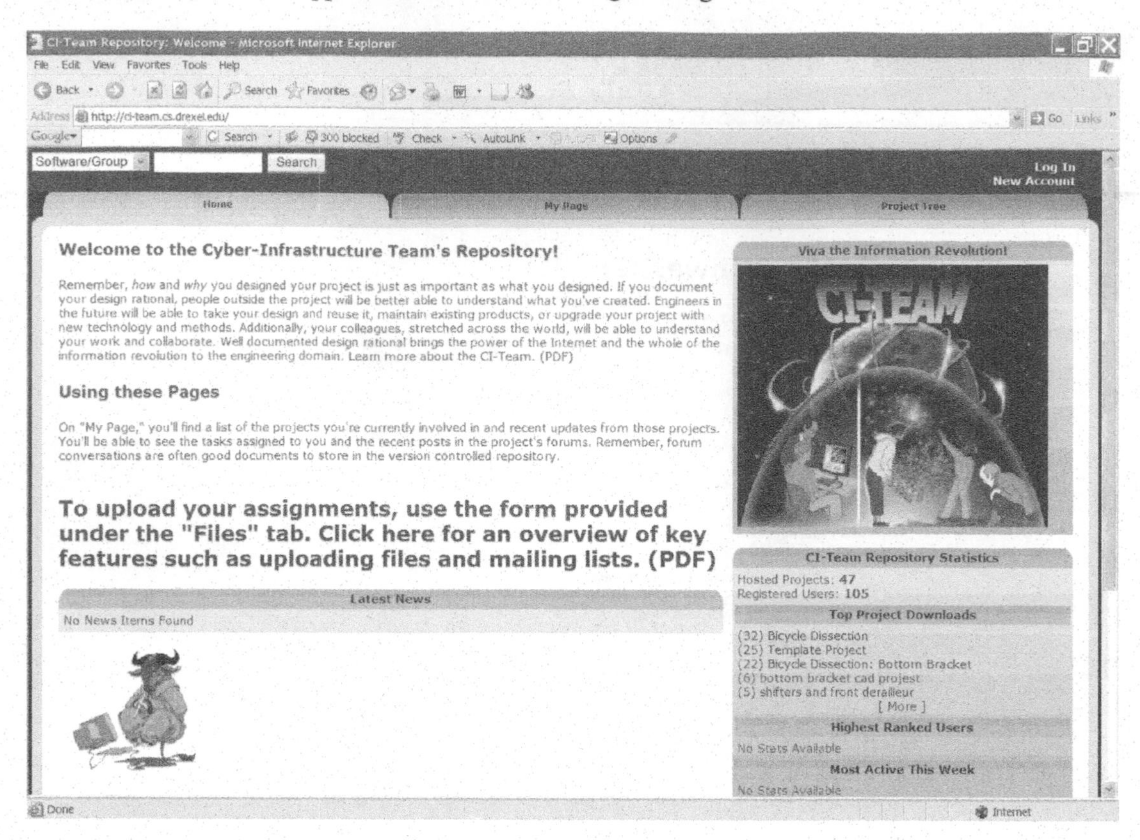

Figure 17 CIBER-U repository screenshot (housed at Drexel University).

Reverse engineering principles will also facilitate the education of a new generation of students on knowledge areas critical to their survival and success as engineers, such as functional modeling, competitive product design, information technology, globalization, and product platforming within an enterprise. However, the proliferation of reverse engineering principles and practice has even further-reaching implications. Consider Table 6, which presents excerpts taken from entry-level engineering positions at General Motors and Boeing. Although the job descriptions are typical for an engineering bachelor's degree, what is enlightening is the list of preferred skills.

A description of the tools and technologies from the two ads and their relevance to reverse engineering are summarized in Table 7. The sustained development of engineering principles, technologies, and tools will continue to shape and influence product development processes in globally competitive markets. Similarly, the teaching of these principles, technologies, and tools will help prepare a wide range of engineering students to enter the workforce with a more effective understanding of how to efficiently develop consumer-driven, cost-effective, and environmentally friendly products in a distributed, technology-mediated environment.

Figure 18 Virtual assembly interface (From Ref. 42).

Table 6 Comparison of Entry-Level Engineering Jobs—General Motors versus Boeing

General Motors	Boeing
Job Description: Work involves project design, research or experimental engineering through the development and testing of engineering projects where decisions are made within the limits of general practices and standards. *Preferred Skills:* – Familiarity with the following tools: Rhapsody, Simulink, Statemate, Magnum – Familiarity with the following technologies: UML, CAN, LIN, FlexRay, TTP, OSEK *Source:* http://www.gm.com/company/careers/job/intern_coop.html	*Job Description:* Engineer is responsible for supporting the Boeing International Space Station (ISS) Mechanical, Structural, and Robotics organization in Houston, TX *Job Requirements:* Requires a Bachelors Degree in mechanical engineering, aeronautical engineering or equivalent from an ABET-accredited institution *Preferred Skills:* Preference will be given to candidates with experience with Mastercam/Verisurf, Geomagic, or Polyworks *Source:* https://jobs.boeing.com/JobSeeker/JobSearch

Table 7 Relevance of Current Tools/Technology to Reverse Engineering

Tool/Technology	Description	Relevance to Reverse Engineering
Rhapsody	Model-driven development environment for systems engineering, allowing both functional oriented and object oriented design techniques to co-exist in one environment	CAD model environments, model manipulation and access
Statemate	Enables engineers to rapidly design and validate complex products through a unique combination of graphic modeling, simulation, code generation, documentation generation, and test plan definition	Model validation, product design, graphic modeling
Magnum	Enables supplier communication and component/service re-use using databases of information and models	Component reuse, model databases, and repositories
OSEK	Open systems and the corresponding computing interfaces for automotive electronics, providing a set of computing architectural standards	Database standards, communication standards
UML	Unified modeling language providing structured classes and model hyper-linking	Model validation and database communication
Mastercam/Verisurf	Computer-aided design (CAD) software for accurate dimensional inspection and manufacturing	Product dissection model inspection, accuracy of CAD models
Geomagic	Software that enables fast, easy-to-understand graphical comparisons between CAD models and as-built parts for reverse engineering, inspection, 2D and 3D geometric dimensioning, and automated reporting	Accuracy of CAD models, reverse engineering, product dissection
Polyworks	Software that creates 3D clouds of points from CAD models for use in digital repositories and part inspection	CAD model development, reverse engineering product models, model accuracy

ACKNOWLEDGMENTS

The National Science Foundation is gratefully acknowledged for their support of Kemper Lewis, Timothy Simpson, William Regli, Rob Stone, and Michael Castellani through the Cyberinfrastructure-TEAM program, grant SCI0537375.

REFERENCES

1. K. Otto and K. Wood, *Product Design: Techniques in Reverse Engineering and New Product Design*, Prentice Hall, Upper Saddle River, NJ, 2001.
2. K. T. Ulrich and S. D. Eppinger, *Product Design and Development*, McGraw-Hill, New York, 1995.
3. C. Hoffman, "The Teardown Artists," *Wired*, **136**(5) (2006).
4. M. Mian, "Modularity, Platforms, and Customization in the Automotive Industry," M.S. Thesis, Department of Industrial and Manufacturing Engineering, The Pennsylvania State University, University Park, PA, 2001.
5. G. Conover and S. Day, "Mining Your Competition," *Automotive Industries*, 32–33 (2002).
6. S. D. Upham, "Automotive Competitive Intelligence 101: A Beginner's Guide," *Automotive Design & Production*, **112**(8), 34 (2000).
7. D. James, "[Web Exclusive] Reverse Engineering Delivers Product Knowledge, Aids Technology Spread," *Electronic Design*, Jan 19, 2006. Accessed May 2006 at http://www.elecdesign.com/, ED Online ID #11966.
8. K. Otto and K. Wood, "A Reverse Engineering and Redesign Methodology for Product Evolution," *Proceedings of the 1996 ASME Design Theory and Methodology Conference*, 96-DETC/DTM-1523, Irvine, CA, 1996.
9. P. Sferro, G. Bolling, and R. Crawford, "Omni-Engineer," *Manufacturing Engineering*, 60–63 (June 1993).
10. D. Sheridan, B. Graman, K. Beck, and J. Harbert, "Improving from the Inside Out," *Appliance Manufacturer* (January 2001).
11. M. Stiller, Michael, "Appliance Teardown Process: Solutions in Steel," *Ceramic Engineering and Science Proceedings*, **21**(5), 127–130 (2000).
12. E. Fisher and S. McGrath, "Analyzing Alternatives," *Appliance Design*, **53**(3), 22–25 (2005).
13. S. Sheppard, "Mechanical Dissection: An Experience in How Things Work," Proceedings of the Engineering Education: Curriculum Innovation & Integration, Santa Barbara, CA, 1992.
14. S. Sheppard, "Dissection as a Learning Tool," Proceedings of the IEEE Frontiers in Education Conference, Nashville, TN, IEEE, 1992.
15. A. M. Agogino, S. Sheppard, and A. Oladipupo, "Making Connections to Engineering During the First Two Years," *22nd Annual Frontiers in Education Conference*, L. P. Grayson, ed., Nashville, TN, IEEE, pp. 563–569, 1992.
16. J. Lamancusa, M. Torres, V. Kumar, and J. Jorgensen, "Learning Engineering by Product Dissection," *ASEE Conference*, Washington, DC, ASEE, 1996.
17. J. S. Lamancusa, J. E. Jorgensen, and J. L. Zayas-Castro, "The Learning Factory—A New Approach to Integrating Design and Manufacturing into the Engineering Curriculum," *Journal of Engineering Education*, **86**(2), 103–112 (1997).
18. D. L. Beaudoin and D. F. Ollis, "A Product and Process Engineering Laboratory for Freshmen," *ASEE Journal of Engineering Education*, **84**(3), 279–284 (1995).

19. R. Felder, R. Beichner, L. Bernold, E. Burniston, P. Dail, and H. Fuller, "Update on IMPEC, An Integrated First-Year Engineering Curriculum at North Carolina State University," Proceedings of ASEE Annual Conference and Exhibition, Milwaukee, WI, ASEE, 1997.
20. S. K. Mickelson, R. Jenison, and N. Swanson, "Teaching Engineering Design through Product Dissection," *Proceedings of the ASEE Annual Conference and Exhibition,* Anaheim, CA, ASEE, 1995.
21. B. Carlson, P. Schoch, M. Kalsher, and B. Racicot, "A Motivational First-Year Electronics Lab Course," *ASEE Journal of Engineering Education,* **86**(4), 357–362 (1997).
22. C. Demetry and J. Groccia, "A Comparative Assessment of Students' Experiences in Two Instructional Formats of an Introductory Materials Science Course," *ASEE Journal of Engineering Education,* **86**(3), 203–210 (1997).
23. R. Barr, P. Schmidt, T. Krueger, and C.-Y. Twu, "An Introduction to Engineering through and Integrated Reverse Engineering and Design Graphics Project," *ASEE Journal of Engineering Education,* **89**(4), 413–418 (2000).
24. C. Fincher, "Trends and Issues in Curricular Development in Higher Education," In *Handbook of Theory and Research*, J. Smart (ed.), Agathon, New York, 1986, pp. 275–308.
25. L. M. Nicolai, "Designing a Better Engineer," *Aerospace America,* **30**(4), 30–33 (1995).
26. M. F. Brereton, "The Role of Hardware in Learning Engineering Fundamentals: An Empirical Study of Engineering Design and Dissection Activity," Ph.D. Dissertation, Mechanical Engineering, Stanford University, Palo Alto, CA, 1998.
27. K. A. Ingle, *Reverse Engineering*, McGraw-Hill, New York, 1994.
28. K. Otto and K. Wood, "Product Evolution: A Reverse Engineering and Redesign Methodology," *Research in Product Development,* **10**(4), 226–243 (1998).
29. D. E. Whitney, *Mechanical Assemblies: Their Design, Manufacture, and Role in Product Development*, Oxford University Press, New York, 2004.
30. K. T. Ulrich and S. Pearson, "Assessing the Importance of Design through Product Archaeology," *Management Science,* **44**(3), 352–369 (1998).
31. T. U. Pimmler and S. D. Eppinger, "Integration Analysis of Product Decompositions," *ASME Design Engineering Technical Conferences—Design Theory & Methodology*, Minneapolis, MN, ASME, **68**, 343–351 (1994).
32. D. M. Sharman and A. A. Yassine, "Characterizing Complex Product Architectures," *Systems Engineering,* **7**(1), 35–60 (2004).
33. G. Pahl and W. Beitz, *Engineering Design: A Systematic Approach*, Springer Verlag, 1996.
34. J. Hirtz, R. Stone, D. McAdams, S. Szykman, and K. Wood, "A Functional Basis for Engineering Design: Reconciling and Evolving Previous Efforts," *Research in Engineering Design,* **13**(2), 65–82 (2002).
35. B. Strawbridge, D. McAdams, and R. Stone, "A Computational Approach to Conceptual Design," *ASME Design Engineering Technical Conferences—Design Theory & Methodology Conference,* Montreal, Quebec, Canada, ASME, Paper No. DETC2002/DTM-34001, 2002.
36. S. Fenves, *A Core Product Model for Representing Design Information*, National Institute of Standards and Technology, Gaithersburg, MD, 2001.

37. M. R. Bohm, R. B. Stone, and S. Szykman, "Enhancing Virtual Product Representations for Advanced Design Repository Systems," *ASME Journal of Computing and Information Science in Engineering,* **5**(4), 360–372 (2005).
38. W. C. Regli, and V. A. Cicirello, "Managing Digital Libraries for Computer-Aided Design," *Computer Aided Design,* **32**(2), 119–132 (2000).
39. W. C. Regli and D. M. Gaines, "A Repository of Designs for Process and Assembly Planning," *Computer Aided Design,* **29**(12), 895–905 (1997).
40. W. C. Regli, and D. M. Gaines, "An Overview of the NIST Repository for Design, Process Planning, and Assembly," *Computer Aided Design,* **29**(12), 895–905 (1997).
41. A. Seth, H. J. Su, and J. M. Vance, "Development of a Dual-Handed Haptic Assembly System:SHARP," under consideration for *ASME Journal of Computing and Information Sciences in Engineering* (2006).
42. C-E. Kim and J. M. Vance, "Collision Detection and Part Interaction Modeling to Facilitate Immersive Virtual Assembly Methods," *ASME Journal of Computing and Information Sciences in Engineering*, **4**, 83–90 (2004).

CHAPTER 5

DESIGN FOR RELIABILITY

B. S. Dhillon
University of Ottawa, Dept. of Mechanical Engineering, Ottawa, Ontario

1 INTRODUCTION

Reliability is an important consideration in the planning, design, and operation of engineering systems because of its many benefits—including reduction in unexpected failures, improved safety, higher reliability, lower maintenance cost, less need for in-process inventory to cover downtime, and lower life cycle cost.

The history of the reliability field goes back to the early years of 1930s, when probability concepts were applied to problems related to electric power systems.[1–4] During World War II, Germans applied the basic reliability concepts to improve reliability of their V1 and V2 rockets. In 1950, the U.S. Department of Defense (DOD) formed an ad hoc committee on reliability. In 1952, this committee was transformed to a permanent body called the Advisory Group on the Reliability of Electronic Equipment (AGREE). In 1957, AGREE released a report that resulted in a specification on the reliability of military electronic equipment.[5,6]

The first book on reliability, *Reliability Factors for Ground Electronic Equipment*, was published in 1956.[7] Over the years, a vast number of publications on reliability have appeared, and today many scientific journals are totally or partially devoted to the field.[8]

A comprehensive listing of publications on various areas of reliability are available at the end of the chapter.[9,10] This chapter presents various important aspects of reliability.

2 COMMON RELIABILITY TERMS AND DEFINITIONS

There are many terms and definitions used in the field of reliability engineering.[11–14] Some of the common ones are as follows:[11–15]

- *Failure.* This is the inability of an item to perform within the initially defined guidelines.
- *Reliability.* This is the probability that an item will carry out its specified mission satisfactorily for the stated time period when used according to the specified conditions.
- *Mean time to failure (exponential distribution).* This is the sum of the operating time of given items divided by the total number of failures, represented as MTTF.
- *Hazard rate.* This is the rate of change of the failed items quantity over the total number of survived items at time t.
- *Useful life.* This is the length of time an item functions within an acceptable level of failure rate.
- *Downtime.* This is the time during which the item is not in a condition to perform its specified mission.
- *Redundancy.* This is the existence of more than one means for accomplishing a required function.
- *Mission time.* This is the time during which the item is performing its stated mission.
- *Repair rate.* This is a figure of merit that measures repair capability. In the case of the exponential distribution, it is the reciprocal of mean time to repair (MTTR).

- *Availability*. This is the probability that an item is available for application or use when required.
- *Active redundancy*. This is a type of redundancy when all redundant units are functioning simultaneously.

3 DESIGN FAILURES AND THEIR COMMON REASONS

History has witnessed many engineering design-related failures.[16,17] Some of the important ones that have occurred over the past 50 years are as follows:[8,16–18]

- In 1963, a U.S. Navy nuclear submarine called the *U.S.S. Thresher* slipped beneath the ocean surface by exceeding its designed maximum test depth and imploded.
- In 1980, in the North Sea, an offshore oil rig known as Alexander L. Kielland broke up under normal weather conditions because of a three-inch crack in a part close to a weld joint.
- In 1986, due to some design problems, the space shuttle *Challenger* exploded seconds after its launch, and all crew members lost their lives.
- In 1986, a reactor at the Chernobyl Nuclear Power Plant, Ukraine, exploded. The explosion killed 31 people and exposed thousands to radiation because of design-related problems.
- In 1988, a Boeing 737-200 lost its cabin roof during a flight because of various metal fatigue cracks emanated from rivet holes in the aluminium skin.

Over the years engineering professionals and others have studied various types of engineering design failures and concluded many different reasons for product/system failures ranging from a disaster to simple malfunction. Some of the common reasons are shown in Figure 1.[16]

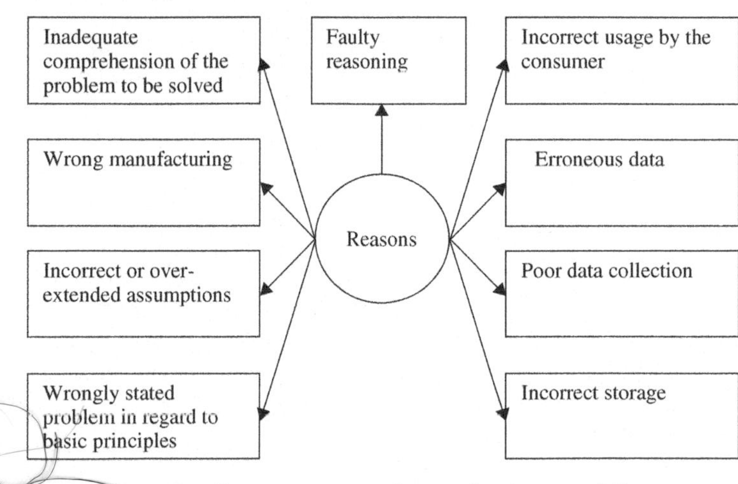

Figure 1 Common reasons for product/system failures.

4 DESIGN PROCESS ASPECTS WITH RESPECT TO PRODUCT RELIABILITY

Design process aspects can have significant impact on product reliability. Thus, they must be considered seriously. Some of these aspects are as follows:[19,20]

- Design simplicity
- Performance of failure mode and effect analysis
- Acquisitions of outside parts with care
- Use of parts with known history
- Simplicity in manufacturability
- Effective understanding of the requirements and the rules
- Confirmation of mechanical design choices mathematically as much as possible (e.g., bearings, load deflections, etc.)
- Making allowances for rough handling, use abuse, and transportation
- Appropriate allowance for possible test house approval requirements in design of mechanical, electrical, safety, and material selection aspects

5 BATHTUB HAZARD RATE CURVE

The *bathtub hazard rate curve*, so called because of its shape, is shown in Figure 2 and is often used to represent the failure rate of many engineering items, particularly the electronic parts. As shown in Figure 2, the curve is divided into three distinct parts: burn-in period, useful-life period, and wear-out period. During the *burn-in period*, the failures are quality-related and they decrease with time.[8,21] During the *useful-life period*, the hazard rate remains constant and the failures occurring during this period are stress-related and are known as *random failures*

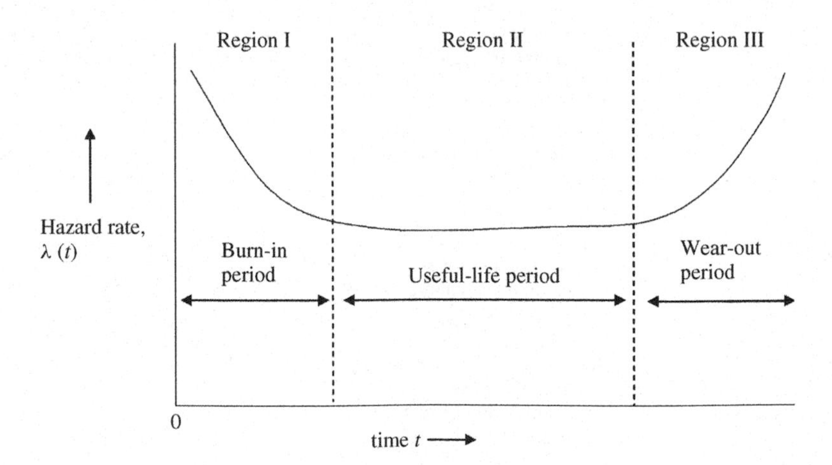

Figure 2 Bathtub hazard rate curve.

Table 1 Causes for Failures for Three Regions of the Bathtub Hazard Rate Curve.

Region	Failure Causes
I: Burn-in period	• Poor quality control • Poor test specifications • Inadequate manufacturing methods • Human error • Substandard materials and workmanship • Poor debugging
II: Useful-life period	• Low safety factors • Undetectable defects • Abuse • Natural failures • Higher random stress than expected • Wrong application
III: Wear-out period	• Wear due to aging • Corrosion and creep • Short designed-in life • Poor maintenance • Wear due to friction • Wrong overhaul practices

because they occur unpredictably. During the *wear-out period*, the hazard rate increases with time, and it starts when the item has bypassed its useful life. Table 1 presents some of the main reasons for failures over these three periods.

6 GENERAL RELIABILITY-RELATED FORMULAS

There are many formulas used in performing reliability analysis. This section presents four general formulas often used in reliability work, all are based on the item reliability function.

6.1 Failure Density Function

The *failure density function* is defined by

$$\frac{dR(t)}{dt} = -f(t) \tag{1}$$

where $f(t)$ is the item probability (failure) density function.
$R(t)$ is the item reliability at time t.

Example 1 Assume that an item's reliability function is defined by

$$R(t) = e^{-\lambda t} \tag{2}$$

where λ is the item constant failure rate.

Obtain an expression for the item failure density function.

Solution: By substituting equation (2) into equation (1), we get

$$\begin{aligned} f(t) &= \frac{de^{-\lambda t}}{dt} \\ &= \lambda e^{-\lambda t} \end{aligned} \tag{3}$$

6.2 Reliability Function

The *reliability function* is defined by

$$R(t) = 1 - \int_0^t f(t)\,dt \tag{4}$$

or

$$R(t) = \int_t^\infty f(t)\,dt \tag{5}$$

or

$$R(t) = e^{-\int_0^t \lambda(t)\,dt} \tag{6}$$

where $\lambda(t)$ is the item hazard or time dependent failure rate.

Example 2

Assume that an item's failure density function is given by equation (3). Obtain an expression for the item reliability function by using equations (4) and (5). Comment on the end result.

Solution: Using equation (3) in equation (4) yields

$$\begin{aligned} R(t) &= 1 - \int_0^t \lambda e^{-\lambda t}\,dt \\ &= e^{-\lambda t} \end{aligned} \tag{7}$$

Similarly, using equation (3) in equation (5), we get

$$\begin{aligned} R(t) &= \int_t^\infty \lambda e^{-\lambda t}\,dt \\ &= e^{-\lambda t} \end{aligned} \tag{8}$$

Equations (7) and (8) are identical. This proves that both equations (4) and (5) give the same result.

6.3 Hazard Rate Function

The *hazard rate function* is defined by

$$\lambda(t) = \frac{f(t)}{R(t)} \tag{9}$$

or

$$\lambda(t) = -\frac{1}{R(t)}\frac{dR(t)}{dt} \tag{10}$$

Example 3 Obtain an expression for hazard rate by using equations (7) and (10).

Solution: Using equation (7) in equation (10), we get

$$\begin{aligned} \lambda(t) &= -\frac{1}{e^{-\lambda t}}\frac{de^{-\lambda t}}{dt} \\ &= \lambda \end{aligned} \tag{11}$$

6.4 Mean Time to Failure

Mean time to failure (MTTF) can be obtained by using any of the following three formulas:

$$MTTF = E(t) = \int_0^\infty tf(t)\,dt \tag{12}$$

or

$$MTTF = \int_0^\infty R(t)\,dt \tag{13}$$

or

$$MTTF = \lim_{s \to 0} R(s) \tag{14}$$

where $E(t)$ is the expected value.
s is the Laplace transform variable.
$R(s)$ is the Laplace transform of the reliability function, $R(t)$.

Example 4 Prove by using equation (7) that equations (13) and (14) yield the same result.

Solution: By substituting equation (7) into equation (13), we get

$$\begin{aligned} MTTF &= \int_0^\infty e^{-\lambda t}\,dt \\ &= \frac{1}{\lambda} \end{aligned} \tag{15}$$

By taking the Laplace transform of equation (7), we get

$$R(s) = \frac{1}{s + \lambda} \tag{16}$$

Using equation (16) in equation (14) yields

$$\begin{aligned} MTTF &= \lim_{s \to 0} \frac{1}{(s + \lambda)} \\ &= \frac{1}{\lambda} \end{aligned} \tag{17}$$

Equations (15) and (17) are identical. Thus, it proves that equations (13) and (14) give the same result.

7 RELIABILITY CONFIGURATIONS

As engineering systems can form various different configurations with respect to reliability, this section presents reliability analysis of three commonly occurring networks or configurations.

7.1 Series Network

This is the simplest configuration, and its block diagram is shown in Figure 3. Each block in the diagram denotes a unit. In this configuration, all the units must work normally for the system to succeed. In other words, if any one of the units fails, the entire system fails.

For independent units, the reliability of the Figure 3 series system is

$$R_s = \prod_{i=1}^{K} R_i \tag{18}$$

where R_s is the series system reliability.
K is the total number of units.
R_i is the unit i reliability; for $i = 1, 2, 3, \ldots, K$.

For constant failure rate of unit i (i.e., for $\lambda_i(t) = \lambda_i$), the unit i reliability from equation (2) is

$$R_i(t) = e^{-\lambda_i t} \tag{19}$$

where λ_i is the unit i constant failure rate.
$R_i(t)$ is the unit i reliability at time t.

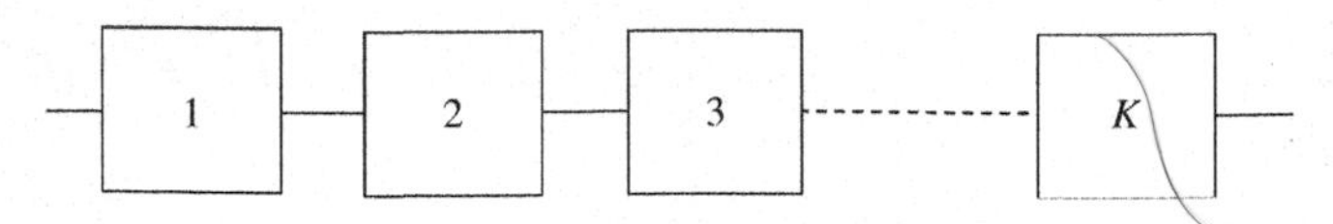

Figure 3 Block diagram representing a K unit series system.

Thus, for constant failure rates of all series units, by substituting equation (19) into equation (18), we get

$$\begin{aligned} R_s(t) &= \prod_{i=1}^{K} e^{-\lambda_i t} \\ &= e^{-\sum_{i=1}^{K} \lambda_i t} \end{aligned} \tag{20}$$

where $R_s(t)$ is the series system reliability at time t; for $i = 1, 2, 3, \ldots.., K$.

Substituting equation (20) into equation (13) yields

$$\begin{aligned} MTTF_s &= \int_0^\infty e^{-\sum_{i=1}^{K} \lambda_i t}\, dt \\ &= \frac{1}{\sum_{i=1}^{K} \lambda_i} \end{aligned} \tag{21}$$

where $MTTF_s$ is the series system mean time to failure.

Using equation (20) in equation (10) yields the following expression for the series system failure rate:

$$\begin{aligned} \lambda_s(t) &= \frac{1}{e^{-\sum_{i=1}^{K} \lambda_i t}} \cdot \frac{de^{-\sum_{i=1}^{K} \lambda_i t}}{dt} \\ &= \sum_{i=1}^{K} \lambda_i \end{aligned} \tag{22}$$

where $\lambda_s(t)$ is the series system failure rate.

7.2 Parallel Network

The block diagram of this network is shown in Figure 4. Each block in the diagram represents a unit. In this configuration, all the units are active or operate simultaneously, and at least one unit must work normally for the system to succeed. More specifically, the system fails when all its units fail.

For independent units, the parallel system reliability is given by

$$R_{ps} = 1 - \prod_{i=1}^{K} (1 - R_i) \tag{23}$$

where R_{ps} is the parallel system or network reliability.
K is the total number of units.
R_i is the unit i reliability; for $i = 1, 2, 3, \ldots., K$.

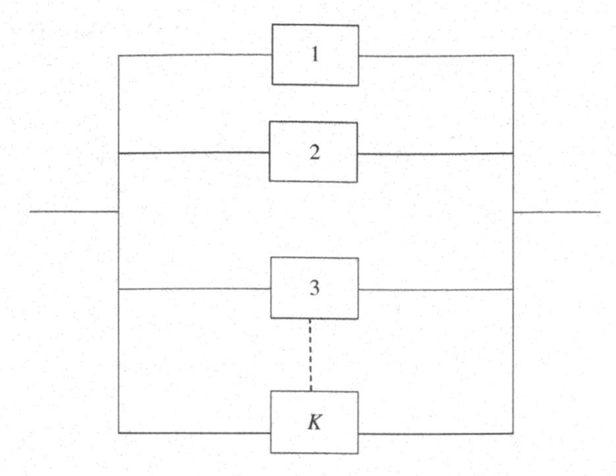

Figure 4 Block diagram representing a K unit parallel system.

For constant failure rates of all parallel units, by substituting equation (19) into equation (23), we get

$$R_{ps}(t) = 1 - \prod_{i=1}^{K}(1 - e^{-\lambda_i t}) \tag{24}$$

where $R_{ps}(t)$ is the parallel system or network reliability at time t.

For identical units, equation (24) becomes

$$R_{ps}(t) = 1 - (1 - e^{-\lambda t})^K \tag{25}$$

where λ is the unit constant failure rate.

Using equation (25) in equation (13) yields

$$\begin{aligned} MTTF_{ps} &= \int_0^{\infty} [1 - (1 - e^{-\lambda t})^K]\, dt \\ &= \frac{1}{\lambda} \sum_{i=1}^{K} \frac{1}{i} \end{aligned} \tag{26}$$

where $MTTF_{ps}$ is the parallel system or network mean time to failure.

7.3 *m*-out-of-*n* Network

In this case, the network or system is composed of n active units, and at least m of them must work normally for the system success. The previous two networks (i.e., series and parallel) are the special cases of this configuration for $m = n$ and $m = 1$, respectively.

Using the binomial distribution, for independent and identical units, the reliability of the m-out-of-n network is given by

$$R_{m/n} = \sum_{j=m}^{n} \binom{n}{j} R^j (1 - R)^{n-j} \tag{27}$$

$$\binom{n}{j} = \frac{n!}{(n-j)!j!} \tag{28}$$

where $R_{m/n}$ is the m-out-of-n network reliability.
R is the unit reliability.

For constant failure rates of all network units, substituting equation (2) into equation (27) yields

$$R_{m/n}(t) = \sum_{j=m}^{n} \binom{n}{j} e^{-j\lambda t}(1 - e^{-\lambda t})^{n-j} \tag{29}$$

where $R_{m/n}(t)$ is the m-out-of-n network reliability at time t.
λ is the unit constant failure rate.

By substituting equation (29) into equation (13), we get

$$\begin{aligned} MTTF_{m/n} &= \int_0^{\infty} \left[\sum_{j=m}^{n} \binom{n}{j} e^{-j\lambda t} \left(1 - e^{-\lambda t}\right)^{n-j} \right] dt \\ &= \frac{1}{\lambda} \sum_{j=m}^{n} \frac{1}{j} \end{aligned} \tag{30}$$

where $MTTF_{m/n}$ is the m-out-of-n network mean time to failure.

8 RELIABILITY EVALUATION METHODS

Over the years many reliability evaluation methods and techniques have been developed. This section describes some of the commonly used methods and techniques.[3,22,23]

8.1 Failure Modes and Effect Analysis (FMEA)

The failure modes and effect analysis (FMEA) is a widely used method in the industrial sector to analyze engineering systems from the reliability aspect. It may simply be described as an approach used to conduct analysis of each potential failure mode in the system in order to examine the results/effects of such failure mode on the system.[24] When criticalities or priorities are assigned to failure mode effects, FMEA is called failure mode, effects, and criticality analysis (FMECA).

The history of FMEA can be traced back to the early 1950s with the development of flight control systems.[25] A comprehensive list of publications on the

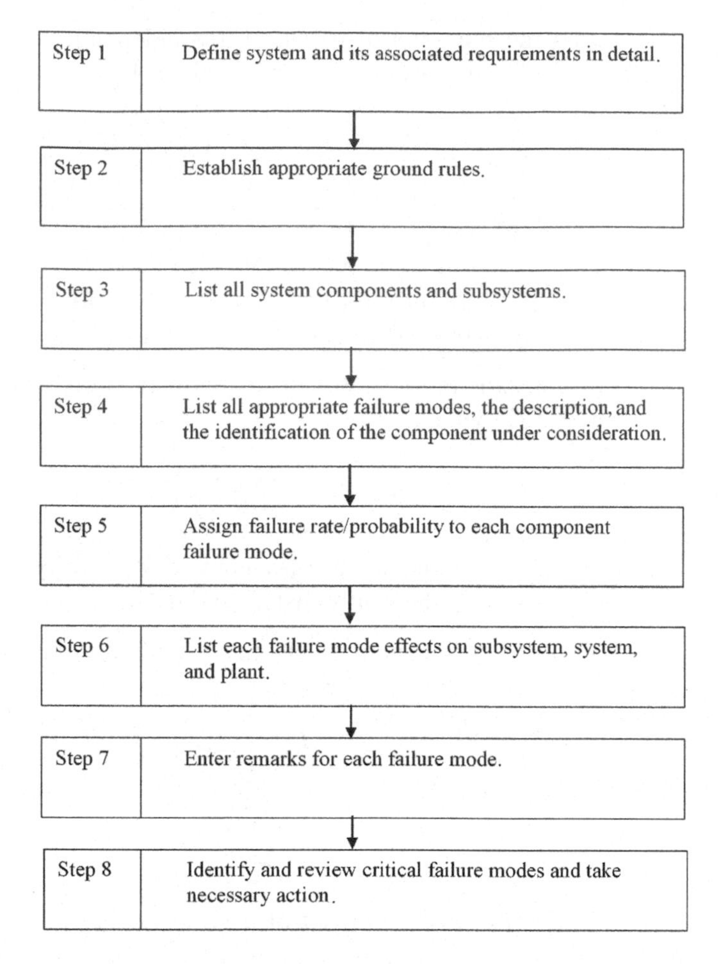

Figure 5 Main steps used in performing FMEA.

method is available in reference 26 at the end of this chapter.[26] The main steps followed to perform FMEA are shown in Figure 5. Additional information on FMEA is available in reference 8 at the end of this chapter.[8]

8.2 Parts Count Method

The parts count method is usually used to estimate equipment/system failure rate during bid proposal and early design phases. The type of information the method uses includes equipment use environment, part quality levels, and generic part types and quantities. Under single use environment, the equipment failure rate is estimated by using the following equation:[27]

$$\lambda_{eq} = \sum_{j=1}^{n} \alpha_j (\lambda_g \theta_q)_j \tag{31}$$

where λ_{eq} is the equipment failure rate, expressed in failures/10^6 hours.
n is the total number of generic part classifications in the equipment.
θ_q is the quality factor of generic parts of classification j.
α_j is the quantity of generic parts of classification j.
λ_g is the generic failure rate of generic part of classification j, expressed in failures/10^6 hours.

The tabulated values for θ_q and λ_g are available in reference 27 at the end of this chapter.[27]

8.3 Fault Tree Analysis

Fault tree analysis is widely used in industry to evaluate reliability of engineering systems—particularly, the ones used in the nuclear power generation. It was developed in the early 1960s at the Bell Telephone Laboratories to analyze the Minuteman launch control system.[22] A *fault tree* may be described as a logical representation of the relationship of primary fault events that result in the occurrence of a given undesirable event called the *top event*. It is depicted using a tree structure with logic gates such as OR and AND.

A comprehensive list of publications on the method is available in reference 9 at the end of this chapter.[9] Four commonly used symbols in performing fault tree analysis (FTA) are shown in Figure 6: Each of these symbols is described below.

1. *OR gate*. This denotes that an output fault event occurs if one or more of the input fault events occur.
2. *AND gate*. This denotes that an output fault event occurs only if all of the input fault events occur.

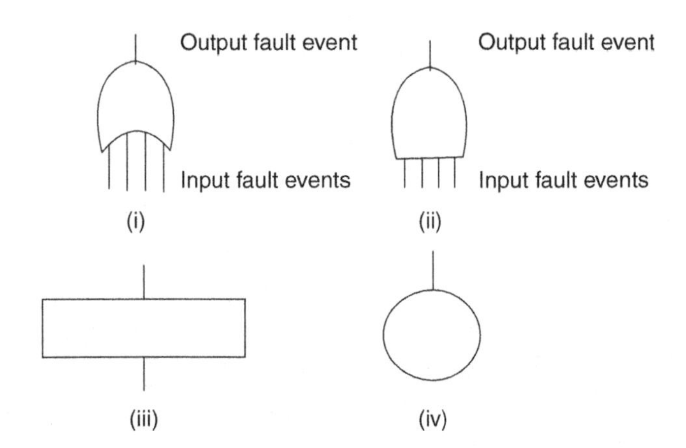

Figure 6 Four basic fault tree symbols: (i) OR gate, (ii) AND gate, (iii) rectangle, (iv) circle.

3. *Rectangle*. This represents a fault event that results from the logical combination of fault events through the input of a logic gate.
4. *Circle*. This denotes a basic fault event or the failure of an elementary component. The fault event's probability of occurrence, failure, and repair rates are normally obtained from field data.

The following example demonstrates the application of the four symbols.

Example 5 A windowless room contains one switch and three light bulbs, and the switch can only fail to close. Develop a fault tree for the top or undesired fault event "dark room" (i.e., room without light). Assume that the room can only be dark if the switch fails to close, all three bulbs burn out, or there is no electricity.

Solution: A fault tree for this example is shown in Figure 7. The fault events are labeled as F_0, F_1, F_2, F_3, F_4, F_5, F_6, F_7, and F_8.

Probability Evaluation of Fault Trees

For independent fault events, the probability of occurrence of the top event can easily be evaluated by applying the basic principles of probability theory.[22] For

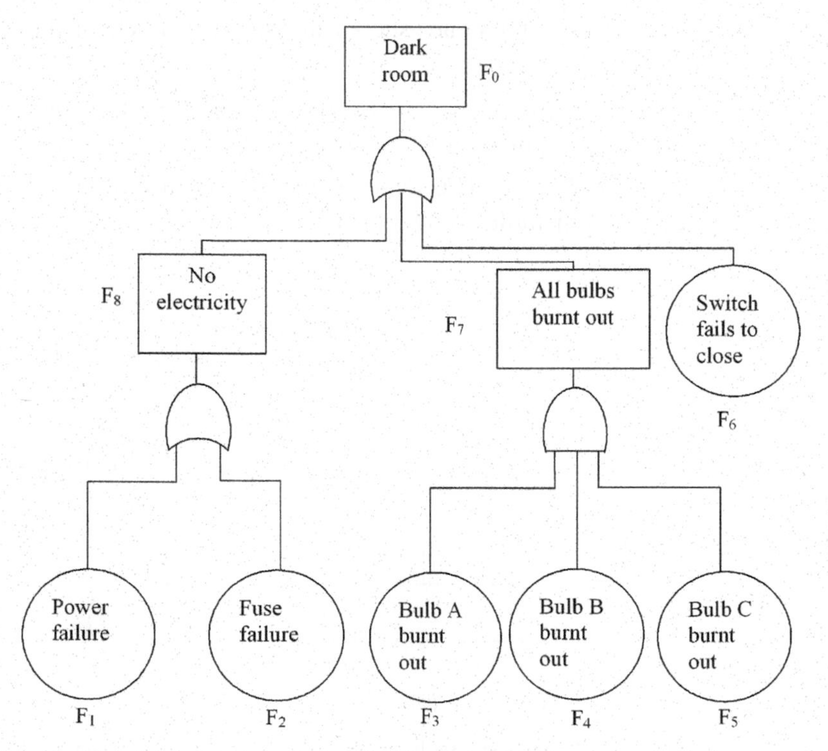

Figure 7 Fault tree for the top event "Dark room".

example, in the case of the fault tree in Figure 7, we have

$$P(F_8) = P(F_1) + P(F_2) - P(F_1)P(F_2) \tag{32}$$

$$P(F_7) = P(F_3)P(F_4)P(F_5) \tag{33}$$

$$P(F_0) = 1 - [1 - P(F_6)][1 - P(F_7)][1 - P(F_8)] \tag{34}$$

where $P(F_i)$ is the probability of occurrence of fault event F_i; for $i = 0, 1, 2, \ldots, 8$.

Example 6 Assume that in Figure 7, the probabilities of occurrence of fault events F_1, F_2, F_3, F_4, F_5, and F_6 are 0.01, 0.02, 0.03, 0.04, 0.05, and 0.06, respectively. Calculate the probability of occurrence of the top event (i.e., "dark room").

Using the specified probability values in equations (32) to (34) we get

$$P(F_8) = 0.01 + 0.02 - (0.01)(0.02)$$
$$= 0.0298$$
$$P(F_7) = (0.03)(0.04)(0.05)$$
$$= 0.00006$$

and

$$P(F_0) = 1 - [1 - 0.06][1 - 0.0006][1 - 0.0298]$$
$$= 0.0881$$

Thus, the probability of occurrence of the top event is 0.0881.

8.4 Markov Method

The Markov method is widely used to perform reliability analysis of repairable and nonrepairable systems. The method is named after a Russian mathematician and is subject to the following assumptions[23]:

- All occurrences are independent of each other.
- The transitional probability from one state to another in the finite time interval Δt is given by $\lambda \Delta t$, where λ is the constant transition rate (i.e., failure or repair rate) from one system state to another.
- The probability of more than one transition occurrence in time interval Δt from one state to another is negligible (e.g., $(\lambda \Delta t)(\lambda \Delta t) \rightarrow 0$).

The method is demonstrated through the following example.

Example 7 Assume that an environmental system can be in either an operating or a failed state. Its constant failure and repair rates are λ_e and μ_e,

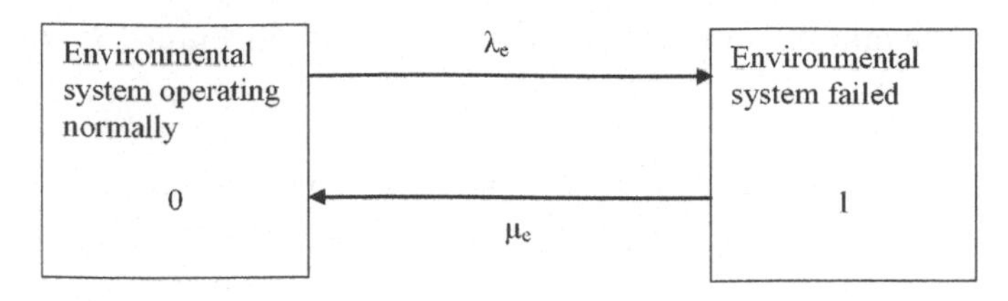

Figure 8 Environmental system state space diagram.

respectively. The system state space diagram is shown in Figure 8. The numerals in boxes denote the system state. Develop expressions for system availability (i.e., the probability of the system being operating at time t) and unavailability (i.e., the probability of the system being failed at time t) by using the Markov method.

Solution:

Using the Markov method, we write down the following two equations for Figure 8 diagram:[23]

$$P_0(t + \Delta t) = P_0(t)(1 - \lambda_e \Delta t) + P_1(t)\mu_e \Delta t \tag{35}$$

$$P_1(t + \Delta t) = P_1(t)(1 - \mu_e \Delta t) + P_0(t)\lambda_e \Delta t \tag{36}$$

where $P_i(t)$ is the probability that the environmental system is in state i at time t; for $i = 0, 1$.

$\lambda_e \Delta t$ is the probability of the environmental system failure in finite time interval Δt.

$\mu_e \Delta t$ is the probability of the environmental system repair in finite time interval Δt.

$(1 - \lambda_e \Delta t)$ is the probability of no failure in finite time interval Δt.

$(1 - \mu_e \Delta t)$ is the probability of no repair in finite time interval Δt.

$P_0(t + \Delta t)$ is the probability of the environmental system being in operating state 0 at time $(t + \Delta t)$.

$P_1(t + \Delta t)$ is the probability of the environmental system being in failed state 1at time $(t + \Delta t)$.

In the limiting case, equations (35) and (36) become

$$\frac{dP_0(t)}{dt} + \lambda_e P_0(t) = \mu_e P_1(t) \tag{37}$$

$$\frac{dP_1(t)}{dt} + \mu_e P_1(t) = \lambda_e P_0(t) \tag{38}$$

At time $t = 0$, $P_0\ (0) = 1$ and $P_1\ (0) = 0$.

Solving equations (37) and (38), we get

$$P_0(t) = A_e(t) = \frac{\mu_e}{(\lambda_e + \mu_e)} + \frac{\lambda_e}{(\lambda_e + \mu_e)} e^{-(\lambda_e + \mu_e)t} \tag{39}$$

$$P_1(t) = UA_e(t) = \frac{\lambda_e}{(\lambda_e + \mu_e)} - \frac{\lambda_e}{(\lambda_e + \mu_e)} e^{-(\lambda_e + \mu_e)t} \tag{40}$$

where $A_e(t)$ is the environmental system availability at time t.
$UA_e(t)$ is the environmental system unavailability at time t.

8.5 Stress-Strength Modeling Method

The stress-strength modeling method is used to predict reliability of an item when probability distributions of the item's stress and strength are known. Thus, *item reliability* is simply the probability that the failure governing stress will not exceed the failure governing strength. Mathematically, it is expressed as follows[8]:

$$\begin{aligned} R &= P(x > z) \\ &= \int_{-\infty}^{\infty} f(z) \left[\int_{z}^{\infty} f(x)\, dx \right] dz \end{aligned} \tag{41}$$

where x is the strength random variable.
z is the stress random variable.
R is the item reliability.
$f(x)$ is the strength probability density function.
$f(z)$ is the stress probability density function.

The application of equation (41) is demonstrated through the following example.

Example 8 Assume that an item's stress and strength probability density functions are defined by the following equations:

$$f(z) = \theta e^{-\theta z}, \qquad 0 \le z < \infty \tag{42}$$

and

$$f(x) = \alpha e^{-\alpha x}, \qquad 0 \le x < \infty \tag{43}$$

where θ and α are the reciprocals of the mean values of stress and strength, respectively.

Obtain an expression for an item's reliability by using equation (41).

By substituting equations (42) and (43) into equation (41), we get

$$R = \int_0^\infty \theta e^{-\theta z} \left[\int_z^\infty \alpha e^{-\alpha x}\, dx \right] dz$$

$$= \int_0^\infty \theta e^{-(\alpha+\theta)z}\, dz \tag{44}$$

$$= \frac{\theta}{\theta + \alpha}$$

For $\alpha = \frac{1}{x_m}$ and $\theta = \frac{1}{z_m}$, equation (44) becomes

$$R = \frac{x_m}{x_m + z_m} \tag{45}$$

where x_m and z_m are the mean strength and stress, respectively.

9 RELIABILITY ALLOCATION METHODS

Reliability allocation is an important factor in producing reliable engineering systems. It may simply be described as the process of assigning reliability requirements to individual parts of components to achieve the required system reliability. Some benefits of reliability allocation are as follows:[8]

- It forces the design engineer to seriously consider reliability equally with other design parameters such as cost, weight, and performance.
- It forces individuals involved in system design and development to fully understand and establish the relationships between reliabilities of system, subsystems, and components.
- It ensures satisfactory design, manufacturing procedures and approaches, and test methods.

There are many reliability allocation methods[8,21–22]. This section examines three of them.

9.1 Similar Familiar Systems Reliability Allocation Method

The similar familiar systems method is based on the familiarity of the design engineer with similar systems or subsystems. More specifically, on the basis of such familiarity, the reliability is allocated to system parts. The main weakness of this method is that in designing new systems, one has to assume that the reliability and life cycle cost of earlier similar designs were adequate.

9.2 Factors of Influence Method

In the factors of influence method, reliability is allocated on the basis of four factors shown in Figure 9: advancements in state-of-the-art, failure criticality,

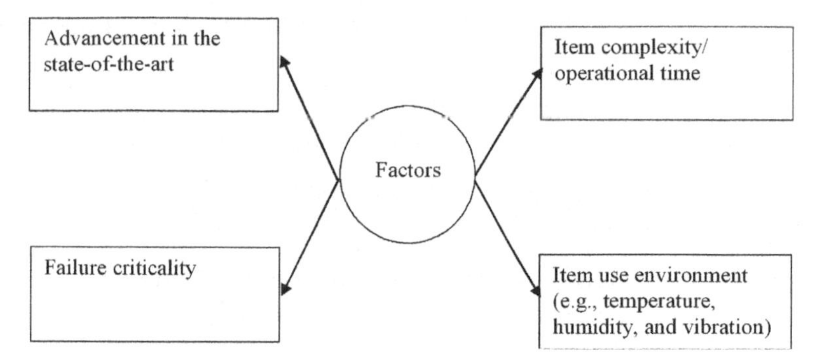

Figure 9 Factors used in allocating reliability to an item.

item complexity, and item use environment. In applying this method, each item is rated with respect to the factors shown in Figure 9 by assigning a number between 1 and 10 (i.e., 1 is allocated to an item least affected by the factor in question and 10 is allocated to an item most affected by the factors of influence). Ultimately, reliability is allocated to an item by using the weight of these assigned numbers for all factors.

9.3 Hybrid Method

The hybrid approach is the result of combining the previous two methods: factors of influence and similar familiar systems. The resulting method is more effective because it incorporates benefits of the other two methods.[8]

10 HUMAN RELIABILITY AND COMMON-CAUSE FAILURES

Engineering systems fail not only due to hardware or software issues but also due to human errors. In fact, the reliability of humans plays a pivotal role in the entire system/product life cycle.

Human errors may be classified under many distinct categories: design errors, operator errors, assembly errors, inspection errors, maintenance errors, handling errors, installation errors, and contributory errors.[28–30] Over the years, many causes for human error have been identified, including the following:[28,31]

- Poor equipment design
- Poor training or skill of concerned personnel (e.g., maintenance, operating, and production personnel)
- Poor lighting in the work area
- Complex tasks
- Inadequate work tools
- Poor work layout

- Poorly written equipment maintenance and operating procedures
- Poor motivation
- High noise level
- High temperature in the work area
- Poor verbal communication
- Poor management

Equations (46) and (47) can be used to estimate human reliability for discrete and time-continuous tasks, respectively[8,31]:

$$R_{\mathrm{hd}} = 1 - \frac{f}{m} \tag{46}$$

and

$$R_{\mathrm{hc}}(t) = e^{-\int_0^t \lambda_{\mathrm{hc}}(t)\,dt} \tag{47}$$

where R_{hd} is the human reliability for discrete tasks.
$R_{\mathrm{hc}}(t)$ is the human reliability at time t for time-continuous tasks.
m is the total number of times the certain task was performed.
f is the number of times the task was performed incorrectly.
$\lambda_{\mathrm{hc}}(t)$ is the time-dependent human error rate.

For $\lambda_{\mathrm{hc}}(t) = \lambda_{\mathrm{hc}}$ (i.e., constant human error rate), equation (47) simplifies to

$$R_{\mathrm{hc}}(t) = e^{-\lambda_{\mathrm{hc}} t} \tag{48}$$

where λ_{hc} is the constant human error rate (e.g., $\lambda_{\mathrm{hc}} = 0.001$ error/hour).

Additional information on human reliability is available in reference 31 at the end of this chapter.[31]

Common-cause failures are receiving increasing attention during reliability analysis of engineering systems because their occurrences result in total system failure. A *common-cause failure* may be defined as any instance where multiple units or items fail due to a single cause.[22] Some reasons for the occurrence of common-cause failures are as follows:[8]

- Deficiency in equipment design
- Operation and maintenance errors
- Common external environment (e.g., humidity, vibrations, temperature, and dust)
- Common manufacturer
- Common external power source
- External catastrophic events such as fire, flood, earthquake, and tornado
- Functional deficiency

Additional information on common-cause failures is available in the references at the end of this chapter.[8,22]

11 FAILURE DATA SOURCES AND RELIABILITY STANDARDS

The following list provides sources and organizations for obtaining failure data:[33]

- MIL-HDBK-217, Reliability Prediction of Electronic Equipment, Department of Defense, Washington, D.C.
- IEEE-STD-500-1977, IEEE Nuclear Reliability Data Manual, Institute of Electrical and Electronic Engineers (IEEE), New York
- ESA Electronic Components Data Bank, Space Documentation Service, European Space Agency (ESA), Via Galileo Galilei, 0044 Frascati, Italy
- M. J. Rossi, Non-electronic Parts Reliability Data, Report No. NPRD-3, Reliability Analysis Center, Rome Air Development Center (RADC), Griffiss Air Force Base, Rome, New York, 1985
- SYREL: Reliability Data Bank, Systems Reliability Service, Safety and Reliability Directorate, UKAEA, Wigshaw Lane, Culcheth, Warrington, Lancashire WA3 4NE, U.K
- Government Industry Data Exchange Program (GIDEP), GIDEP Operations Center, Department of the Navy, Naval Weapons Station, Seal Beach, Corona, California
- National Aeronautics and Space Administration (NASA) Parts Reliability Information Center, George C. Marshall Flight Center, Huntsville, Alabama
- National Technical Information Center (NTIS), 5285 Port Royal Road, Springfield, Virginia
- Defense Technical Information Center, DTIC-FDAC, 8725 John J. Kingman Road, Suite 0944, Fort Belvoir, Virginia
- Reliability Analysis Center (RAC), Rome Air Development Center (RADC), Griffiss Air Force Base, Rome, New York

There are many standards concerned with different aspects of reliability. Some of these are listed as follows.[8]

- *MIL-STD-785*. Reliability Program for Systems and Equipment, Development and Production, Department of Defense, Washington, D.C.
- *MIL-STD-721*. Definitions of Terms for Reliability and Maintainability, Department of Defense, Washington, D.C.
- *MIL-STD-756*. Reliability Modeling and Prediction, Department of Defense, Washington, D.C.
- *MIL-STD-472*. Maintainability Prediction, Department of Defense, Washington, D.C.
- *MIL-STD-1629*. Procedures for Performing a Failure Mode, Effects and Criticality Analysis, Department of Defense, Washington, D.C.

- *MIL-STD-781*. Reliability Design, Qualification and Production Acceptance Tests: Exponential Distribution, Department of Defense, Washington, D.C.
- *MIL-STD-2155*. Failure Reporting, Analysis and Corrective Action System (FRACAS), Department of Defense, Washington, D.C.
- *MIL-STD-52779*. Software Quality Assurance Program Requirements, Department of Defense, Washington, D.C.
- *MIL-STD-189*. Reliability Growth Management, Department of Defense, Washington, D.C.
- *MIL-STD-2074*. Failure Classification for Reliability Testing, Department of Defense, Washington, D.C.

REFERENCES

1. S. A. Smith, "Service Reliability Measured by Probabilities of Outage", *Electrical World*, **103**, 371–374 (1934).
2. W. J., Layman, "Fundamental Considerations in Preparing a Master System Plan," *Electrical World*, **101**, 778–792 (1933).
3. P. E. Benner, "The Use of the Theory of Probability to Determine Spare Capacity," *General Electric Rev.*, **37**, 345–348 (1934).
4. S. A. Smith, "Probability Theory and Spare Equipment", *Edison Electric Inst. Bulletin.*, 310–314 (March 1934).
5. AGREE Report, *Reliability of Military Electronic Equipment*, Department of Defense, Washington, DC, 1957.
6. MIL-R-25717 (USAF), *Reliability Assurance Program for Electronic Equipment*, Department of Defense, Washington, DC, 1957.
7. K. Henney, ed., *Reliability Factors for Ground Electronic Equipment*, McGraw-Hill, New York, 1956.
8. B. S. Dhillon, *Design Reliability: Fundamentals and Applications*, CRC Press, Boca Raton, FL, 1999.
9. B. S. Dhillon, *Reliability and Quality Control: Bibliography on General and Specialized Areas*, Beta Publishers, Gloucester, Ontario, Canada, 1992.
10. B. S. Dhillon, *Reliability Engineering Applications: Bibliography on Important Application Areas*, Beta Publishers, Gloucester, Ontario, Canada, 1992.
11. MIL-STD-721, *Definitions of Terms for Reliability and Maintainability*, Department of Defense, Washington, DC.
12. J. J. Naresky, "Reliability Definitions", *IEEE Transactions on Reliability*, **19**, 198–200 (1970).
13. T. McKenna and R. Oliverson, *Glossary of Reliability and Maintenance,* Gulf Publishing Company, Houston, Texas, 1997.
14. T. P. Omdahl, ed., *Reliability, Availability, and Maintainability (RAM) Dictionary*, ASQC Quality Press, Milwaukee, Wisconsin, 1988.
15. W. H. Von Alven, ed., *Reliability Engineering*, Prentice-Hall, Englewood Cliffs, NJ, 1964.

16. J. W. Walton, *Engineering Design*, West Publishing Company, New York, 1991.
17. E. A. Elsayed, *Reliability Engineering*, Addison Wesley Longman, Reading, Massachusetts, 1996.
18. B. S. Dhillon, *Engineering Design: A Modern Approach*, Richard D. Irwin, Chicago, 1996.
19. P. L. Hurricks, *Handbook of Electromechanical Product Design*, Longman Scientific and Technical, Long Group UK Limited, London, 1994.
20. A.D.S. Carter, *Mechanical Reliability*, MacMillan, London, 1986.
21. W. Grant Ireson and C. F. Coombs, eds., *Handbook of Reliability Engineering and Management*, McGraw-Hill, New York, 1988.
22. B. S. Dhillon, and C. Singh, *Engineering Reliability: New Techniques and Applications,* John Wiley, New York, 1981.
23. Shooman, M.L., Probabilistic Reliability: An Engineering Approach, McGraw Hill Book Company, New York, 1968.
24. T. P. Omdahl, ed., *Reliability, Availability, and Maintainability (RAM) Dictionary,* ASQC Quality Press, Milwaukee, WI, 1988.
25. MIL-F-18372 (Aer), *General Specification for Design, Installation, and Test of Aircraft Flight Control Systems*, Bureau of Naval Weapons, Department of the Navy, Washington, DC, 1954.
26. B. S. Dhillon, "Failure Modes and Effect Analysis: Bibliography," *Microelectronics and Reliability*, **32**, 729–731 (1992).
27. MIL-HDBK-217, *Reliability Prediction of Electronic Equipment*, Department of Defense, Washington, DC, 1998.
28. D. Meister, "The Problem of Human-Initiated Failures", Proceedings of the 8th National Symposium on Reliability and Quality Control, 234–239 1962.
29. J. I. Cooper, "Human-Initiated Failures and Man-Function Reporting", *IRE Trans. on Human Factors*, **10**, 104–109 (1961).
30. H. L. Williams, "Reliability Evaluation of Human Component in Man-Machine Systems", *Electrical Manufacturing*, 78–82 (April 1958).
31. B. S. Dhillon, *Human Reliability: With Human Factors*, Pergamon Press, New York, 1986.
32. W. C. Gangloff, "Common Mode Failure Analysis", *IEEE Transactions on Power Apparatus and Systems*, **94**, 27–30 (1975).
33. B. S. Dhillon, *Mechanical Reliability: Theory, Models, and Applications,* American Institute of Aeronautics and Astronautics, Inc., Washington, DC, 1988.

CHAPTER 6

DESIGN FOR MAINTAINABILITY

O. Geoffrey Okogbaa, Wilkistar Otieno
University of South Florida, Tampa, Florida

1 INTRODUCTION

Environmental concerns have created the need for sustainable development and have brought a global understanding that there is economic benefit in promoting products that are environmentally responsible and profitable, with reduced health risks for the consumers. The idea, according to 1987 UN report, is to meet current needs without jeopardizing the needs of future generations. This has led to a renewed emphasis on *design for environment* (DfE).

Life-cycle analysis (LCA) is a development platform upon which to anchor the formalisms for evaluating environmental effects and economic impact of the different stages of a product life cycle under the overarching umbrella of DfE. Included in these different stages are product and process designs, material selection, product manufacture and demanufacture, assembly and disassembly, reliability, maintenance, recycling, material recovery, and disposal. Since the sequential analyses of these different product stages do not provide a comprehensive, picture of the total effect in the context of product design efficiency and efficacy, a *concurrent* dynamism is implored in which the different stages are explored using the concept of *concurrent* engineering.

Thus, there are now different tools for addressing the different stages of LCA in support of design for environment, and these are typically referred to as design for X (DfX), such as design for manufacture (DfM) and design for assembly (DfA), design for product assurance (DfPA), and design for remanufacture (DfR), among others.

The need for self-diagnoses and fault tolerance and the need to conserve dwindling natural resources have led to the increased emphasis on design for product assurance (DfPA), especially with respect to reusability, survivability, and maintainability. Thus, new frontiers in engineering design, beyond the issues of concurrent design and design for manufacturability, may well depend on a paradigm shift that encompasses maintenance intervention practices to sustain fault tolerance and reusability. The cornerstone of such a paradigm shift hinges on the ability to develop analytical techniques that realistically characterize the failure and renewal process distributions for a complex multiunit repairable system, with particular emphasis on the system's transient phenomenon, which characterizes its aging processes. This problem is especially amplified in autonomous systems where human interaction and intervention are minimal. For such systems, the issue of maintenance intervention has taken on renewed importance because these systems should, in general, experience little to no failures.

Maintenance has been long recognized as necessary to increase a system's reliability, availability, and safety. Maintenance is necessary not only for systems where high reliability is required or where the failure may result in a catastrophe such as in the case of airplane or space shuttle, but also for any system where availability is of concern. In fact, relatively few systems are designed to operate without maintenance of any kind.[1]

For small systems with very few components, maintenance can be implemented by using the operator's experience. However, for large, complex systems that are common today, personal experiences are no longer enough in maintenance planning. As a matter of fact, maintenance planning based solely on personal experience may lead to significant loss to the systems. As an example, components may be neglected until they fail, thereby causing significant losses. At the opposite extreme, brand-new components may be replaced for no reason. This is especially true for systems that consist of large numbers of complex and expensive components that experience random or gradual failures. For such systems, the issue of maintenance planning has taken on renewed importance because these systems generally represent large capital expenditures and should experience few or no failures.

In the past few decades, extensive research has been done with regard to maintenance planning. However, only a few of these have addressed the problem of multiunit systems with economic dependency. In a few of the existing research studies the authors conducted *steady-state analyses* based on two major assumptions:[2–4]

1. The planning horizon is infinite.
2. The long-run expected replacement rate is constant.

For systems with long lives, the steady-state results are appropriate approximation of system behavior. However, the steady-state analysis does not consider the inherent transient characteristics of the system failure behavior. After all, no system really has an infinite planning horizon, and most systems have only short life spans. Consequently, steady-state analysis, while useful, is biased and limited in its applications.

Currently, no unified approach exists in the literature which considers the realistic failure characteristics of a complex multiunit system under transient response (system with a finite horizon). In addition, there are no maintenance strategies that incorporate the system's renewal and potential aging process. These problems limit efforts to provide a framework necessary to integrate the intervention functions into the overall system availability as well as assurance estimates for complex systems. Very little research has been done on how the system transient behavior impacts its remaining life. Such analyses are important if effective preventive maintenance for complex systems is to be implemented. In the long run this would help in eventual realization of the completely autonomous systems where human intervention is reduced to the minimum.

1.1 Maintainability

Maintainability and reliability considerations are playing increasingly vital roles in virtually all engineering disciplines. As the demand for systems that perform better and cost less increase, there is a corresponding demand or requirement to minimize the probability of failures and the need to quickly bring the system back to normal operations when unavoidable failures occur.

Terms and Definitions

Maintainability is the probability that when operating under stated environmental conditions, the system (facility or device or component) having failed or in order to prevent expected failure will be returned to an operating condition (repaired) within a given interval of downtime. Maintainability is a characteristic of equipment design that is expressed in terms of ease and cost of maintenance, availability of equipment, safety, and accuracy in the performance of maintenance actions. Other related terms include the following:

- *Maintenance.* All actions necessary for retaining a system in, or restoring it to, a serviceable condition. Systems could be maintained by repair or replacement.
- *Corrective maintenance.* This is performed to restore an item to satisfactory condition after a failure.
- *Mean time to failure (MTTF).* The mean time interval between two consecutive system failures is the MTTF.
- *Mean time between maintenance (MTBM).* This is the mean time between two consecutive maintenance procedures (replacement or repair).
- *Maintenance downtime.* This is the portion of downtime attributed to both corrective and preventive maintenance.

Classes of Maintenance Policies

Preventive Maintenance (PM). Preventive maintenance (PM) is performed to optimize the reliability and availability of a system prior to failure. Such maintenance involves overhaul, inspection, and condition verifications. It is widely considered as an effective strategy for reducing the number of system failures, thus lowering the overall life-cycle development cost.[5] PM is becoming increasingly beneficial, due to the need to increase the life cycle and performance of assets. Significant cost savings are achieved by predicting an impending failure in machinery. This might be in the form of savings achieved on spare parts, labor costs, downtime avoided, or prevention of damage due to accidents. Several parameters of the machinery are monitored to ensure their normal functioning. Such parameters might include vibration analysis, contaminant analysis, and monitoring of energy consumption, temperature, and noise levels.

A major effort in the development of a PM program is to determine the optimal replacement intervals. Operationally, maintenance management involves a transient and often uncertain environment with little data and proven tools that can assist in decision making. In particular, it is very difficult to develop an optimal maintenance schedule for a complex (multiunit) system with components that experience increasing failure rates in a dynamic environment and strong stochastic or economic dependencies.[2,6] Several approaches to this problem have fallen short because of the intractability of the problem, unrealistic assumptions about complex system behavior, and unnecessary restrictions on the maintenance model formulation and policy specification.

PM can be classified as condition-based and time-based policies. Time-based PM can be justified if and only if the following conditions hold:

1. The component under consideration has a significant aging behavior, which is characterized by increasing failure rate (IFR).

2. Preventive maintenance cost is much higher than the corrective maintenance cost.
3. Before failure, the component does not exhibit any abnormal behavior. In other words, it is not possible to detect any abnormal signal if such exists before the component fails.

Condition-based PM heavily depends on the available technology for early detection of imminent failures. Condition-monitoring techniques and equipment that detect failure effects are used. These are widely known as predictive techniques, and can be categorized as dynamic, particle, chemical, physical, or electrical in nature.[7] Some of them include lubricant analysis, vibration analysis, thermography, penetrating liquids, radiography, ultrasound, and corrosion controls.

Corrective Maintenance (CM). This is performed to restore an equipment to satisfactory condition after a failure. It is assumed that in most cases the cost associated with CM is much higher than that associated with PM.

Opportunistic Maintenance (OM). The opportunistic maintenance concept originates from the fact that there can be economic and stochastic dependency between various components of a multiunit system. *Opportunism* here refers to the idea of jointly replacing several components at some point in time based on three conditions:

1. A component will be replaced when it fails.
2. A component will be actively replaced when it reaches age T_a (active replacement age).
3. A component will be replaced if it reaches age T_p (passive replacement age) and there is another component in the system that is being replaced, either due to failure or due to active preventive maintenance. In this case, $(T_p < T_a)$.

The Unit, System or Component

In actual practical situations or considerations, maintainability may be viewed or defined differently for a system, components, and so on. However, the system or unit of interest normally determines what is being studied, and there is usually no ambiguity. From the system modeling point of view, a system can be considered either as a single-unit system or as a multiunit system. In a single-unit model, the entire system is viewed as one component, and its failure distribution, failure process, maintenance activities, and effects are well defined.

Most real-world systems are complex in nature and may consist of hundreds of different components. Thus, the assumption that such a system would follow a single failure distribution is too limiting and unrealistic. Hence, it would be of little use in maintenance program development. Rather, the first step should be to decompose the system into subsystems or components for which the failure distributions are more traceable, and the maintenance activities and associated costs and effects are well defined. For this reason, we define a complex system as a multiunit system.

Whether a group of components should be considered as several individual single-unit systems or as an integrated multiunit system depends on whether there exists economic or stochastic dependency between the components. *Stochastic dependency* implies that each component's transition probability depends on the status of other components in the systems and the notion that the failure of one component may increase the failure probability of the other components. *Economic dependency* implies an opportunity for group replacement of several components during a replacement event. This is justified by the fact that the joint replacement of several components will cost less than separate replacements of the individual components.[6]

On the one hand, weak economic and stochastic dependency means that decisions can be independently made for each component in the system. On the other hand, if stochastic and economic dependency between components is very strong, then an optimal decision on the repair or replacement of one component is not necessarily optimal for the whole system. Thus, for complex systems, maintenance intervention plans should be for the whole system rather than for each individual component.

Definition of Failure

A system or unit is commonly referred to as having *failed* when it ceases to perform its intended function. When there is total cessation of function, engines stop running, structures collapse, and so on—the system has clearly failed. However, a system can also be considered to be in a failed state when its deterioration function is within certain critical limits or boundaries. Such subtle forms of failure make it necessary to define or determine quantitatively what is meant by failure.

Time Element. The way in which time is specified can also vary with the nature of the system under consideration:

- In an intermittent system, one must specify whether calendar time or number of hours of operation is to be used in measuring time (car, shoes, etc.).
- If the system operation is cyclic (switch, etc.), then time is likely to be specified in terms of number of operations.
- If the maintainability is to be specified in calendar time, it may also be necessary to indicate the number or frequency of stops and gos.

Operating Condition.

- Operating parameters include loads, weight, and electrical load.
- Environmental conditions include temperature, extremes, dust, salt, vibrations, and similar factors.

1.2 Steady State versus Transient State

Steady State

If the system planning horizon is much longer than the life of the components, then it would be appropriate to develop a predictive maintenance program based on the long-run stable condition, or the steady state. In this regard, the expected replacement rate (ERR) would provide important information for maintenance management on spare parts inventory, size of maintenance work force, and maintenance equipment. However, the reason for maintenance intervention is the inherent failure characteristics of a system that is transient by nature.[8] Thus, while the steady-state results are useful, the transient response provides the information that reflects the system's useful life profiles for planning, maintenance, and supportability requirements.

Transient State

A system is said to be in a transient state before it stabilizes into equilibrium of steady state, often due to exogenous environmental conditions or internal control factors. Transient state analysis is most important in systems whose homogenous stochastic behaviors converge very slowly to a steady state. In such cases the steady state would not be indicative of a system's actual behavior. Thus the transient state represents the true system behavior during the useful life. Analytical models that are based on the transient behavior of the system and that take the form of differential and integral equations can be used to model maintenance policies. Numerical methods can then be used to solve the resulting equations in an attempt to obtain a clear picture of the system's transient behavior. Several numerical methods have been proposed and have been used successfully in attacking this type of problem. However, the complex nature of manufacturing systems and the characteristics of the proposed transient models preclude the use of just any of the existing methods. The Runge-Kutta and the Runge-Kutta-Gill methods are best to use because of their robustness relative to linear/nonlinear models. Afterward, simulation should be used to validate the developed transient state models.

1.3 Basic Preventive Maintenance Strategy

Considering an existing bottling system environment, the objective of preventive maintenance is to identify the right time and right subsystems with optimized schedule to perform the maintenance before the occurrence of failures. Failures of large systems or complex systems usually follow certain types of probability distribution, which are mostly determined by the system design reliability/availability and BOM (bill of materials). Practically before a failure happens, there often are some symptoms such as abnormal changes of system mechanic vibration, temperature, and so on.

The basic strategy of preventive maintenance is to decide on and execute the optimal maintenance plan, considering the designed system reliability features and real-time field-collected system data, as well as the constraint of maintenance cost.

1.4 Implementation of Preventive Maintenance

Considering the bottling process, the following four aspects should facilitate implementation of the optimal preventive maintenance:

1. System operation monitoring
2. Design maintainability/availability performance
3. Empirical performance history
4. Comprehensive evaluation and analysis that lead to decision making

Each element of the preventive maintenance procedure is covered in detail in this section.

System Operation Monitoring

This involves keeping track of the key system parameters during the bottling operation. It includes vibration analysis, infrared technology, and fluid analysis and tribology. These aspects of system operation monitoring can be explained as follows.

Vibration Measurement and Analysis. A vibration analyzer and some operator experience can be used to identify the causes of abnormal machinery conditions such as rotor imbalance and misalignment. Other problems, including bearing wear and the severity of the wear, can also be identified and quantified. Vibration monitoring provides earlier warning of machine deterioration than temperature monitoring. Small and gradual machine deterioration show up as significant vibration increases. Early detection usually permits continued operation until a scheduled shutdown. The vibration measurement and analysis can be implemented by deploying a handheld vibration meter PT908, solid-state vibration switch VS101, or TM101.

PT908 vibration meter is one of the most useful tools in checking the overall condition on all types of rotating machinery. PredicTech's PT908 is portable and is economical, reliable, and rugged. The vibration meter can measure acceleration, velocity, and displacement in peak or RMS. Field operators don't need any vibration analysis knowledge to use this device. The displayed digital vibration level is an indicator of machine running condition.

VS101 solid-state vibration switch is a direct upgrade of the mechanical vibration switch. This solid vibration switch measures excessive machine vibration and machine failure, with additional dry-contact relay for machine alarm. VS101 responds to destructive vibration by shutting down the machine when the vibration trip level is exceeded to preventing catastrophic damage and extensive repairs and downtime. VS101 can be used in rotation machines like electric motors, blowers, pumps, agitators, gear boxes, small compressors, and fans.

The TM101 transmitter monitor is used to measure any rotation machinery vibration such as case vibration, bearing housing vibration, or structural vibration. Output is in acceleration, velocity, or displacement. This system can be used in hazardous areas.

Infrared Thermography. Abnormal operation of mechanical systems usually results in excessive heat, which can be directly caused by friction, cooling degrading, material loss, or blockages. Infrared thermograph allows monitoring of the temperatures and thermal patterns while the equipment is online and running with full load. Unlike many other test methods, infrared can be used on a wide variety of equipment, including pumps, motors, bearings, generators, blower systems, pulleys, fans, drives, and conveyors. Infrared thermography should be deployed in key system components, such as motors and generators, blower systems, and bearings.

Identifying thermal pattern doesn't necessarily locate a problem. In mechanical applications, a thermograph is more useful in locating a problem area other than determining the root cause of the overheating, which usually is produced inside a component and is not visible to the camera. Other approaches such as vibration analysis, oil analysis, and ultrasound can be employed to determine the problem.

Fluid Analysis and Tribology. Tribology studies the interdisciplinary field of friction, lubrication, wear, and surface durability of materials and mechanical systems. Typical tribological applications include bearings, gears, bushings, brakes, clutches, chains, human body implants, floor tiles, seals, piston engine parts, and sports equipment.

Machine fluid (usually oil) analysis is used to evaluate two common contamination problems—namely, water and dust—which dramatically increase wear rates and inadequate lubrication. By monitoring, reporting, and recommending the correction of contamination problems, oil analysis is of a very proactive condition monitoring technology. It can also effectively predict impending catastrophic failure of mechanical and electrical systems through which the oil flows.

There are four typical categories of abnormal wear for mechanical systems: abrasive wear, adhesive wear, fatigue wear, and corrosive wear:

1. *Abrasive wear* particles are normally indicative of excessive dirt or other hard particles in the oil that are cutting away at the load-bearing surfaces.

2. *Adhesive wear* particles reveal problems with lubricant starvation as a result of low viscosity, high load, high temperature, slow speed, or inadequate lubricant delivery.
3. *Fatigue wear* particles are often associated with mechanical problems, such as improper assembly, improper fit, misalignment, imbalance, or other conditions.
4. *Corrosive wear* particles are the result of corrosive fluid such as water or process materials contacting metal surfaces.

Typical tribometers used for this purpose include the pin-on-disk tribometer, linear reciprocating tribometer, and high temperature tribometer.

Design Maintainability/Availability Performance

These system parameters are used to calculate the expected the mean time to failure (MTTF) and the mean time to repair (MTTR), which are fundamental theoretic reference metric of system maintenance plan.

To add maintainability into a design, designers need an intensive understanding of the system. This includes the system's configuration, topology, component interdependency, and failure distribution. Design for maintainability is well hinged in the ability to develop techniques that realistically characterize the failure and renewal process distributions for complex, multiunit repairable systems, with particular emphasis on a system's transient phenomenon as well as its aging process.

Empirical Performance History

System maintainability/availability performance varies when deployed in different physical operation environment such as at various humidity and voltage stability levels, with different load, under different human operators. Therefore, the empirical performance data also serve as a very important reference to predict early failures and schedule preventive maintenance for specific individual systems.

Comprehensive Evaluation and Analysis

To determine a preventive maintenance schedule, all the foregoing design and empirical reliability data and parameters must be comprehensively evaluated with more detailed scientific analysis. There are numerous statistical process control (SPC) methodologies and tools available to perform the comprehensive evaluation and analysis. This section covers several of the proven methods.

Design of Experiment (DOE)/ANOVA. This is used as a mechanism to determine the significant factors and their interactions affecting the system reliability performance.

Control Charts. Control charts are useful as a tool to determine the boundary limits of process performance and related nonconformance levels.

Automation of Preventive Maintenance—CMIS. The automation of scheduling the optimal preventive maintenance can be achieved by implementing the CMIS (computer manufacturing integrated systems), which is efficient yet costly. In the specific case of a bottling system, all the monitoring devices/terminals must be connected to a central station equipped with high-performance servers. The servers are programmed to trigger at different system thresholds the performance levels or status of the system and thus alert the decision maker to take action, based on comparisons of real-time data with preestablished thresholds of operation/maintainability performance. Typically, the software algorithms and protocols created from the system design reliability and historical failure data—as well as other customer-specified requirements—generate these limits or thresholds.

The rest of the chapter is sequenced as follows. Section 2 introduces the probability theory and probability distributions. The third section gives an introduction into system reliability functions, its measurement parameters, and how system maintenance is related to system reliability. The fourth section delves into maintenance, definitions and maintenance classifications, and system maintenance modeling. Finally, system availability is discussed as a way to measure the integrity of any maintenance policy of a reparable system.

2 REVIEW OF PROBABILITY AND RANDOM VARIABLES

Each realization of a manufacturing activity represents a random experimental trial—in other words, such an activity may be looked at as a process or operation that generates raw data, the nature or outcome of which cannot be predicted with certainty. Associated with each experiment or its realization is the sample space, which is the set of all possible outcomes of the experiment. An *outcome* of an experiment is defined as one of the set of possible observations which results from the experiment. One and only one outcome results from one realization of the experiment. Most quantities occurring in a manufacturing environment (a realization of a manufacturing experiment) are subject to random fluctuations. Because of the fluctuation and the randomness of the experiments, the outcomes are *random variables*. In order to characterize these random variables so that they can become useful tools in describing a manufacturing system, it is important to understand what random variables are.

2.1 Random Variables

An intuitive definition of a *random variable* is that it is a quantity that takes on real values randomly. It follows from this definition that a random vector is an n-tuple of real valued random variables. An operating definition of a random variable is that it is a function that assigns a real value (a number on the real line) to each sample point in the sample space S. This assignment or mapping is one-to-one and is not bidirectional. In other words, each event in the sample space can only take on one value at a time. As an example, let the height of males in a region or country be a variable of interest in a study. Due to its nature, such a variable can be described as random. Thus, the random variable can be expressed as the function $f(x) =$ a numerical value equal to the height of a unique individual male named x. That is,

$$f\ (\text{John1}) = 6\ \text{feet} \quad f\ (\text{Paul20}) = 7\ \text{feet} \quad f\ (\text{Don10}) = 6\ \text{feet}$$

Note, however, that two elements from the sample space can have the same real values assigned to them. In our example, John1 and Don10 have the same height. However, John1 cannot have heights of 6 feet and 7 feet at the same time; hence the mapping is unique.

The domain of the random variable is the sample space (S) and the range of the random variable is the real line R. It is the range of the random variable that determines the types of values that are assigned to the random variable under consideration.

Random variables occurring in manufacturing situations or in industry in general are subject to random fluctuations that exhibit certain regularities, and they sometimes have well-defined forms. In some cases, they are also of a given form or belong to some class or family. Thus, depending on the type of random experiment that generated the domain of the random variable, the mapping can be generalized into closed form expressions, formula, equations, rules or graphs that describe how the values are assigned. These rules or equations are known as probability density functions (for continuous random variables) and probability mass functions (for discrete random variables). In other words, probability density or mass functions are simply closed-form expressions or rules that indicate how assignments to values on the real line are made from the domain of the random variable. The nature of the experiment ultimately determines the type of equations or rules that apply.

One of the problems of using the density function to characterize a manufacturing process is that in some cases such closed-form expressions or equations are difficult to come by. Hence, we are often left to examine the moments resulting from the data at hand using the moment-generating functions. Usually the first and second moments (the mean and variance) could provide useful insight as to the behavior of the random variable. However, in some cases these parameters are not enough to completely characterize the underlying distribution, and so we have to look to other approaches that would provide the confidence needed to ensure that indeed the distribution assumed is the appropriate one.

The following are some density or mass functions that are often used to describe random variables that typically occur in a manufacturing setting. These distributions can be categorized into discrete and continuous. The discrete distributions the include binomial, negative binomial, geometric, hypergeometric, and Poisson distributions, while the continuous distributions that will be examined include normal, exponential, and Weibull distributions.

2.2 Discrete Distributions

Binomial

Consider a random experiment with the following conditions:

1. Each trial has only two possible outcomes; namely, the occurrence or nonoccurrence of an event (e.g., conforming/nonconforming, defective/nondefective, success/failure).
2. The probability (p) of occurrence of an event is constant and is the same for the each trial.
3. There are n trials (n is integer).
4. The trials are statistically independent.

Such an experiment is called a binomial experiment, which follows the Bernoulli sequence. The random variable of interest is the number of occurrences of a given outcome or event.

Example Let x be the number of occurrences of an event (where n is known). The probability of having exactly x_0 occurrences in n trials, where p is the probability of an occurrence is given by

$$P(x = x_0) = {}^nC_{x_0} p^{x_0}(1 - p)^{n-x_0}, \quad x_0 = 0, 1, \ldots n \tag{1}$$

where ${}^nC_{x_0} = \frac{n!}{(n-x_0)!x_o!}$

The mean and variance of the binomial are $\mu = np$, and $\sigma^2 = np(1 - p)$, respectively, where $0 \leq p \leq 1$.

The probability that a certain wide column will fail under study is 0.05. If there are 16 such columns, what is the probability that the following will be true?

a. At most two will fail.

b. Between two and four will fail.

c. At least four will fail.

Solution:

a. $$P(X \leq 2|n = 16, p = 0.05) = \sum_{0}^{2} \binom{16}{x} (0.05)^x (0.95)^{16-x}$$

$$x = 0, \binom{16}{0} (0.05)^0(0.95)^{16-0} = 0.440$$

$$x = 1, \binom{16}{1}(0.05)^1(0.95)^{16-1} = 0.371$$

$$x = 2, \binom{16}{2}(0.05)^2(0.95)^{16-2} = 0.146$$

$$P(X \leq 2) = P(X = 0) + P(x = 1) + P(X = 2)$$

$$\therefore \quad P(X \leq 2) = 0.957$$

b. $P(X \leq 30) = \Phi(-1.81) = 0.035$

$$= \sum_0^3 \binom{16}{x}(0.05)^x(0.95)^{16-x} - 0.957$$

$$= 0.957 + \binom{16}{3}(0.05)^3(0.95)^{16-3} - 0.975$$

$$= 0.036$$

c. $P(X \geq 4|n = 16, p = 0.05) = 1 - P(X \leq 3)$

$$= 1 - \sum_0^2 \binom{16}{x}(0.05)^x(0.95)^{16-x}$$

$$= 1 - (0.036 + 0.957)$$

$$= 0.007$$

Negative Binomial

In this case, the random variable is the number of trials. For a sequence of independent trials with probability of the occurrence of an event equal to p, the number of trials x before exactly the rth occurrence is known as the negative binomial, or the Pascal distribution. The probability of exactly x_0 trials before the rth occurrences is given by

$$P(X = x_0) = \binom{x-1}{r-1} p^r(1-p)^{x_0-r}, \quad x_0 = r, r+1, \ldots n \tag{2}$$

where the mean and variance is given, respectively, by

$$\mu = \frac{r}{p}, \quad \sigma^2 = \frac{r(1-p)}{p^2}$$

Example The probability that on production, a critical defect is found is 0.3. Find the probability that the tenth item inspected on the line is the fifth critical defect.

Solution: x = sample size = 10, r = the outcome = 5

$$P(x = 5) = \binom{10-1}{5-1}(0.3)^5(0.7)^{10-5}$$

$$= \binom{9}{4}(0.3)^5(0.7)^5 = 0.05145$$

Geometric Distribution

The random variable is the number of trials. In a Bernoulli sequence, the number of trials until a specified event occurs for the first time is governed by the geometric distribution. Thus, for a sequence of independent trials with probability of occurrence p, the number of trials x before the first success is the geometric distribution, which is also a member of the family of Pascal distributions. If the occurrence of the event is realized on the xth trial, then there must have been no occurrence of such event in any of the prior $(x - 1)$ trials. Hence, it is same as the negative binomial distribution with $r = 1$:

$$P(X = x) = p(1 - p)^{x-1}$$

The mean and variance, respectively, are

$$E(X) = \mu = \frac{1}{p}, \quad \sigma^2 = \frac{(1-p)}{p^2}$$

In a time or space problem that can be modeled as a Bernoulli sequence, the number of time (or) space intervals until the first occurrence is called the *first occurrence time*. If the individual trials or intervals in the sequence are statistically independent, then the first occurrence time is also the time between any two consecutive occurrences of the same events—that is, the recurrence time is equal to the first occurrence time. The mean recurrence time, which is commonly known in engineering as the *average return period* or the *average run length* (ARL), is equal to the reciprocal of p, the probability of occurrence of the event within one time unit.

Example A system maintainability analysis procedure requires an experiment to determine the number of defective parts before the system is shut down for corrective maintenance. The probability of a defect is 0.1. Let X be a random variable denoting the number of parts to be tested until the first defective part is found. Assuming that the trials are independent, what is the probability that the fifth test will result in a defective part?

Solution:

$$P(X = 5) = 0.1(1 - 0.1)^{5-1}$$

$$= 0.066$$

Hypergeometric Distribution

The *hypergeometric distribution* is used to model events in a finite population of size N when a sample of size n is taken at random from the population without replacement and where the elements of the population can be dichotomized as belonging to one of two disjoint categories. Thus, in a finite population N with different categories of items (e.g., conforming/nonconforming, success/failure, defective/nondefective), if a sample is drawn in such a way that each successive drawing is not independent (i.e., the items are not replaced), then the underlying distribution of such an experiment is the hypergeometric. The random variable of interest is the number of occurrences X of a particular outcome for a classification or category a, with the sample size of n. The probability of exactly x_0 occurrences is given by:

$$h(x = x_0|n, a, N) = \frac{\binom{a}{x_0}\binom{N-a}{n-x_0}}{\binom{N}{n}} x_0 = 0, 1, \ldots n \quad (3)$$

The hypergeometric satisfies all the conditions of the binomial except for independence in trials and constant p.

where X = the random variable representing the number of occurrences of a given outcome

a = category or classification of N

N = population size

n = sample size

$\binom{N}{n}$ = the number of samples of size n.

$\binom{N}{n}\binom{N-a}{n-x_0}$ = the number of samples having x_0 outcomes out of a.

The mean $\mu = n(a/N)$, and the variance $\sigma^2 = n(a/N)(1 - a/N)$.
Let $p = a/N$; then $\mu = n\ p$, and $\sigma^2 = n\ p\ (1 - p)$.

Example To test the reliability of an existing machine, a company is interested in evaluating a batch of 50 identical products. The procedure calls for taking a sample of 5 items from the lot of 50 and passing the batch if no more than 2 are found to be defective, therefore dimming the machine fit. Assuming that the batch is 20 percent defective, what is the probability of accepting the batch?

Solution: Given: $N = 50$, $n = 5$, $a = 20\%$ of $50 = 10$

$$= P(x \leq 2) = \sum_{x=0}^{2} \left(\frac{\binom{10}{5}\binom{50-10}{5-x}}{\binom{50}{5}} \right)$$

For $x = 0$, $P(x = 0) = 0.3106$
For $x = 1$, $P(x = 1) = 0.4313$
For $x = 2$, $P(x = 2) = 0.2093$

$$P(x \leq 2) = P(x = 0) + P(x = 1) + P(x = 2) = 0.9517$$

Poisson Distribution

Many physical problems of interest to engineers involve the occurrences of events in a continuum of time or space. A Poisson process involves observing discrete events in a continuum of time, length, or space, with μ as the average number of occurrence of the event. For example in the manufacture of an aircraft frame, cracks could occur anywhere in the joint or on the surface of the frame. Also in the construction of a pipeline, cracks could occur along continuous welds. In the manufacture of carpets, defects can occur anywhere in a given area of carpet. The light bulb in a machine tool could burn out at any time. Examples abound of the types of situations where the occurrence rather than the nonoccurrence of events in a continuum is of interest. Such time–space problems can be modeled with the Bernoulli sequence by dividing the time or space into small time intervals, assuming that the event will either occur or not occur (only two possibilities). In the case of the Poisson, it is usually assumed that the event will occur at any time interval or any point in space and also that the event may occur no more than once at a given time or space interval. Four of the assumptions of the Poisson follow:

1. An event can occur at random and at any time or point in space.
2. The occurrence of an event in a given time or space interval is independent of that in any other nonoverlapping intervals.
3. The probability of occurrence of an event in a small interval Δt is proportional to Δt and is given by $\mu \Delta t$, where μ is the mean rate of occurrence of the event (μ is assumed a constant).
4. The probability of two or more occurrences in the interval Δt is negligible and numerically equal to zero (higher orders of Δt are negligible).

A random variable X is said to have a Poisson distribution with parameter μ if its density is given by

$$f(X = x) = \frac{\mu^x e^{\mu}}{x!} \qquad \mu > 0, \; x = 0, 1, 2, \ldots \tag{4}$$

with mean $= \mu$, and variance $= \sigma^2 = \mu$

Example 1 A compressor is known to fail on average 0.2 times per hour.

a. What is the probability of a failure in 3 hours?
b. What is the probability of at least two failures in 5 hours?

Solution: Note: For discrete distributions;
$P(X = x_0) = P[X \leq x_0] - P[X \leq (x_0 - 1)]$
Where: $P(X \leq x_0|\mu) = F(x_0, \mu)$
Thus: $P(X = x_0) = F(x_0, \mu) - F(x_0 - 1, \mu)$

a. $\mu = (0.2)(3) = 0.6$

$$P(x = 1) = F(1, 0.6) - F(0, 0.6) = 0.878 - 0.549 = 0.324$$

b. $\mu = (0.2)(5) = 1.0$

$$P(x \geq 2) = 1 - P(x \leq 1) = 1 - F(1, 1) = 1 - 0.736$$

The F values can be found in the cumulative Poisson table available in most texts on statistics.

Example 2 An aircraft's landing gear has a probability of 10^{-5} per landing of being damaged from excessive impact. What is the probability that the landing gear will survive a 10,000 landing design life without damage?

Solution: Considering the problem as being Poisson in nature, $\lambda = np = 10^{-5} \times (10,000) = 0.1$

$$P(x = 0) = \frac{0.1^0 e^{-0.1}}{0!} = 0.904837$$

2.3 Continuous Distributions

Normal Distribution

The normal probability is the perhaps the most frequently used of all probability densities and some of the most important statistical techniques are based on it. The normal random variable appears frequently in practical problems, and it provides a good approximation to a large number of other probability laws. The density function for a normal random variable is symmetric and bell-shaped. Due to the bell-shaped nature, a normally distributed random variable has a very high probability of taking on a value close to μ and correspondingly a lower probability of taking on values that are further away on either side of μ. Thus, a random variable X is normally distributed if its density function is of the form

$$f_X(x) = \frac{1}{\sigma\sqrt{2\pi}} e^{\frac{(x-\mu)^2}{2\sigma^2}} \quad (5)$$

where x is real, μ is any real number, and σ is positive.

Computing the probability associated with the occurrence of the normal probability event requires integrating the density function. Unfortunately, this integration cannot be carried out in closed form. However, numerical techniques can be used to evaluate the integral for specific values of μ and σ^2 after properly transforming the normal random variable with mean μ and variance σ^2 to a standard normal variable with mean $\mu = 0$, and $\sigma^2 = 1$. Tables of the standard normal are available in most statistics and probability texts. For the standard normal, the probability of the normal random variable X taking on a value less than x_0 is the cumulative distribution function defined as $P(X < x_0) = z$, where $z = (x_0 - \mu)/\sigma$ is the number of standard deviates between the mean and x_0.

The actual area represented by the value of z is given by $\Phi(z)$, where $\Phi(z)$ is the integral of the density function with boundaries from $-\infty$ to z. Most standardized normal tables have a sketch that shows the boundaries and the areas that result in values indicated on the tables. Because the normal is symmetric, the tail values of the areas are identical. The same is true of the corresponding standard deviation, except for a change in sign. Deviations above the mean are denoted as positive, while those below the mean are denoted as negative. This makes it possible to evaluate different values on both sides of the mean with just the table value for lower half of the normal function ($-\infty$ to z) or the upper half of the function (z to $+\infty$).

Example The time to wear out of a cutting-tool edge is distributed normally with $\mu = 2.8$ hours and $\sigma = 0.6$ hour.

a. What is the probability that the tool will wear out in less than 1.5 hours?
b. How often should the cutting edge be replaced to keep the failure rate less than 10 percent of the tools?

Solution:

a. $P\{t < 1.5\} = F_t(1.5) = \Phi(z)$

Where $z = \dfrac{(t - \mu)}{\sigma} = \dfrac{1.5 - 2.8}{0.6} = -2.1667$

$\Phi(-2.1667) = 0.0151$

b. $P\{T < t\} = 0.10$; thus $\Phi(z) = 0.1$

Interpolating from tables shows that $z \approx -1.28$

Thus $t - u = -1.28\sigma$ and $t = 2.03$ hours

Lognormal Distribution

Situations frequently arise where a random variable Y is a product of other independent random variables y_i—that is, $y = y_1 y_2 y_3 \cdots y_N$. For instance, the failure of a shaft may be proportional to the product of forces of different magnitude. Taking the natural logarithm of the above equation,

$$\ln y = \ln y_1 + \ln y_2 + \cdots + \ln y_N.$$

Let $\ln y$ be distributed normally, then y will be lognormally distributed. The lognormal density function for y is given by:

$$f(y) = \frac{1}{y\sigma\sqrt{2\pi}} \exp\left(-\frac{1}{2\sigma^2}(\ln y - u)\right) \tag{6a}$$

Let $\omega = \sigma$, and let $\mu = \ln y_0$

Then we can rewrite lognormal density function in a more common form as

$$f_y(y) = \frac{1}{\sqrt{2\pi}\omega y} \exp\left\{-\frac{1}{2\omega^2}\left[\ln\left(\frac{y}{y_0}\right)\right]^2\right\}, \tag{6b}$$

and

$$F_y(y) = \Phi\left[\frac{1}{\omega}\ln\left(\frac{y}{y_0}\right)\right]$$

The parameters ω and y_0 are known for any scenario.

Example Fatigue life data for a shaft are fit to a lognormal distribution with the following parameters: $y_0 = 2 \times 107$ cycles, and $\omega = 2.3$. To what value should the design life be set if the probability of failure is not to exceed 1.0 percent?

Solution: From the normal distribution tables, z such that $\Phi(z) = 0.01$ is -2.32.

$$-2.32 = \left[\frac{1}{2.3}\ln\left(\frac{y}{2 \times 10^7}\right)\right]$$

Thus

$$y = 9.63 \times 10^4 \text{ cycles}$$

Exponential Distribution

If events occur according to a Poisson process, then the time T between consecutive occurrences has an exponential distribution. Thus, in a Poisson process, if the rate of occurrence of the events in a continuum is λ, then time between occurrence is exponentially distributed with mean time to occur equal to θ and the mean occurrence rate $\lambda = 1/\theta$. The exponential distribution has been used to study life distributions in the physical and biological sciences as well as in engineering. If T is defined as the exponential random variable, then the density and cumulative distribution functions are given as follows:

$$f(t, \lambda) = \lambda exp(-\lambda\tau) = \frac{1}{\theta}\exp\left(-\frac{t}{\theta}\right), \quad where\ \lambda = \frac{1}{\theta} \tag{7}$$

$$F(T < t) = \int_0^t \frac{1}{\theta}\exp\left(-\frac{1}{\theta}\tau\right) d\tau = 1 - \exp\left(-\frac{1}{\theta}t\right) \tag{8}$$

The average value $\mu = \theta = $ s, the standard deviation.

Example The life in years of a certain kind of electrical switch has an exponential distribution with $\lambda = 1/2$ ($\theta = 2$). If 100 switches are installed in a system, find the probability that at most 30 will fail during the first year.

Solution: First try to find the probability that one switch will fail in the first year ($t = 1$). $F(t) = (1 - e^{-\lambda t}) = 0.3935$ = probability that one will fail during the first year. Now try to find $P(x \leq 30)$, with $n = 100$, $p = 0.3935$, using the binomial:

$$P(x \leq 30) = \sum_{0}^{30} \binom{100}{x} (0.3935)^x (0.6065)^{100-x}$$

Since $np > 5$ and $p > 0.1$, we can use normal approximation, where

$$\mu = np = (100)(0.3905) = 39.35$$

$$\sigma^2 = np(1 - p) = 23.86, \sigma = 4.485$$

Using the continuity for correction 30, where 30 lies between 29.5 and 30.5 such that:
For "at most 30" (30 or less), the appropriate value to use is 30.5.
For "at least 30" (30 or more), the appropriate value to use is 29.5.

$$Z = \left(\frac{30.5 - 39.35}{4.485}\right) = -1.81$$

$$P(X \leq 30) = \Phi(-1.81) = 0.035$$

Hence, the probability that at most 30 will fail during the first year is about 0.04.

Weibull Distribution

Advances in technology have made possible the design and manufacture of complex systems whose operation depends on the reliability and availability of the subsystems and components that make up such systems. The time to failure of the life of component measured from a specified point or time interval is a random variable. In 1951, W. Weibull introduced a distribution that has been found to be very useful in the study of reliability and maintenance of physical systems.

The most general form of the Weibull is the three-parameter form presented in equation (9), (10), and (11):

$$f(x) = \left(\frac{\beta}{\theta}\right)\left(\frac{\zeta - \gamma}{\theta - \gamma}\right)^{\beta} e\left(\frac{-(x - \gamma)}{(\theta - \gamma)}\right)^{\beta} \tag{9}$$

$$f(x) = (\alpha\beta)(x - \gamma)^{\beta-1}e^{-(\alpha x)^{\beta}}, \quad \text{where } \alpha = \frac{1}{\theta} \tag{10}$$

$$f(x) = 1 - \exp[-(\alpha x)^{\beta}] \tag{11}$$

where $\theta > 0$ and $\alpha > 0$

$$x > \gamma$$

$$\beta > 0$$

where θ = the characteristic life of the distribution. It is also referred to as the scale parameter and is used to locate the distribution on the x-axis.

γ = the location parameter or the minimum life (useful for warranty. specification).

β = the shape parameter or the slope.

For the Weibull distribution, substituting $x = \theta$ in the cumulative distribution, gives:

$$f(x = \theta) = 1 - e^{-1} = 0.0632$$

So for any Weibull distribution, the probability of failure prior to θ is 0.632. Thus, θ will always divide the area under the probability density function (pdf) into 0.632 and 0.368 for all values of the slope β.

For the two-parameter Weibull, the minimum life γ is zero. A major benefit of modeling life distributions with the Weibull is that the distribution is robust enough that for different values of β it is possible to accommodate a host of other distribution types. For example, when $\beta = 1$, the Weibull becomes the exponential. Also when $\beta = 4$, the Weibull starts to resemble the normal distribution. It is also the distribution of choice to model the different regions of a component's or system's life such as the early, constant, and wear-out regions. The Weibull is particularly useful when the failure rate of the system is not constant but increases (IFR = increasing failure rate) or decreases (DFR = decreasing failure rate). The mean and variance are given as follows:

$$\mu = \theta\Gamma\left(1 + \frac{1}{\beta}\right) \tag{12}$$

$$\sigma^2 = \theta^2\left[\Gamma\left(1 + \frac{2}{\beta}\right)\Gamma^2\left(1 + \frac{1}{\beta}\right)\right] \tag{13}$$

where $\Gamma(n) = (n - 1)!$ is defined as the Gamma function on an integer n.

Example The life of a magnetic resonance imaging (MRI) machine is modeled by a Weibull distribution with parameters $\beta = 2$ and $\theta = 500$ hours. Determine the mean life and the variance of the MRI. Also, what is the probability that the MRI will fail before 250 hours?

Solution:

a. $$\mu = 500\Gamma\left(1 + \frac{1}{2}\right) = 443.11\text{hours}$$

$$\sigma^2 = 500^2\left[\Gamma\left(1 + \frac{2}{2}\right)\Gamma^2\left(1 + \frac{1}{2}\right)\right] = 53650.5$$

b. $$F(x) = 1 - \exp[-(250/500)^2] = 0.2212$$

The gamma values can be found in the gamma tables available in most statistics texts.

2.4 Pareto Analysis for Data Segregation

Any systematic maintenance strategy is aimed at preventing the dominant causes of failure of critical equipment, and, in turn, toward achieving acceptable system availability. In the natural scheme of things, all the major components of a system rarely have the same effect, magnitude wise, on system performance or output. In social system theory, it is usually assumed that 80 percent of society's wealth is held by about only 20percent of the people. The Pareto chart is a bar chart, or histogram, that demonstrates that the intensity of a given phenomenon (such as failure or nonconformance) due to one of the system's components is relative to failure caused by the other components in the system. Hence, the chart shows the contribution of each member to the total system failure. Pareto charts are powerful in identifying the critical failure contributors. Thus, in terms of evaluation, the available resources can be directed to the component that causes failure the most so as to decrease system failure.

Pareto chart analysis requires a statistical approach of measurement or aggregation for identifying the most important problems through different measuring scales or combination of scales. Typical scales or units include frequency of occurrence, cost, labor use, and exposure factor. The idea is that while the sum total of a certain type of occurrence may be due to a number of different sources (or components), the vast majority of those occurrences are due to a small percentage of the sources.

The Pareto rule, also called the 80/20 rule (i.e., 80 percent of the occurrences are due to 20 percent of the sources of the occurrences) has been used very successfully in nonconformance analysis for quality assurance. Through the analyses of scrap information, the current number of significant nonconforming items and their causes are identified.

As an example, consider the problem of poor lathe performance for a certain operation. The following causes based on performance history and process experience have been identified:

DESCRIPTION	CODE
Operator effect	A
Power surges	B
Material mixtures	C
Tool (wear, age)	D
Miscellaneous	E

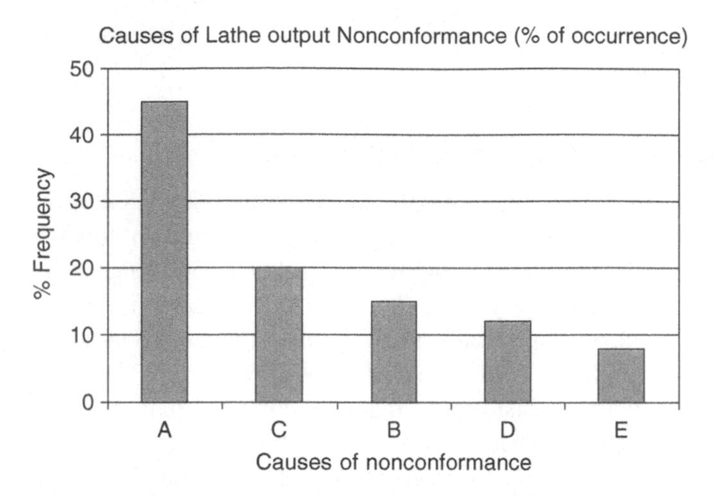

Figure 1 Pareto chart of various causes of lathe output nonconformance (% of occurrence).

Suppose a sampling system was set up whereby over the next week, parts coming out the lathe were examined and the defects were classified according to the categories identified earlier. The frequency of occurrence may be displayed as shown in Figure 1. The plot shows the defect type on the horizontal axis and the frequency on the vertical axis. From this Pareto chart, it is obvious that the major contributors to nonconformance were operator and material problems. As possible solution options, operator training or retraining and vendor evaluation/qualification should be instituted as a means of reducing the problem of poor performance.

3 REVIEW OF SYSTEM RELIABILITY AND AVAILABILITY

System reliability is defined as the probability that a system will perform properly for a specified period of time under a given set of operating conditions. This definition implies that both the loading under which the system operates and the environmental conditions must be taken into consideration when modeling system reliability. However, perhaps the most important factor that must be considered is time, and it is often used to define the rate of failure.

3.1 Reliability Models

Define t = time to failure (random variable), and let T equal the age of the system. Then $F(t) = P(t \leq T)$ = Distribution function of failure process $R(t)$ = Reliability function $= 1 - F(t)$,

Assume $F(t)$ is differentiable.

$$F'(t) = \text{the failure density} f(t) = \frac{d}{dt}F(t) \tag{14}$$

$$R(t) = 1 - F(t) = 1 - \int_0^t f(s)\,ds = \int_t^\infty f(s)\,ds \tag{15}$$

If t is a negative exponential random variable, then:

$$f(t) = \frac{1}{\theta}\exp(-t/\theta)$$

$$F(t) = 1 - \exp\left(-\frac{t}{\theta}\right)$$

$$R(t) = 1 - F(t) = \exp\left(-\frac{t}{\theta}\right)$$

Figure 2 gives a relative frequency of failure from the viewpoint of initial operation at time $t = 0$. The failure distribution function $F(t)$ is the special case when $t_1 = 0$ and t_2 is the argument of $F(t)$—that is, $F(t_2)$.

Failure Probability in the Interval (t_1,t_2)

$$\begin{aligned}\int_{t_1}^{t_2} f(t)\,dt &= \int_0^{t_2} f(t)\,dt - \int_0^{t_1} f(t)\,dt \\ &= F(t_2) - F(t_1) \\ &= [1 - R(t_2)] - [1 - R(t_1)] \\ &= R(t_1) - R(t_2)\end{aligned} \tag{16}$$

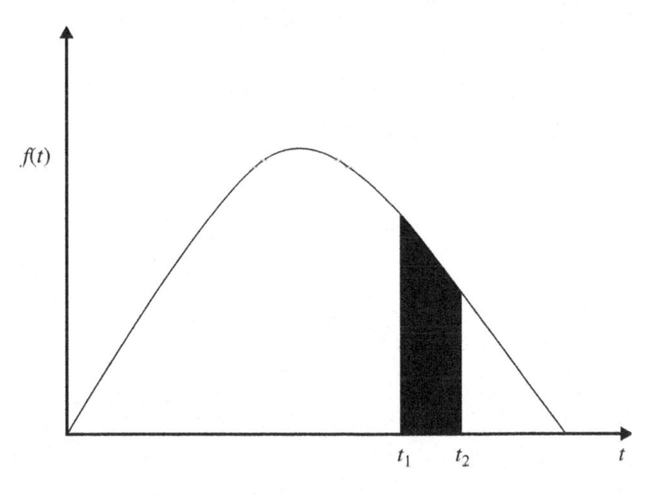

Figure 2 Failure density function.

Reliability of Component of Age t

The reliability (or survival probability) of a fresh unit corresponding to a mission of duration x is by definition $R(x) = \overline{F}(x) = 1 - F(x)$, where $F(x)$ is the life distribution of the unit. The corresponding conditional reliability of the unit of age t for an additional time duration x is

$$\overline{F}(x|t) = \frac{\overline{F}(t \cap x)}{\overline{F}(t)}, \quad \overline{F}(x) > 0 \tag{17}$$

$$\text{But} \quad \overline{F}(t \cap x) = \overline{F}(t + x) = \text{Total life of the unit up to time}(t + x)$$

$$\therefore \overline{F}(x|t) = \frac{\overline{F}(t + x)}{\overline{F}(t)}$$

Similarly, the conditional probability of failure during the interval of duration x is $F(x|t)$ where: $F(x|t) = 1 - \overline{F}(x|t)$ by definition.

$$\text{But} \quad \overline{F}(x|t) = \frac{\overline{F}(t + x)}{\overline{F}(t)}$$

$$\text{Thus,} \quad F(x|t) = 1 - \frac{\overline{F}(t + x)}{\overline{F}(t)} = \frac{\overline{F}(t) - \overline{F}(t + x)}{\overline{F}(t)}$$

Conditional Failure Rate (Hazard or Intensity Function v(t))

The conditional failure probability is given by $F(x|t)$. Hence, the conditional failure rate is given by $\frac{F(x|t)}{x}$

$$\text{where} \quad \frac{F(x|t)}{x} = \frac{1}{x}\left[\frac{\overline{F}(t) - \overline{F}(t + x)}{\overline{F}(t)}\right] = \left(\frac{1}{x}\right)\frac{R(t) - R(t + x)}{R(t)} \tag{18}$$

The hazard or intensity function is the limit of the failure rate as the interval (x, in this case) approaches zero. The hazard function is also referred to as instantaneous failure rate because the interval in question is very small. That is,

$$\begin{aligned} v(t) &= \lim_{x \to 0} \frac{R(t) - R(t + x)}{x}\left(\frac{1}{R(t)}\right) \\ &= -\lim_{x \to 0} \frac{R(t + x) - R(t)}{x}\left(\frac{1}{R(t)}\right) \\ &= -\frac{1}{R(t)}\frac{dR(t)}{dt} \end{aligned} \tag{19}$$

Note: $$R(t) = 1 - \int_0^t f(s)\,ds = 1 - F(t)$$

Hence $$\frac{d}{dt}R(t) = -\frac{d}{dt}F(t) = -f(t)$$

$$\therefore\ v(t) = \frac{1}{R(t)}[f(t)] = \frac{f(t)}{R(t)}$$

The quantity $v(t)$ represents the probability that a device of age t will fail in the small interval t to $t + x$.

Estimation of the Intensity Function Using Binomial Distribution Approach

Consider a population with N identical items and failure distribution function $F(t)$. If $N_S(t)$ is a random variable denoting the number of items functioning successfully at time t, then $N_S(t)$ is represented by

$$P[N_S(t) = n] = \frac{N!}{n!(N-n)!}(R(t)^n)(1 - R(t))^{N-n}$$

where R(t) is the probability of success.

The expected value of $N_S(t)$ is $E(N_S(t))$ if x is binomially distributed. Then, $np = NR(t)$, where $E(N_S(t)) = NR(t) = \overline{N}(t)$.

$$\text{Hence:}\quad R(t) = \frac{E(N_S(t))}{N} = \frac{\overline{N}(t)}{N}$$

$$\text{But}\quad F(t) = 1 - R(t) = 1 - \frac{\overline{N}(t)}{N} = \frac{N - \overline{N}(t)}{N}$$

$$f(t) = \frac{d}{dt}F(t) = -\frac{d}{Ndt}\overline{N}(t)$$

$$= \lim_{\Delta t \to 0} \frac{N(t) - \overline{N}(t + \Delta t)}{\Delta t}(1/N)$$

Replacing N with $\overline{N}(t)$ where $\overline{N}(t)$ = average number of successfully functioning at time t yields, leads to

$$v(t) = \lim_{\Delta t \to 0} \frac{\overline{N}(t) - \overline{N}(t + \Delta t)}{\overline{N}(t)\Delta t} = -\frac{1}{\overline{N}(t)}\frac{d}{dt}\overline{N}(t) = \frac{N}{\overline{N}(t)}f(t) = \frac{f(t)}{R(t)} \qquad (20)$$

Relationship between the Failure Density, the Intensity, and the Reliability Functions

$$v(t) = \frac{f(t)}{R(t)} = -\frac{1}{\overline{N}(t)}\frac{d}{dt}\overline{N}(t) = -\frac{d}{dt}(\ln \overline{N}(t)) = -\frac{d}{dt}(\ln R(t))$$

$$R(t) = \exp\left(-\int_0^t v(\tau)\,d\tau\right)$$

$$f(t) = v(t)\exp\left(-\int_0^t v(\tau)\,d\tau\right)$$

Mean Time to Failure (MTTF and MTBF)

The mean time to failure is the expected value of the time to failure. By definition, the expected value of a density function $y = f(x)\{(f(x)$is continuous$\}$ is the following:

$$E(x) = \int_{-\infty}^{\infty} xf(x)\,dx, \quad -\infty < x < \infty \tag{21}$$

For the mean time to failure T, $(T > 0)$,

$$E(T) = \int_{0}^{\infty} R(s)\,ds, \quad 0 \le T \le \infty \tag{22}$$

By proper transformation and integration (integration by parts), the mean time to failure is equal to $\int_0^\infty sf(s)ds$. How?

$$E(T) = \int_{0}^{\infty} R(s)\,ds, 0 \le T \le \infty$$

$$Note: \int udv = uv - \int vdu$$

Let $u = R(s)$, and let $dv = ds$

then $v = s$, and

$$du = d(R(s)) = -f(s)\,ds$$

$$\text{Thus: } \int_{0}^{\infty} R(s)\,ds, 0 \le T \le \infty = sR(s)\Big|_0^\infty + \int sf(s)\,ds$$

At $s = 0, sR(s) = 0$; at $s = \infty, sR(s) = 0$.

$$\therefore \quad \int_{0}^{\infty} R(s)\,ds = \int sf(s)\,ds = \text{the expected value}$$

Example Let us use the exponential density function to determine the MTTE:

$$f(t) = \frac{1}{\theta}\exp(-t/\theta)$$

$$E(t) = \int_{0}^{\infty} s\left(\frac{1}{\theta}\right)\exp(-s/\theta)\,ds$$

$$\text{Let} \quad s = u, \quad \exp(-t/\theta)\,ds = dv$$

$$v = -\frac{\theta}{\theta}\exp(-t/\theta)$$

Using integration by parts, we get the following:

$$E(t) = \theta, MTTF = \theta$$

Therefore,

$$R(\theta) = \exp(-\theta/\theta) = \exp(-1)$$

$$R(\text{MTTF}) = \exp(-1) = 0.368$$

For normal density function:

$$Z = \frac{t-\mu}{\sigma} = \frac{t - MTTF}{\sigma}$$

when $t = \text{MTTF}$

$$Z_z = 0, \quad \Phi(0) = 0.5$$

Thus, even if MTTF is the same, reliability could change, depending on the underlying probability structure that governs the failure.

3.2 Intensity Functions for Some Commonly Used Distributions

From Section 3.1.5, the intensity functions of some commonly used distributions can be computed as follows:

Exponential Distribution

$$f(t) = 1/\theta e^{-t/\theta}, \; R(t) = e^{-t/\theta} \tag{23}$$

$$v(t) = \frac{f(t)}{R(t)} = \frac{1}{\theta} \tag{24}$$

Standardized Normal

$$f(t) = \frac{1}{\sqrt{2\pi}} \exp\left(-\frac{z^2}{2}\right) = \frac{\phi(z)}{\sigma} \tag{25}$$

$$R(t) = 1 - \int_{-\infty}^{Z} \frac{1}{\sqrt{2\pi}} \exp\left(-\frac{\tau^2}{2}\right) d\tau = 1 - F(t) = 1 - \phi(z) \tag{26}$$

$$v(t) = \frac{f(t)}{R(t)} = \frac{\left(z = \dfrac{t-\mu}{\sigma}\right)/\sigma}{R(t)}$$

$$= \frac{\phi(z)}{\sigma(1-\phi(z))} \tag{27}$$

where $\phi(z)$ = pdf for standard normal random variable
$\varphi(z)$ = cdf (cumulative density function) for standard normal random variable.

Log Normal

$$f(t) = \frac{1}{\sigma t\sqrt{2\pi}} \exp\left[-\frac{1}{2}\left(\frac{\ln\tau - \mu}{\sigma}\right)^2\right] \tag{28}$$

$$R(t) = 1 - \int_0^t \frac{1}{\sigma t\sqrt{2\pi}} \exp\left[-\frac{1}{2}\left(\frac{\ln\tau - \mu}{\sigma}\right)^2\right] d\tau \tag{29}$$

$$= 1 - F(t) = 1 - P(T \leq t) = 1 - P\left[Z \leq \frac{\ln t - \mu}{\sigma}\right]$$

$$R(t) = 1 - P\left[z \leq \frac{\ln t - \mu}{\sigma}\right]$$

$$v(t) = \frac{f(t)}{R(t)} = \frac{\phi\left(\dfrac{\ln t - \mu)}{\sigma}\right)/\sigma}{1 - F(t)} = \frac{\phi\left(\dfrac{\ln t - \mu)}{\sigma}\right)}{t\sigma(1 - \phi(z))} \tag{30}$$

Gamma Distribution

$$v(t) = \frac{\frac{\lambda^n}{\Gamma(n)} t^{n-1} e^{-\lambda t}}{\dfrac{\sum_{k=0}^{n-1} (\lambda t)^{k} \exp(-\lambda t)}{k!}} \tag{31}$$

Weibull Distribution

$$f(t) = \frac{\beta(t-\delta)^{\beta-1}}{(\theta-\delta)^\beta} \exp\left[\left(\frac{t-\delta}{\theta-\delta}\right)^\beta\right], \quad t \geq \delta \geq 0 \tag{32}$$

$$R(t) = 1 - F(t) = 1 - \int_0^t f(\tau)\delta\tau = \exp\left[\left(\frac{t-\delta}{\theta-\delta}\right)^\beta\right] \tag{33}$$

$$v(t) = \frac{\beta(t-\delta)^{\beta-1}}{(\theta-\delta)^\beta} \tag{34}$$

3.3 The Three-Parameter Weibull Probability Distribution

The Weibull probability density function is an extremely important distribution to characterize the probabilistic behavior of a large number of real-world phenomena. It is especially useful as a failure model in analyzing the reliability of different types of systems. The complete three-parameter Weibull probability distribution, $W(a, b, c)$ is given by

$$f(x; a, b, c) = \frac{c(x-a)^{c-1}}{b^c} \exp\left\{-\left(\frac{x-a}{b}\right)^c\right\}$$

where $x \geq a, b > 0, c > 0$ and a, b, and c are the location, scale, and shape parameters, respectively.

Successful application of the Weibull distribution depends on having acceptable statistical estimates of the three parameters. There is no existing method in obtaining closed-form estimates of the Weibull distribution parameters. Because of this difficulty, the three-parameter Weibull distribution is seldom used. Even the popular two-parameter model does not offer closed-form estimates of the parameters and relies on numerical procedures.

The likelihood Function

The log likelihood function of three-parameter model is given by

$$\log L(a, b, c) = n \log c - nc \log b + (c - 1) \sum_{i=1}^{n} \log(x_i - a) - \sum_{i=1}^{n} \left(\frac{x_i - a}{b} \right)^c \tag{35}$$

The objective is to obtain estimates of a, b, and c so as to maximize $\log L(a, b, c)$ with respect to the constraints:

$$a \leq \min_{1 \leq i \leq n} x_i, \; b > 0, \text{ and } c > 0$$

The following systems of likelihood equations are obtained:

$$\frac{\partial \log L}{\partial a} = -(c - 1) \sum_{i=1}^{n} \log(x_i - a) + \frac{c}{b} \sum_{i=1}^{n} (x_i - a)^{c-1} = 0 \tag{36}$$

$$\frac{\partial \log L}{\partial b} = -\frac{nc}{b} + \frac{c}{b^{c+1}} \sum_{i=1}^{n} (x_i - a)^c = 0 \tag{37}$$

$$\frac{\partial \log L}{\partial c} = \frac{n}{c} - n \log b + \sum_{i=1}^{n} \log(x_i - a) - \sum_{i=1}^{n} \left(\frac{x_i - a}{b} \right)^c \{\log(x_i - a) - \log b\} = 0 \tag{38}$$

Equations (36), (37), and (38) constitute a system of nonlinear equations. There are no existing methods to solve them analytically. Moreover, to our knowledge, there are no effective numerical methods available. Based on this concern and the importance of the three-parameter Weibull distribution in practice, Tsokos and Qiao extended their converging two-parameter numerical solution, to the three-parameter model.[9,10] They showed that the three-parameter solution also converges rapidly, and it does not depend on any other conditions. They briefly summarized their findings in the following theorem:

Theorem 1. *The Simple Iterative Procedure (SIP) always converges for any positive starting point, that is, c_k generated by SIP converges to the unique fixed point c^* of $q(c)$ as $k \to \infty$, for any starting point $c_o > 0$. And the convergence is at least a geometric rate of 1/2 : Where $c = q(c)$, $c_{k+1} = c_k + q(c_k)/2$ and $c = c_o$ for $k = 0, 1, 2 \ldots$*

3.4 Estimating $R(t)$, $v(t)$, $f(t)$ Using Empirical Data

Small sample size ($n \leq 10$)

Consider the following ordered failure times:

$$_{\mathrm{o}}T_1, {}_{\mathrm{o}}T_2, {}_{\mathrm{o}}T_3, {}_{\mathrm{o}}T_4, \ldots, {}_{\mathrm{o}}T_n$$

Where: $_{\mathrm{o}}T_1 \leq {}_{\mathrm{o}}T_2 < {}_{\mathrm{o}}T_3 \leq \cdots \leq_{\mathrm{o}} T_n,$ and the subscript o denotes ordered.

$$_nP_j = \hat{F}(_{\mathrm{O}}T_J)$$

$_nP_j$ is the fraction of the population failing prior to the jth observation in a sample of size n.

The best estimate for $_nP_j$ is the median value. That is,

$$_nP_j = \hat{F}(_{\mathrm{o}}T_J) = \frac{j - 0.3}{n + 0.4} \tag{39}$$

Hence, the cumulative distribution at the jth ordered failure time t_j is estimated as:

$$\hat{F}(_oT_J) = \frac{j - 0.3}{n + 0.4} \tag{40}$$

$$\hat{R}(_{\mathrm{O}}\mathrm{T_J}) = 1 - \hat{F}(_{\mathrm{O}}T_J)$$

$$= 1 - \frac{j - 0.3}{n + 0.4} = \frac{n + 0.4 - j + 0.3}{n + 0.4} = \frac{n - j + 0.7}{n + 0.4} \tag{41}$$

$$\hat{v}(_OT_j) = \frac{\hat{R}(_OT_j) - \hat{R}(_OT_{j+1})}{(_OT_{j+1} - {}_OT_j)\hat{R}(_OT_j)} \tag{42}$$

$$\hat{v}(_OT_j) = \frac{1}{(_OT_{j+1} - {}_OT_j)(n - j + 0.7)} \tag{43}$$

$$\hat{f}(_OT_j) = \frac{\hat{R}(_OT_j) - \hat{R}(_OT_{j+1})}{(_OT_{j+1} - {}_OT_j)\hat{R}(_OT_j)} = \frac{1}{(n + 0.4)(_OT_{j+1} - {}_OT_j)} \tag{44}$$

Large Sample Size

$$R(t) = \frac{\overline{N}(t)}{N} \tag{45}$$

$$v(t) = \frac{\overline{N}(t) - \overline{N}(t + x)}{\overline{N}(t)x} \tag{46}$$

$$f(t) = \frac{\overline{N}(t) - \overline{N}(t + x)}{N.x} \tag{47}$$

where $x = \Delta t$

3.5 Failure Process Modeling

The growing importance of maintenance has generated considerable interest in the development of maintenance policy models. The mathematical sophistication of these models has increased with the growth in the complexity of modern systems. If maintenance is performed too frequently or less frequently, the cost or system availability would become prohibitive. Thus, a major effort in the development of maintenance programs is to determine the optimal intervention intervals.

In order to develop any maintenance decision model, it is very important to understand and characterize the system's failure-repair process. The most important maintenance decision is to determine when a component needs to be replaced. To model a reparable system, the first question to be answered is, "What would the system look like after a maintenance activity?" In other words, what is the characteristic of the system failure process?

In making mathematical models for real-world phenomena, it is always necessary to make assumptions so as to make the calculations more tractable. However, too many simplifications and assumptions may render the models inapplicable. One simplifying assumption often made is that component failures are independent and identically distributed. From the stochastic process modeling point of view, the system failure process is generally classified as one of the three typical stochastic counting processes—homogenous Poisson process (HPP), nonhomogeneous Poisson process (NHPP), and renewal process (RP). A process $\{N(t), t \geq 0\}$ is said to a counting process if $N(t)$ represents the total number of events that occur by time t.[11] In addition, a counting process must fulfill the following conditions:

1. $N(t) \geq 0$.
2. $N(t)$ is an integer.
3. If $s < t$, then $N(t) \leq N(s)$.
4. For $s < t$, then $N(t) - N(s)$ equals the number of events that occur in the interval (s, t).

Specifically, a counting process is said to be Poisson in nature if, in addition to the above conditions, the following are also met:

1. $N(0) = 0$

2. The process has independent increments.
3. The number of events in any interval of length t is Poisson distributed with a mean of λt:

$$P\{N(t+s) - N(s) = n\} = e^{\lambda t}\frac{(\lambda t)^n}{n!}, \quad n = 0, 1, \ldots \text{and } E[N(t)] = \lambda t. \quad (48)$$

Independent increments mean that the numbers of events that occur in disjoint time intervals are independent.

A homogenous Poisson process, also known as stationary Poisson process, is a counting process in which the inter-arrival rate of two consecutive events is constant. This implies that the failure distribution of such a system is the exponential and that the failure rate or the hazard function is constant over time. Such an assumption on a failure process would mean that the system life is memoriless, and that a system that has already been in use for a time t is as good as new regarding the time it has until it fails. For such a process, preventive maintenance (PM) is ineffective since the failure process is random.

NHPP, also referred to as a nonstationary process, is a generalized Poisson process. The difference is that in NHPP the inter-arrival rate is a function of time. The following conditions must be fulfilled by an NHPP in addition to those of a Poisson process:

1. $P\{N(t+h) - N(t) \geq 2\} = o(h)$. That is, the probability that the number of events that occur in the interval $[t, t+h]$ is a function of the time elapsed.
2. $P\{N(t+h) - N(t) = 1\} = \lambda(t)h + o(h)$.

The major drawback of applying NHPP in maintenance decisions is that a single failure distribution would be used to describe the failure behavior of a complex system. As a result, it would be difficult to evaluate maintenance effects on different components.

Nonhomogeneous Poisson Process

Consider a complex system in the development and testing process. The system is tested until it fails, then it is repaired or redesigned if necessary, and then it is tested again until it fails. This process continues until we reach a desirable reliability, which would reflect the quality of the final design. This process of testing a system has been referred to as *reliability growth.* Likewise, the reliability of a repairable system will improve with time as component defects and flaws are detected, repaired, or removed. Consider a nonhomogeneous Poisson process, NHPP, with a failure intensity function given by the following:

$$v(t) = \frac{\beta}{\theta}\left(\frac{t}{\theta}\right)^{\beta-1}, t > 0 \quad (49)$$

This failure intensity function corresponds to the hazard rate function of the Weibull distribution in equation (34). NHPP is an effective approach to analyzing the reliability growth and predicting the failure behavior of a given system. In addition to tracking the reliability growth of a system, we can utilize such a modeling scheme for predictions, since it is quite important to be able to determine the next failure time after the system has experienced some failures during the developmental process. Being able to have a good estimate as to when the system will fail again is important in strategically structuring maintenance policies. The probability of achieving n failures of a given system in the time interval $(0, t)$ can be written as

$$P(x = n; t) = \frac{\exp\left\{-\int_0^t v(x)\,dx\right\}\left\{\int_0^t v(x)\,dx\right\}^n}{n!} \qquad \text{for } t > 0 \tag{50}$$

When the failure intensity function $V(t) = \lambda$, equation (50) reduces into a homogeneous

$$\text{Poisson process,} \qquad P(x = n; t) = \frac{e^{-\lambda t}(\lambda t)^n}{n!}, \qquad 0 < t. \tag{51}$$

For tracking reliability growth of the system with the Weibull failure intensity,

$$V(t) = \frac{\beta}{\theta}\left(\frac{t}{\theta}\right)^{\beta-1}, 0 < t, 0 < \beta, \theta \tag{52}$$

Where β, θ are scale parameters, respectively, the Poisson density function reduces to

$$P(x = n; t) = \frac{1}{n!}\exp\left\{-\frac{t^\beta}{\theta^\beta}\right\}\left(\frac{t}{\theta}\right)^{n\beta} \tag{53}$$

Equation (53) is the nonhomogeneous Poisson (NHPP) or Weibull process. In reliability growth analysis it is important to be able to determine the next failure time after some failures have already occurred. With respect to this aim, the time difference between the expected failure time and the current failure time or the mean time between failures (MTBF) is of significant interest. The maximum likelihood estimate (MLE) for the shape and scale parameters β and θ are

$$\hat{\beta} = \frac{n}{\sum_{i=1}^{n} \log\left(\frac{t_n}{t_i}\right)} \qquad \text{and} \qquad \hat{\theta} = \frac{t_n}{n^{1/\beta}}$$

The MLE of the intensity function and its reciprocal can be approximated using the above estimates of β and θ. In reliability growth modeling, we would expect the failure intensity $v(t)$ to be decreasing with time; thus, $\beta < 1$. However, values of $\beta > 1$ indicate that the system is wearing out rapidly and would require intervention. Thus, since our goal is to improve reliability, we would need to establish the relationship between MTBF and $v(t)$:

$$MTBF = \frac{1}{v(t)} \qquad \text{for} \quad \beta < 1$$

Nonparametric Kernel Density Estimate

Let $f(x)$ be the unknown probability density function (failure model). We shall nonparametrically estimate $f(x)$ with the following:

$$\hat{f}_n(x) = \frac{1}{nh}\sum_{i=1}^{n} K\left(\frac{x - X_i}{h}\right) \tag{54}$$

The effectiveness of $\hat{f}_n(x)$ depends on the selection of the kernel function K, and bandwidth h. The kernel density estimate is probably the most commonly used estimate and is certainly the most studied mathematically. It does, however, suffer from a slight drawback when applied to data from long-tailed distributions. The kernel density estimate was first proposed by Rosenblatt in 1956. Since then, intensive work has been done to study its various properties, along with how to choose the bandwidth.[12–17] The kernel function is usually required to be a symmetric probability density function. This means that K satisfies the following conditions:

$$\int_{-\infty}^{\infty} K(u)\,du = 1, \quad \int_{-\infty}^{\infty} uK(u)\,du = 0, \quad \int_{-\infty}^{\infty} u^2K(u)\,du = k_2 > 0$$

Commonly used kernel functions are listed in Table 1.

These kernel functions can be unified and put in a general framework, namely, symmetric beta family, which is defined by the following:

$$K(u) = \frac{1}{Beta(1/2, \gamma + 1)}(1 - u^2)_+^{\gamma} \quad \gamma = 0, 1 \ldots \tag{55}$$

where the subscript + denotes the positive part, which is assumed to be taken before the exponentiation. The choices $\gamma = 0$, 1, 2, and 3 lead to, respectively, the uniform, the Epanechnikov, the Biweight, and Triweight Kernel functions.

Table 1 Intensity Functions for Some Commonly Used Distributions.

Kernel	1. Form	2. Inefficiency
Uniform	$\frac{1}{2}I(\lvert u\rvert \le 1)$	1.0758
Epanechnikov	$\frac{3}{4}(1 - u^2)I(\lvert u\rvert \le 1)$	1.0000
Biweight	$\frac{15}{16}(1 - u^2)^2I(\lvert u\rvert \le 1)$	1.0061
Triweight	$\frac{35}{32}(1 - u^2)^3I(\lvert u\rvert \le 1)$	1.0135
Gaussian	$\sqrt{\frac{1}{2\pi}}\exp\left(-\frac{1}{2}u^2\right)$	1.0513

As noted in Marron and Nolan,[17] this family includes the Gaussian kernel in the limit as $\gamma \to +\infty$.

Properties of the kernel function K determine the properties of the resulting kernel estimates, such as continuity and differentiability. From the definition of kernel estimate, we can see that if K is a density function, that is, K is positive and its integral over the entire line is 1, then $\hat{f}_n(x)$ is also a probability density function. If K is n times differentiable, then also $\hat{f}_n(x)$ is n times differentiable. From Table 1, we observe that only the Gaussian kernel will result in everywhere differentiable kernel estimate. This is one reason why the Gaussian kernel is most popular.

If $h \to 0$ with $nh \to \infty$, and the underlying density is sufficiently smooth (f'' absolutely continuous and f''' being squarely integrable), then we have the following:

$$bias[\hat{f}_n(x)] = \frac{1}{2}h^2 k_2 f''(x) + o(h^2) \tag{56}$$

and

$$Var[\hat{f}_n(x)] = \frac{f(x)R(K)}{nh} + o((nh)^{-1}). \tag{57}$$

where $R(K) = \int_{-\infty}^{\infty} K^2(u)du$ is used. The derivation of the conclusions can be found in Silverman.[12] Combining the two expressions and integrating over the entire line, we have the asymptotic mean integrated squared error (AMISE):

$$\text{AMISE}[\hat{f}_n(x)] = \frac{1}{4}h^4 k_2^2 R(f'') + \frac{R(K)}{nh} \tag{58}$$

It can be seen from equation (58) that it depends on four basic quantities: namely, the underlying probability density function $f(x)$, the sample size n, the bandwidth h, and the kernel function K. The first two quantities are usually out of our control. But we can select the other quantities, the bandwidth and the kernel function, to make the AMISE$[\hat{f}_n(x)]$minimal.

First we fix the kernel function and find the best bandwidth. That is, for any fixed K, setting $\frac{\partial \text{AMISE}(\hat{f}_n(x))}{\partial h}$ equal to 0 yields the optimal bandwidth:

$$h_o = \left[\frac{R(K)}{k_2^2 R(f'')}\right]^{\frac{1}{5}} n^{\frac{-1}{5}}$$

The corresponding minimal AMISE is given by

$$\text{AMISE}_o = \frac{5}{4}[\sqrt{k_2}R(K)]^{4/5} R(f'')^{1/5} n^{-4/5}$$

To determine the optimal kernel function, we proceed by minimizing AMISE$_o$ with respect to K, which is equivalent to minimizing $\sqrt{k_2}R(K)$. The optimal kernel function obtained was the Epanechnikov kernel given by $\frac{3}{4}(1 - u^2)I(|u| \le 1)$. The value of $\sqrt{k_2}R(K)$ for the Epanechnikov is $\frac{3}{5\sqrt{5}}$. Thus, the ratio $\sqrt{k_2}R(K)/\frac{3}{5\sqrt{5}}$ provides a measure of relative inefficiency of using other kernel functions. For example, Table 1 gives such values of this ratio for the common kernel functions.

Thus, we can conclude that we can select any of these kernels and obtain almost equal effective results. Thus, K should be chosen based on other issues, such as ease of computation and properties of $\hat{f}_n$. The Gaussian kernel possess these properties and thus is commonly used.

4 REPAIRABLE SYSTEMS AND AVAILABILITY ANALYSIS

In many classes of systems where corrective maintenance plays a central role, reliability is no longer the central focus. In the case of repairable systems (as a result of corrective maintenance), we are interested in

- The probability of failure
- The number of failures
- The time required to make repairs

Under such considerations, two measures (parameters) of system performance in the context of system effectiveness become the focus—namely, *availability* and *maintainability*. Other related effectiveness measures of significance include the following:

- Serviceability
- Reparability
- Operational readiness
- Intrinsic availability

Definitions:

O_t = operating time
I_t = idle time (scheduled system free time)
d_t = downtime (includes administrative and logistics time needed to marshal resources, active repair time, and additional administrative time need to complete the repair process)
a_t = active repair time
m_d = mean maintenance downtime resulting from both corrective and preventive maintenance times
MTTF = mean time to failure
MTBF = mean time between failures
MTBM = mean time between repair or maintenance
CM = corrective maintenance
PM = preventive maintenance

In general, most practitioners consider MTTF and MTBF to be identical. In a strict reliability sense, MTTF is used in reference nonreparable system such as satellites. By contrast, MTBF is used in reference to reparable systems that can entertain multiple failures and repairs in the system's life cycle. For practical purposes, both measures of system performance are identical and the use of the MTTF and MTBF is indistinguishable from that viewpoint.

4.1 Definition of Systems Effectiveness Measures

Serviceability

This is the ease with which a system can be repaired. It is a characteristic of the system design and must be planned at the design phase. It is difficult to measure on a numeric scale.

Reparability

This is the probability that a system will be restored to a satisfactory condition in a specified interval of active repair time. This measure is very valuable to management since it helps quantify workload for the repair crew.

Operational Readiness (OR)

The probability that a system is operating or can operate satisfactorily when the system is used under stated conditions. This includes free (idle) time. Define the following then: $OR = \frac{O_t+I_t}{O_t+I_t+d_t}$

Maintainability

This is the probability that a system can be repaired in a given interval of downtime.

Availability

Availability $A(t)$ is defined as the probability that a system is available when needed or the probability that a system is available for use at a given time. It is simply the proportion of time that the system is in an operating state and it considers only operating time and down time.

$$A(t) = \frac{O_t}{O_t + d_t} \tag{59}$$

Intrinsic Availability

Intrinsic availability is defined as the probability that a system is operating in a satisfactory manner at any point in time. In this context, time is limited to operating and active repair time. Intrinsic availability, A_I, is more restrictive than availability and hence is always less than availability. It excludes free or idle time. It is defines as follows:

$$A_I = \frac{O_t}{O_t + a_t} \tag{60}$$

Inherent Availability

Inherent availability, A_{IN}, is defined as the probability that the system is operating properly given corrective maintenance activities. It excludes preventive maintenance times, administratively mandated free or idle time, logistics support times, and administrative time needed to inspect and ready the system after repair. Inherent availability is defined as follows:

$$A_{IN} = \frac{MTBF}{MTBF = MTTR} \tag{61}$$

Achieved Availability

According to Elsayed[19], achieved availability is defined as the measure of availability that considers both corrective maintenance and preventive maintenance times.

It is a function of the frequency of maintenance (CM or PM) as well as the actual repair or maintenance times. Functionally, it is defines as:

$$A_a = \frac{MTBM}{MTBM + m_{\mathrm{d}}} \tag{62}$$

As indicated earlier, availability $A(t)$ = probability that a system is performing satisfactorily at a given time. It considers only operating time and downtime. This definition refers to point availability and is often not a true measure of the system performance. By definition, point availability is

$$A(T) = \frac{1}{T}\int_0^T A(t)\,dt \tag{63}$$

This is the value of the point availability averaged over some interval of time T. This time interval may represent the design life of the system or the time to accomplish some mission. It is often found that after some initial transient effects, the point availability assumes some time-independent value. This steady state or asymptotic availability is given by

$$A^*(\infty) = \lim_{T\to\infty} \frac{1}{T}\int_0^T A(t)\,dt \tag{64}$$

If the system or its components cannot be repaired, then the point availability at time t is simply the probability that it has not failed between time 0 and t. In this case:

$$A(t) = R(t)$$

Substituting for $A(t) = R(t)$ in equation (64), we get the following:

$$A^*(\infty) = \lim_{T\to\infty} \frac{1}{T}\int_0^T R(t)\,dt$$

$$As\ T \to \infty,\ \int_0^T R(t)\,dt = MTTF$$

$$\text{Hence:} \quad A^*(\infty) = \frac{MTTF}{\infty} = 0$$

$$\therefore \quad A^*(\infty) = 0$$

This result is quite intuitive given our assumption. Since all systems eventually fail, and if there is no repair, then the availability averaged over an infinitely long time is zero. This is the same reasoning for reliability at time infinity being zero.

Example A constant failure rate system is being examined to determine its availability, given a desired maximum design life. As in most constant failure rate systems, the design life is being measured against its average life, or as a function of the system's mean time to failure (MTTF). The design question is to determine the system availability in the case where the design life is specified in terms of the MTTF for a nonreparable system whose MTTF is characterized by a system intensity function that is constant: $v(t) = c$.

Assuming that the design life $T = 1.2$ (MTTF), what system availability will sustain this design life?

Solution: For a constant failure rate system, with no repairs,

$$v(t) = c = \lambda, \qquad \text{hence } R(t) = e^{-\lambda t}$$

$$A^*(T) = \frac{1}{T}\int_0^T e^{-\lambda t}\,dt = \frac{1}{\lambda T}(1 - e^{-\lambda T})$$

Expanding by Taylor series:

$$A^*(T) = \frac{1}{\lambda T}\left(1 - 1 + \lambda T - \frac{1}{2}(\lambda T)^2 + \cdots\right) \quad \text{for } \lambda T <<< 1$$

Hence

$$A^*(T) \approx 1 - \frac{1}{2}(\lambda T) = 1 - p$$

$$p = \frac{1}{2}(\lambda T) \Rightarrow \quad \text{hence } 2p = \lambda T$$

But for $v(t) =$ constant, $\lambda = \frac{1}{MTTF}$

$$T = 1.2MTTF$$

$$\text{Thus: } 2p = \frac{1.2MTTF}{MTTF} \Rightarrow p = 0.6$$

$$\text{Hence} \quad A^*(T) = 0.4$$

This says that the maximum availability that can be expected in the system for a design life that is 20 percent more than the expected average life of the system is about 40 percent.

4.2 Repair Rate and Failure Rate

In order to estimate availability, one must take into account the repair rate, which is typically larger than the failure rate for system stability. For example, a repair time of 5 hours is equal to a rate of $(1/5) = 0.2$, whereas an MTTF of 400 hours is equal to a rate of $(1/400) = 0.0025$.

If the underlying probability structure of the repair process can be characterized as having a constant intensity function, then the pdf of the repair process is the exponential, by definition. Thus, if

$$v(t) = \mu$$

$$m(t) = f(t) = \mu e^{\mu t}$$

Where $m(t)$ is the maintenance density function, then the MTTR $= \frac{1}{\mu}$.

4.3 Modeling Maintainability

Using the definition of the maintenance density function $m(t)$ as our base, we can define the maintainability function $M(t)$ as follows:[10]

$$M(t) = \int_0^t m(\tau)\, d\tau \tag{65}$$

The corresponding mean repair time, or mean time to repair (MTTR), is given by

$$MTTR = \int_0^\infty tm(t)\, dt \tag{66}$$

Earlier in the analysis and development of the intensity function (see equation (17)) we observed that the corresponding conditional probability that a unit of age t will survive for an additional time duration Δt is given by

$$\overline{F}(\Delta t|t) = \frac{\overline{F}(t \cap \Delta t)}{\overline{F}(t)}, \overline{F}(t) > 0$$

where $\overline{F}(t) = 1 - F(t)$, given that $F(t)$ is the probability of failure in time t. But by the definition of conditional probability:

$$\overline{F}(t \cap \Delta t) = \overline{F}(t + \Delta t) = \text{Total life of a unit up to and including } \Delta\text{t,}$$
$$\text{that is } (\text{t} + \Delta\text{t})$$

$$\therefore \overline{F}(\Delta t/t) = \frac{\overline{F}(t + \Delta t)}{\overline{F}(t)} \tag{67}$$

Similarly, the conditional probability of failure during the interval of duration Δt is $F(\Delta t|t)$ where $F(\Delta t|t) = 1 - \overline{F}(\Delta t|t)$.

$$\text{But } \overline{F}(\Delta t|t) = \frac{\overline{F}(t + \Delta t)}{\overline{F}(t)} \Rightarrow F(t) = 1 - \frac{\overline{F}(t + \Delta t)}{\overline{F}(t)} = \frac{\overline{F}(t) - \overline{F}(t + \Delta t)}{\overline{F}(t)} \tag{68}$$

$$v(t) = -\lim_{\Delta t \to 0} \left[\frac{\overline{F}(t+\Delta t) - \overline{F}(t)}{\Delta t} \right] \left(\frac{1}{\overline{F}(t)} \right)$$

$$= \left(\frac{-1}{\overline{F}(t)} \right) \frac{d}{dt} \overline{F}(t) = -\frac{1}{R(t)} \frac{d}{dt} R(t) \tag{69}$$

Analogously, let the time to repair a system, from the point of failure, be given by t'. Then for a small interval Δt, the probability that repair will be completed in the time interval $t + \Delta t$ is given by :

$$m(t)\Delta t = P(t \le t' \le t + \Delta t) \Rightarrow m(t) = \frac{P(t \le t' \le t + \Delta t)}{\Delta t} \tag{70}$$

The conditional probability of repair in the interval $t + \Delta t$ is given by

$$P(t \cap \Delta t) = \frac{P(t \le t' \le t + \Delta t)}{P(t' \ge t)} \tag{71}$$

The conditional repair rate is given by

$$\frac{P(t \cap \Delta t)}{\Delta t} = \frac{P(t \le t' \le t + \Delta t)}{P(t' \ge t)\Delta t} \tag{72}$$

Hence

$$w(t) = \frac{P(t \le t' \le t + \Delta t)}{P(t' \ge t)\Delta t} = \frac{m(t)}{1 - M(t)}$$

$$w(t) = \frac{m(t)}{1 - M(t)} \Rightarrow m(t) = w(t)[1 - M(t)]$$

but $\quad m(t) = \frac{d}{dt} M(t)$

$$w(t) = \lambda(t) = \frac{\frac{d}{dt} M(t)}{1 - M(t)} \tag{73}$$

Integrating:

$$\int_0^t w(\tau)\, d\tau = \int_0^{M(t)} \frac{dM}{(1 - M)} \tag{74}$$

Note that $\int \frac{1}{(1-x)}\, dx = -\ln(1 - x)$

Thus: $\int_0^{M(t)} \frac{dM}{(1 - M)} = -\ln[1 - M(t)]$

$$\ln[1 - M(t)] = \int_0^t w(\tau)\, d\tau \Rightarrow 1 - M(t) = e^{-\left(\int_0^t w(\tau)\, d\tau\right)}$$

Hence, $M(t) = 1 - \exp\left(-\int_0^t w(\tau)\,dt\right)$ (75)

Also, since $m(t) = w(t)[1 - M(t)]$

Then $m(t) = w(t)\left[1 - \left(1 - \exp\left(-\int_0^t w(\tau)\,dt\right)\right)\right] = w(t)\exp\left[-\int_0^t w(\tau)\,dt\right]$

Hence:

$$m(t) = w(t)\exp\left[-\int_0^t w(\tau)\,dt\right] \quad (76)$$

Example Most human performance and human factors experts believe the nature of the maintenance activity, as well as all other activities performed by humans, makes them amenable to be modeled by the lognormal probability structure. Consider an aluminum hot-roll line whose maintenance density function from equation (6a) is given by

$$m(t) = \frac{1}{t\sigma\sqrt{2\pi}}\exp\left[-\frac{1}{2\sigma^2}(\ln t - \mu)^2\right]$$

where t is the downtime delay in days, $\mu = 2$ months, and $\sigma = 1$ month. Also, the time to failure (in months) is exponentially distributed as follows:

$$f(t) = \frac{1}{\theta}e^{-\left(\frac{t}{\theta}\right)},$$

where $\theta = MBTF = 30$ days

a. Find the maintainability function M(t).
b. Find the steady state availability.

Solution:

a.

$$\text{Given: } m(t) = \frac{1}{t\sigma\sqrt{2\pi}}\exp\left[-\frac{1}{2\sigma^2}(\ln t - \mu)^2\right]$$

Let $y = \ln(t)$

Then:

$$dy = \frac{1}{t}\,dt$$

Hence:

$$f(y) = m(y) = \frac{1}{\sigma\sqrt{2\pi}}\exp\left[-\frac{1}{2\sigma^2}(y - \mu)^2\right]$$

$$M(t) = \int_0^t m(\tau)\,d\tau = \int_0^t \frac{1}{\sigma\sqrt{2\pi}} \exp\left[-\frac{1}{2\sigma^2}(y-\mu)^2\right] dt = \Phi\left(\frac{y-\mu}{\sigma}\right)$$

$$= \Phi\left(\frac{\ln t - \mu}{\sigma}\right)$$

$$M(t) = \Phi\left(\frac{\ln t - 2}{1}\right) = \Phi(\ln t - 2)$$

$$MTTR = E(t) = E(e^y) = \int_{-\infty}^{\infty} e^y f(y)$$

$$= \int_{-\infty}^{\infty} \frac{1}{\sigma\sqrt{2\pi}} e^y \exp\left\{-\left(\frac{1}{2}\right)\left(\frac{y-\mu}{\sigma}\right)^2\right\} dy$$

$$E(t) = \exp\left(\mu + \frac{\sigma^2}{2}\right) \int_{-\infty}^{\infty} \frac{1}{\sigma\sqrt{2\pi}} \exp\left\{-\left(\frac{\{y-(\mu+\sigma^2)\}^2}{2\sigma^2}\right)\right\} dy$$

But: $\displaystyle\int_{-\infty}^{\infty} \frac{1}{\sigma\sqrt{2\pi}} \exp\left\{-\left(\frac{\{y-(\mu+\sigma^2)\}^2}{2\sigma^2}\right)\right\} dy = 1 \Rightarrow$ Std normal

Hence:

$$E(t) = \exp\left(\mu + \frac{\sigma^2}{2}\right) \Rightarrow MTTR) = \exp\left(\mu + \frac{\sigma^2}{2}\right) = \exp\left(\mu + \frac{\sigma^2}{2}\right)$$

with $\mu = 2$ days, $\sigma = 1$day

$$MTTR = \exp\left(2 + \frac{1}{2}\right) = 12.812\text{days}$$

b. For ease of calculation, inherent availability is asymptotically equal to the steady state availability. Using equation (61), we can find the steady state availability as follows:

$$A^*(t) = \frac{MTBF}{MTBF + MTTR} = \frac{30}{12.812 + 30} = 0.707 \approx 71\%$$

Example Let the maintenance density function of a system be given by

$$m(t) = te^{-\left(\frac{t^2}{2}\right)}$$

Also, let the underlying failure probability structure for the system be given by the Weibull distribution with characteristic life $\theta = 72$ months, and the slope $\beta = 2$.

a. Find the maintainability function, $M(t)$.
b. Find the MTTR.
c. What is the system steady state availability?

Solution:

a. From equation (65), we can find the maintainability function.

$$M(t) = \int_0^t m(t)\,dt$$

$$= \int_0^t te^{-\frac{t^2}{2}}\,dt$$

$$let\ \ u = \frac{t^2}{2} then\ du = -t\,dt$$

$$M(t) = \int_0^t -e^{-u}\,du = 1 - e^{-\frac{t^2}{2}}$$

b. Equation (66) gives the following:

$$MTTR = \int_0^\infty tm(t)\,dt = \int_0^t t^2e^{-\frac{t^2}{2}}\,dt$$

$$\text{Let } u = te^{-\frac{t^2}{2}}$$

Applying integration by part gives:

$$MTTR = \left[-te^{-\frac{t^2}{2}} + \int e^{-\frac{t^2}{2}}\,dt\right]_0^\infty = \int_0^\infty e^{-\frac{t^2}{2}}\,dt$$

$$\text{Since } \int e^{a^2t^2}\,dt = \frac{\sqrt{\pi}}{2a} \text{ for } a > 0,$$

$$Let\ a = \frac{1}{\sqrt{2}} \text{ then } a^2 = \frac{1}{2}$$

$$\int_0^\infty e^{-\frac{t^2}{2}}\,dt = \frac{\sqrt{\pi}}{2\frac{1}{\sqrt{2}}} = \sqrt{\frac{\pi}{2}} = 1.254 \text{ hours}$$

c. Equation (61) gives us the steady state availability:

$$A^*(s) = \frac{MTTF}{MTTF + MTTR}$$

Given $\theta = 72$ months and $\beta = 2$,

$$\text{MTTF} = \theta\Gamma\left(1 + \frac{1}{\beta}\right) = 72\Gamma\left(1 + \frac{1}{2}\right)$$

From the gamma tables:

$$\Gamma\left(1 + \frac{1}{2}\right) = 0.8862$$

Thus $MTTF = 72 \times 0.8862 = 63.82$ months

$$A^*(s) = \frac{63.82}{63.82 + 1.254} = 0.98$$

4.4 Preventive versus Corrective Maintenance

Unscheduled downtime due to system failure can render a line inoperable for some time, leading to reduced production output. Both preventive and corrective maintenance (PM and CM, respectively) strategies can be employed to reduce system downtime. Preventive maintenance (with added predictive capability) is a proactive strategy that attempts to keep the system in operation with minimal or limited number of breakdowns. Its implementation, when combined with predictive maintenance, which depends on a good understanding of the probability structure of the underlying failure process, is employed in order to reduce the episodic occurrences of the failure event. By contrast, corrective maintenance is employed as a tool to return the system to an operating state once it has experience a catastrophic event leading to system failure.

Corrective maintenance (CM) programs are designed to reduce equipment repair cycle time by ensuring that a trained technician is available to address equipment maintenance requirements, with emphasis on lowest mean time to repair. Ordinarily, PM is scheduled periodically to coincide with specific and equal time intervals. This is consistent with the idea of constant failure rate for the intensity function. Under such circumstance, PM is of little value in enhancing system performance. Thus, it is common (as we shall show later) to argue that *preventive* maintenance alone does not lead to improvement in system availability or reliability. By contrast, *predictive* maintenance provides the capability to extend preventive maintenance to the realm of intervention under the condition of nonconstant intensity function. In this domain, the intervention intervals are no longer constant but are determined through a predictive platform that models the underlying probability structure of the failure process to provide the intervention epochs. The better the predictive capability, the better is the efficacy of preventive interventions, and the lesser the episodic occurrences that require corrective maintenance actions.

4.5 Reliability and Idealized Maintenance

Typically, reliability is an appropriate measure of system performance when a component, system, or unit does not undergo repair or maintenance. In the case where there is repair or maintenance, the system effectiveness measure used is availability. However, in the case of idealized PM, where maintenance returns the system to an as-good-as-new condition, then the reliability of the maintained system can be considered the same as typical system reliability, for obvious reasons. This, in essence, is the same as the reliability measure of a system where the failure rate or the intensity function is a constant. Thus, assume that $v(t) = c = \lambda, t > 0$. *Then, by definition*, we can find the reliability function:

$$R(t) = e^{-\int_0^t v(\tau)\,d\tau} = e^{-\lambda t} \tag{77}$$

Now assume that for preventive maintenance, the intervention interval is denoted by T and that t is the running time of the system less any downtime. Now if we perform preventive maintenance to the system at time T as called for by the PM schedule, then the question becomes, "What is the effect of this idealized maintenance on the intensity function and hence the system reliability?" In other words, does it lead to improved reliability or does the reliability decrease? The answer to that question depends on our assumption about the nature of the maintenance activity.

The reason we want to frame this question in the context of the intensity function is that most of the data collected for the purpose of intervention are typically in terms of the intensity function or the failure rate. For idealized maintenance, which brings the system to as-good-as-new condition, the system does not have any memory of any accumulated wear effect because the condition of the system is as good as new. Thus, it is assumed that intervention does not fundamentally change the underlying structure of the failure process.

The graphs of the intensity function of an unmaintained system $v(t)$ and that of a maintained system $v_m(t)$ are shown on Figure 3.

That is, for the maintained system at $t > T$

$$v_M(t) = \lambda(t)$$

$$R_M(T) = e^{-\int_0^T \lambda(t)\,dt} = R(t), 0 < T < t$$

Define each maintenance epoch as an integer N, where $N = 0, 1, \ldots \infty$

We will use the cumulative damage effect model to develop the resulting intensity function and hence the reliability function for the maintenance intervals.[1,20] The next maintenance epoch, beyond the first one at interval T, is denoted by

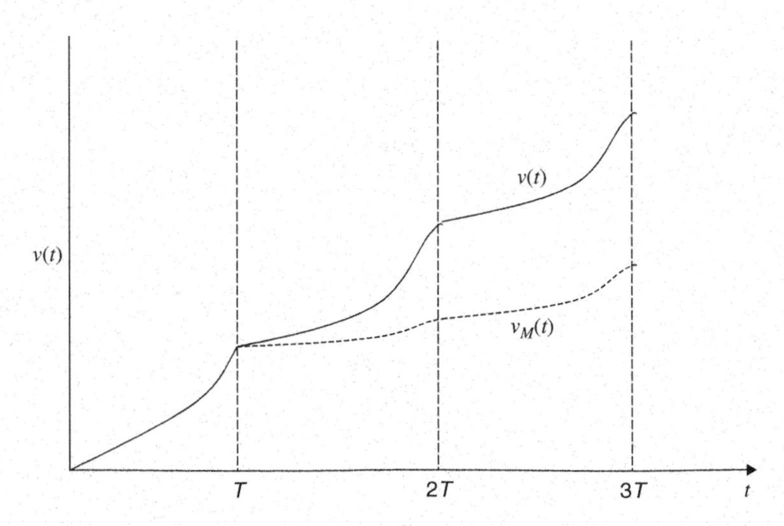

Figure 3 The effect of preventive maintenance on the failure rate/intensity function.

the interval $T \leq t \leq 2T$. The intensity function $v_M(t)$ is given by

$$v_M(t) = v(T) + v(t - T), T \leq t \leq 2T \tag{78}$$

This is graphed in Figure 3. Hence for all subsequent maintenance epochs, the intensity function will change to reflect the new system status. Thus, for system life cycle t, and assuming K equals the number of maintenance epochs during the life cycle, then the cumulative life cycle intensity function is given by:

$$v_M(t) = \sum_{K=0}^{\infty} (v(t - KT) + v(KT)), KT \leq t \leq (K + 1)T \tag{79}$$

Correspondingly, we can determine the reliability of each maintenance epoch and hence the reliability of the maintained system using the cumulative intensity functions as follows:

$$R_M(t) = \exp\left(-\int_0^t v_M(t)\,dt\right) = \exp\left(-\int_{KT}^{(K+1)T} \sum_{k=0}^{\infty} [v(t - KT) + v(KT)]\right) \tag{80}$$

According to,[1] the reliability of a maintained system with equal intervention intervals T is given by the following:

$$R_M(t) = R(T)^K R(t - KT) \tag{81}$$

where $KT \leq t \leq (K + 1)T$, and $N = 0, 1 \ldots \infty$

As shown previously, the MTTF for a system is defined as the average life of that system. The system reliability is really a measure of the life performance of the system. The average life of a system or random variable is the *expected value* of the life of that system. This means that the MTTF can be looked at as the expected value of the system. Thus, if we denote t as the random variable defining the time to failure of the system, then the expected valuc or the expected life of a system is given as:

$$E(t) = MTTF = \int_0^{\infty} R(t)\,dt \tag{82}$$

Using the same argument, we can determine the average life, or MTTF, of a maintained system by replacing $R(t)$ of a regular system by $R_M(t)$ of a maintained system.[1] Hence from equations (81) and (82), the average life, or MTTF, of a maintained system is given by the following:

$$E(t) = MTTF_M = \int_0^{\infty} R_M(t)\,dt$$

But $R_M(t) = R(T)^K R(t - KT),\ KT \leq t \leq (K + 1)T$ from equation (81)

$$\text{Thus: } MTTF = \int_0^{\infty} R(T)^K R(t - KT)\,dt, \text{ where } KT \leq t \leq (K+1)T, K = 0, 1, 2 \ldots \infty$$

Let $t' = t - KT$

when $t = KT, t' = 0$

when $t = (K+1)T, t' = KT + T - KT = T$

$$\therefore\ MTTF = \sum_{K=0}^{\infty} R(T)^K \int_0^T R(t')\,dt'$$

$$\text{But } \sum_0^{\infty} R(T)^K = \frac{1}{1 - R(t)} = \text{sum of an infinite series}$$

How?

$$\text{Let } \sum_0^{\infty} R(T)^K = Q$$

$$\text{Then: } Q = 1 + R(t) + R(t)^2 + R(t)^3 + \cdots + R(t)^K \tag{83}$$

$$R(t)Q = R(t) + R(t)^2 + R(t)^3 + \cdots + R(t)^{K+1} \tag{84}$$

Subtracting equation (84) from (83), we get the following:

$$Q - R(t)Q = 1 - R(t)^{K+1}$$

$$Q(1 - R(t)) = 1 - R(t)^{K+1}$$

$$\text{Thus: } Q = \frac{1 - R(t)^{K+1}}{1 - R(t)}$$

$$As\ K \to \infty \Rightarrow Q = \frac{1 - 0}{1 - R(t)} \Rightarrow \sum_0^{\infty} R(T)^K = \frac{1}{1 - R(t)}$$

$$\text{Thus, } MTTF = \frac{\int_0^T R(t)\,dt}{1 - R(t)} \tag{85}$$

The MTTF for two important lifetime distributions, namely, the exponential and the Weibull, will be examined next to provide some insight as to how to manage the system under maintenance.

Exponential Life

For the exponential, the average life in the presence of maintenance intervention is given by

$$MTTF_{\text{Exponential}} = \frac{\int_0^T R(t)\,dt}{1 - R(T)} = \frac{\int_0^T e^{-\left(\frac{t}{\theta}\right)}\,dt}{\left[1 - e^{-\left(\frac{T}{\theta}\right)}\right]} \tag{86}$$

$$\int_0^T e^{-\left(\frac{t}{\theta}\right)}\,dt = -\theta\left[e^{-\left(\frac{t}{\theta}\right)}\right]\Big|_0^T = -\theta\left[e^{-\left(\frac{T}{\theta}\right)} - 1\right] = \theta\left[1 - e^{-\left(\frac{T}{\theta}\right)}\right]$$

$$MTTF_{\text{PM}}\ (\text{Exponential}) = \theta\left[\frac{1 - e^{-\left(\frac{T}{\theta}\right)}}{1 - e^{-\left(\frac{T}{\theta}\right)}}\right] = \theta = MTTF$$

Thus, in the case of the exponential, the MTTF is the same regardless of whether there is preventive maintenance. This is in agreement with our earlier notion that in the case of constant intensity function or failure rate, PM does not lead to any improvement in system performance.

Weibull Life

Considering a hot-roller with underlying failure distribution as the Weibull and life cycle of 15 years (i.e., 180 months, characteristic life $\theta = 72$ months, and slope $\beta = 2$), we want to compute the MTTF_{PM}(Weibull) given by the following equation:

$$MTTF_{\text{PM}}(\text{Weibull}) = \frac{\int_0^T R(T)}{1 - R(T)} = \frac{\int_0^T e^{-\left(\frac{t}{\theta}\right)}\,dt}{1 - R(T)} = \frac{R_{MTTF}}{1 - R(T)} \tag{87}$$

Since the numerator (or R_{MTTF}) cannot easily be computed in closed form, numerical integration based on the Simpson's rule is used. The maintenance epochs (T) were considered to be fractions of the hot-roller's life cycle. The MTTF was then calculated by estimating the numerator using the numerical approximation for different planned maintenance epochs, as shown in Table 2.

The result presented in Figure 4 is intuitive, and that is, as the maintenance epochs' planning horizon increases, the MTTF exponentially decreases.

4.6 Imperfect Maintenance

Imperfect maintenance occurs mainly due to maintenance flaws induced by the human operator or, in the case of autonomous maintenance, by defective material or system architecture. The impact of imperfect maintenance is that the system could fail right after it has been repaired or it can fail during the process of repair. In recognition of the fact that this possibility exists, we will model the system reliability under imperfect maintenance by introducing the probability of system breakdown caused by a minute probability of failure.

Again, define the system reliability in the presence of maintenance as given in equation (81) as:

$$R_M(t) = R(T)^K R(t - KT), \quad KT < t < (K + 1)T$$

Table 2 MTTF for Different Maintenance Epochs

LC	f (fraction)	T = f(LC)	$\int_0^T e^{(\frac{t}{\theta})^{beta}} dt$	MTTF
180	0.033	6	4.9266	726.3031
180	0.067	12	10.7731	401.1182
180	0.100	18	16.4628	277.0683
180	0.133	24	21.925	212.4933
180	0.167	30	27.98	173.1876
180	0.200	36	31.9312	146.9352
180	0.233	42	36.3857	128.3027
180	0.267	48	40.4357	114.5087
180	0.300	54	44.0683	103.9829
180	0.333	60	47.2825	95.7718
180	0.367	66	50.0881	89.2625
180	0.400	72	52.5039	84.043
180	0.433	78	54.556	79.8251
180	0.467	84	56.2756	76.4007
180	0.500	90	57.6972	73.6151
180	0.533	96	58.8564	71.3498
180	0.567	102	59.7891	69.5121
180	0.600	108	60.5292	68.0276
180	0.633	114	61.1087	66.8356
180	0.667	120	61.5562	65.8856
180	0.700	126	61.8972	65.135
180	0.733	132	62.1535	64.548
180	0.767	138	62.3435	64.0939
180	0.800	144	62.4825	63.7468
180	0.833	150	62.5828	63.4848
180	0.867	156	62.6541	63.2897
180	0.900	162	62.7043	63.1263
180	0.933	168	62.739	63.0425
180	0.967	174	62.7627	62.9684
180	1.000	180	62.7787	62.9163

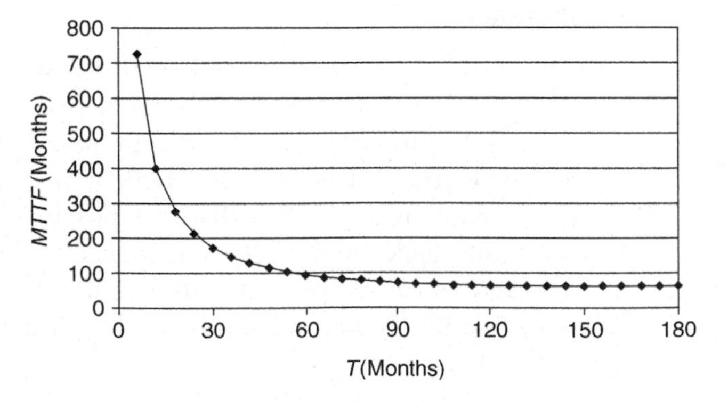

Figure 4 Graph of *MTTF* vs. *T* (maintenance epochs)

Now with the added possibility of failure, we define the probability of such minute failure occurrence as p, then the reliability of nonfailure is $(1 - p)$. Thus, for the system under imperfect maintenance, the reliability is given by the probability that the system will survive multiplied by the probability that the repair will not lead to failure:

$$R_M(t) = R(T)^K R(t - KT)(1 - p)^K, \quad KT < t < (K + 1)T, p << 1 \quad (88)$$

Thus, the MTTF is given by the following:

$$MTTF = \int_0^\infty R_M(t) = \sum_{K=0}^{\infty} \int_{KT}^{(K+1)T} R(T)^K R(t - KT)(1 - p)^K,$$
$$KT < t < (K + 1)T, \; p << 1 \quad (89)$$

But

$$(1 - p)^K \approx e^{-KP}, \text{ for } p <<< 1$$

Thus

$$MTTF = \sum_0^\infty e^{-Kp} R(T)^K \int_0^T R(\tau)\, d\tau$$

where $\tau = t - KT$

The series $\sum_0^\infty e^{-Kp} R(T)^K \approx \dfrac{e^{-Kp}}{1 - R(T)}$

$$\text{Thus: } MTTF = \frac{e^{-Kp} \int_0^T R(t)\, dt}{1 - R(T)} \quad (90)$$

For the Exponential

$$MTTF = e^{-Kp}\theta$$

Since the constant e^{-Kp} is always less than 1, the MTTF in the case of imperfect maintenance is less than for perfect maintenance. Although this is not a great revelation, it is important to note that every effort should be made to mitigate the errors of omission or commission that typically drive up the value of p. Table 3 demonstrates the dominant effect of p, given both p and K on the value of MTTF for imperfect maintenance.

4.7 Modeling Availability

Consider a two-state system classified as working (operational/available) or not working (nonoperational/failed). In terms of availability, the probabilities of the two states can be defined as $A(t)$, and $\tilde{A}(t)$, respectively, at any given time t,

Table 3 Effect of K and p on the MTTF for Imperfect Maintenance

K	p	Kp	e^{-Kp}
1	0.001	0.001	0.9990005
2	0.001	0.002	0.998002
3	0.001	0.003	0.9970045
10	0.001	0.01	0.99004983
40	0.001	0.04	0.96078944
1	0.005	0.005	0.99501248
2	0.005	0.01	0.99004983
3	0.005	0.015	0.98511194
10	0.005	0.05	0.95122942
40	0.005	0.2	0.81873075
1	0.1	0.1	0.90483742
2	0.1	0.2	0.81873075
3	0.1	0.3	0.74081822
10	0.1	1	0.36787944
40	0.1	4	0.01831564
1	0.5	0.5	0.60653066
2	0.5	1	0.36787944
3	0.5	1.5	0.22313016
10	0.5	5	0.00673795
40	0.5	20	2.0612E-09

and where λ is defined as the failure rate and μ is the repair rate. The system initial conditions are given thus:

$$A(t) = 1, \tilde{A}(t) = 0, \text{ and } A(t) + \tilde{A}(t) = 1$$

By using the differential equation approach, it is fairly straightforward to develop the equation for availability.

Assume that we are interested in looking at the system operation during the small time interval Δt given that the system has been in operation during time t. Now consider the change in system availability $A(t)$ between the time t and $t + \Delta t$. There are two possibilities that affect the system state:

1. $\lambda\Delta t$ is the conditional probability of failure during Δt, given that the system was available at time t.
2. $\mu\Delta t$ is the conditional probability that the system has been repaired in the time interval of Δt given system failure.

Thus: $P[\text{system failure during } \Delta t] = \lambda\Delta t$
$P[\text{repair during } \Delta t|\text{system failure}] = \mu\Delta t$

Thus probability that the system is available at time $t + \Delta t$ is given by:

$$A(t + \Delta t) = \{A(t)[1 - \lambda\Delta t]\} + \{[1 - A(t)]\mu\Delta \text{t}\} \tag{91}$$

This, in essence, means that the availability of the system in the interval $t + \Delta t$ is made of two components:

- The system was available in the time t and did not fail in the interval $t + \Delta t$.

  or

- It failed during Δt with probability $[1 - A(t)]$ and was thus repaired with probability $\mu\Delta t$.

Hence:

$$A(t + \Delta t) = A(t)[1 - \lambda\Delta t] + [1 - A(t)]\mu\Delta t$$
$$= A(t) - \lambda A(t)\Delta t + \mu\Delta t - \mu A(t)\Delta t$$

Thus

$$\frac{A(t + \Delta t) - A(t)}{\Delta t} = -(\lambda + \mu)A(t) + \mu$$

$$\text{As} \quad \Delta t \to 0, \frac{A(t + \Delta t) - A(t)}{\Delta t} = \frac{d}{dt}[A(t)]$$

$$\frac{d}{dt}A(t) = -(\lambda + \mu)A(t) + \mu$$

$$\frac{d}{dt}A(t) + (\lambda + \mu)A(t) = \mu$$

Using the method of integration factor (IF), where

$$IF = e^{\int(\mu+\lambda)dt} = e^{(\mu+\lambda)t}$$

$$\frac{d}{dt}[A(t)e^{(\mu+\lambda)t}] = \mu e^{(\mu+\lambda)t} \Rightarrow A(t)e^{(\mu+\lambda)t} = \mu \int e^{(\mu+\lambda)t}$$

$$A(t)e^{(\mu+\lambda)t} = \frac{\mu}{(\mu + \lambda)}e^{(\mu+\lambda)t} + C, \quad \text{where C is the integration constant.}$$

At $t = 0$, $A(0) = 1$

$$\text{Hence: } C = 1 - \frac{\mu}{(\mu + \lambda)} = \frac{\lambda}{(\mu + \lambda)}$$

$$\therefore A(t) = \frac{\mu}{(\mu + \lambda)} + \frac{\lambda}{(\mu + \lambda)}e^{(\mu+\lambda)t}$$

Thus the solution is given by:

$$A(t) = \frac{\mu}{\mu + \lambda} + \frac{\lambda}{\mu + \lambda}e^{-(\mu+\lambda)t} \tag{92}$$

$$A^*(t \to \infty) = \lim_{t\to\infty} A(t) = \frac{\mu}{\lambda + \mu} \tag{93}$$

Define the following:

Mean repair rate = MRR = μ

Mean failure rate = MFR = λ

Also MTTF $= 1/\lambda$, MTTR $= 1/\mu$

$$\text{Then:}\quad A^*(t \to \infty) = \frac{MRR}{MRR + MFR} = A(t), \text{ where } t \text{ is the time interval.} \tag{94}$$

Rewriting:

$$\frac{\mu}{\lambda + \mu} \quad \text{as} \quad \frac{\frac{1}{\lambda}}{\left(\frac{1}{\mu} + \frac{1}{\lambda}\right)} = \frac{1}{\lambda}\left(\frac{1}{\left(\frac{1}{\mu} + \frac{1}{\lambda}\right)}\right)$$

$$\frac{1}{\lambda}\left(\frac{1}{\left(\frac{1}{\mu} + \frac{1}{\lambda}\right)}\right) = \frac{1}{\lambda}\left(\frac{1}{\frac{\lambda+\mu}{\lambda\mu}}\right) = \frac{\mu\lambda}{\lambda}\left[\frac{1}{\lambda + \mu}\right] = \frac{\mu}{\lambda + \mu}$$

$$\text{Thus :}\quad A^*(t \to \infty) = \frac{\mu}{\lambda + \mu} = \frac{\frac{1}{\lambda}}{\left(\frac{1}{\mu} + \frac{1}{\lambda}\right)}$$

$$= \frac{MTTF}{MTTR + MTTF} = \textit{interval availability} \tag{95}$$

Availability Computation Using Empirical Data

Example An aluminum hot-roll line is used to reduce aluminum ingot through several processes to a final configuration for making aluminum cans. The line is prone to failure and the plant engineer is interested in determining the availability of the line versus the design recommendations, as per the EOM. Typically, the EOM stipulations aimed at value of $A(t)$ should be about 70 percent of the specified value, based on the complexity of the line and the fragile nature of the final product. Table 4 represents the times (in hours) over a 30-day period during which aluminum hot-roll line was down at time (td_i) and up at time (tu_i)—that is, repaired and put back into operation. The repair process is sensitive to the line content and so must be done rather quickly, or the product will have to be discarded. Determine the following:

a. MTTF
b. MTTR
c. The 30-day (720-hour) availability
d. Interval availability, A(t)

Solution: The main assumption is that the hot roller runs for 24 hours a day, seven days a week.

Total number of hours of operation for the line considering a 30-day horizon = $24 \times 30 = 720$ hours. During this time, there were 30 occurrences (n) of failures and repairs. The times (td_i) and (tu_i) will be used to estimate MTTR and MTTF.

Table 4 Availability Estimates Using Empirical Estimates of MTTF, MTTR, and Downtimes

i	t_{d_i}	CM downtime $d_i = (t_{u_i} - t_{d_i})$	t_{ui}	Uptime before failure $u_i = t_{d_i} - t_{u_i}$
0	0	0	0	0
1	80.80	5.24	86.04	80.80
2	99.00	2.07	101.06	12.95
3	107.02	5.34	112.36	5.96
4	124.89	6.07	130.96	12.53
5	136.82	5.89	142.71	5.86
6	155.11	7.93	163.05	12.40
7	176.06	4.53	180.59	13.01
8	203.99	5.75	209.74	23.40
9	225.51	1.70	227.20	15.77
10	239.53	1.99	241.52	12.33
11	249.82	4.21	254.03	8.30
12	299.66	4.45	304.11	45.63
13	322.67	3.77	326.44	18.56
14	390.34	3.09	393.43	63.90
15	424.93	3.46	428.39	31.50
16	430.75	3.69	434.45	2.36
17	456.36	5.29	461.65	21.92
18	474.25	6.21	480.46	12.60
19	509.19	4.69	513.87	28.73
20	516.86	3.55	520.41	2.98
21	534.10	2.85	536.95	13.69
22	544.78	4.05	548.82	7.82
23	557.20	2.33	559.53	8.37
24	565.27	5.56	570.83	5.74
25	596.56	5.90	602.46	25.73
26	610.47	3.87	614.34	8.00
27	631.04	4.33	635.37	16.70
28	653.52	3.95	657.47	18.15
29	679.21	4.64	683.86	21.74
30	715.50	4.50	720.00	31.64
TOTAL		**130.92**		**589.08**

Uptime $u_i = (t_{d_i} - t_{u_i})$, where $tu_o = 0$
Downtime $d_i = (t_{u_i} - t_{d_i})$

a. $MTTF_{CM} = \frac{1}{n}\sum_{i=1}^{n}(td_i - tu_{i-1}) = \frac{1}{n}\sum_{i=1}^{n} \mathrm{u_i} = \frac{589.08}{30} = 19.65$ hours

b. $MTTR_{CM} = \frac{1}{n}\sum_{i=1}^{n}(tu_i - td_i) = \frac{1}{n}\sum_{i=1}^{n} \mathrm{d_i} = \frac{130.92}{30} = 4.36$ hours

Mission Length $= T = 720$ hours

c. $A(t) = \frac{1}{T}\sum_{i=1}^{n}(td_i - tu_{i-1}) = \frac{1}{T}\sum_{i=1}^{n} \mathrm{u_i} = \left(\frac{589.08}{720}\right) \times 100 = 81.82\%$

On the other hand,

$$\tilde{A}(t) = \text{unavailability} = 1 - A(t) = 100 - 81.82\% = 18.18\%$$

d. $A(T) =$ Interval availability

$$A(T) = \frac{MTTF}{MTTF + MTTR} = \frac{\lambda}{\lambda + \mu} = \frac{1}{1 + \frac{\mu}{\lambda}} = \frac{1}{1 + \frac{MTTR}{MTTF}} = \frac{1}{1 + \frac{4.36}{19.65}}$$

$$= 81.81\%$$

4.8 Coherent Functions and System Availability

A coherent system structure ensures that only those system structures function that are sensible or reasonable, so that there are no irrelevant components. We can use the idea of coherent system structures to analyze system reliability or availability. For example, the reliability of a pure parallel configuration can be defined as:

$$R = P(x_1 + x_2 + \cdots + x_n)$$

Where the x_i represents the event that path i is successful or operational. Given that

$P(A_1 + A_2) = P(A_1) + P(A_2) - P(A_1A_2)$ for no mutually exclusive events, then,

$$R = [P(x_1) + P(x_2) + \cdots + P(x_n)) \\ - [P(x_1x_2) + P(x_1x_3) + \cdots + P_{i \neq j}(x_ix_j)] + \cdots + (-1)^{n-1}P(x_1x_2 \cdots x_n) \tag{96}$$

This can also be expressed first in terms of system unreliability, where

$$R(t) = 1 - \Pi_{i=1}^{n}(1 - R_i(t))$$

Thus, one can define the reliability of the system in terms of a coherent function using the system structure function developed in equation (96) earlier. For example, for an n component system, the system reliability using the coherent function structure approach can be defined as

$$R(t) = \phi\{R_1(t) + R_2(t) \cdots + R_n(t)\} \tag{97}$$

where ϕ is the coherent operator of the system. In the case of a three-component system, this is equal to

$$\phi(x) = 1 - [(1 - x_1)(1 - x_2)(1 - x_3)] \\ = x_1 + x_2 + x_3 - x_1x_2 - x_1x_3 - x_2x_3 + x_1x_2x_3 \tag{98}$$

Similarly, the system availability for the system can be defines as a coherent function defined as

$$A(t) = \phi\{A_1(t) + A_2(t) + \cdots + A_n(t)\} \tag{99}$$

Example Suppose the hot-roll line has three identical conveyor belts that operate in parallel such that they each feed into a washer. Each conveyor fails independently. The underlying probability structure for the intensity function for the time to failure of the conveyor belts is constant with an average life of 60 days. The system is run such that the probability of all three belts failing at the same time is very negligible. In addition, when a conveyor belt fails, the other belts will continue to feed the washers while the failed one(s) is (are) repaired. Assuming that the maintenance density function is given by:

$$m(t) = \mu e^{-\mu t}, \text{ where}$$

$$\mu = \text{mean repair rate}$$

$$\text{and} \quad MTTR = 2\text{days}$$

a. What is the steady state availability of this conveyor system?
b. What is the probability that the conveyor system will fail during the production period of 30 days?

Solution

a. $A^*(t) = \frac{\mu}{\lambda+\mu}$
   Since the conveyor belts are identical, then

$$A_i = \frac{\mu_i}{\lambda_i + \mu_i} \Rightarrow A_1 = A_2 = \frac{\mu}{\lambda + \mu}$$

But $A_{\text{System}}(t) = x_1 + x_2 + x_3 - x_1x_2 - x_1x_3 - x_2x_3 + x_1x_2x_3$
For steady state:

$$A_{\text{System}} = \left(\frac{3\mu}{\lambda + \mu}\right) - 3\left(\frac{\mu}{\lambda + \mu}\right)^2 + \left(\frac{\mu}{\lambda + \mu}\right)^3$$

$$= \frac{3\mu}{(\lambda + \mu)} - \frac{3\mu^2}{(\lambda + \mu)^2} + \frac{\mu^3}{(\lambda + \mu)^3}$$

$$A_{\text{System}} = \frac{\mu}{(\lambda + \mu)^3}[3(\lambda + \mu)^2 - 3\mu(\lambda + \mu) + \mu^2]$$

$$\mu = \frac{1}{2}, \lambda = \frac{1}{60} \Rightarrow \mu = 0.5, \lambda = 0.01667$$

$$A_{\text{System}} = \frac{0.5}{(0.5 + 0.01667)^3}[3(0.516667)^2 - 3(0.5)(0.51667) + (0.5)^2]$$

$$A_{\text{System}}(t) = 0.999$$

b.

$$A_i(t) = \frac{\mu}{\lambda+\mu} + \frac{\lambda}{\lambda+\mu}e^{-(\lambda+\mu)t}$$

$$= A_1(t) = A_2(t) = A_3(t)$$

$$A_i(30) = \frac{0.5}{0.01667 + 0.5} + \frac{0.01667}{0.01667 + 0.5}e^{-(0.51667)(30)}$$

$$A_i(30) = \frac{0.5}{0.516667} + \frac{0.01667}{0.516667}e^{-(15.5)}$$

$$= 0.9677$$

$$A_{\text{System}}(30) = 1 - [1 - 0.9677]^3$$

$$= 0.9999$$

REFERENCES CITED

1. E. E. Lewis, *Introduction to Reliability Engineering*, 2nd ed., John Wiley & Sons, New York, 1994.
2. J. Huang, "Preventive Maintenance Program Development for Multi-unit System with Economic Dependency—Stochastic Modeling and Simulation Study," Ph.D. Dissertation, University of South Florida, Tampa, Florida, 1993.
3. X. Zheng and N. Fard, "A Maintenance Policy for Repairable Systems based on Opportunistic Failure-rate Tolerance," *IEEE Transactions on Reliability*, **40,** 237–244 (June 1991).
4. X. Zheng, and N. Fard, "Hazard-Rate Tolerance Method for an Opportunistic Replacement Policy," *IEEE Transactions on Reliability*, **41**(1), 13–20 (March 1992).
5. O. G. Okogbaa and X. Peng, "A Methodology for Preventive Maintenance Analysis under Transient Response," Proceedings of 1996 Annual Reliability and Maintenance Symposium, Las Vegas, Nevada, USA, pp. 335–340, 1996.
6. D. P. S. Sethi, "Opportunistic Replacement Policies," in *The Theory and Application of Reliability*, C. P. Tsokos and I. N. Shimi (ed.), Academic Press, New York, 1977, pp. 433–447.
7. J. Moubray, *Reliability-Centered Maintenance*, 2nd ed., Industrial Press, New Jersey, 1997.
8. Watenhost, C. W. I., and Groenendijk, W. P., "Transient Failure Behavior of Systems," *IMA Journal of Mathematics Applied in Business and Industry*, **3**(4), 1992.
9. C. P. Tsokos and H. Qiao, "Parameter Estimation of the Weibull Probability Distribution," *Journal of Mathematics and Computers in Simulation*, **37,** 47–55 (1994).
10. C. P. Tsokos and H. Qiao, "Best Efficient Estimates of the Intensity Function of the Weibull Process," *Journal of Applied Statistics*, **25,** 110–120 (1998).
11. M. S. Ross, *Introduction to Probability Models,* 8th ed., Academic Press, New York, (2003).
12. B. W. Silverman, *Density Function Estimation for Statistics and Data Analysis*, Chapman & Hall, London, 1986.
13. W. Hädle, *Smoothing Techniques with Implementation in S*, Springer-Verlag, New York, 1991.
14. D. W. Scott, *Multivariate Density Estimation: Theory, Practice, and Visualization,* John Wiley, New York, 1992.
15. M. P. Wand and M. C. Jones, *Kernel Smoothing*, Chapman and Hall, London, 1995.

16. J. S. Simonoff, *Smoothing Methods in Statistics*, Springer-Verlag, New York, 1986.
17. J. S. Marron and D. Nolan, "Automatic Smoothing Parameter Selection: A survey", *Empirical Economic*, **13**, 187–208; "Canonical Kernels for Density Function, "Statistic Prob. Letters, 7, 195–199 (1988).
18. A. E. Elsayed, *Reliability Engineering*, Addison Wesley Longman, New York, 1996.

REFERENCES

T. Aven, "Optimal Replacement under a Minimal Repair Strategy—A General Failure Model," *Advances in Applied Probability*, **15**, 198–211 (1983).

R. Barlow and L. C. Hunter, "Optimum Preventive Maintenance Policies," *Operations Research*, **8**, 90–100 (1960).

F. Beichelt, "A Generalized Block-Replacement Policy," *IEEE Transactions on Reliability*, **30**, 171–172 (1981).

M. Berg and B. Epstein, "A Modified Block Replacement Policy," *Naval Research Logistics Quarterly*, **23**, 15–24 (1976).

M. Berg, "Optimal Replacement Policies for Two-unit Machines with Increasing Running Costs—I," *Stochastic Processes and Applications*, **4**, 89–106 (1976).

M. Berg, "Optimal Replacement Policies for Two-unit Machines with Increasing Running Costs—II," *Stochastic Processes and Applications*, **5**, 315–322 (1977).

M. Berg, and B. Epstein, "Comparison of Age, Block, and Failure Replacement Policies," *IEEE Transactions on Reliability*, **27**, 25–29 (1978).

M. Berg and B. Epstein, "A Note on a Modified Block Replacement Policy for Units with Increasing Marginal Running Costs," *Naval Research Logistics Quarterly*, **26**, 157–160 (1979).

P. J. Boland, "Periodical Replacement with Minimal Repair Costs Vary with Time," *Naval Research Logistics Quarterly*, **29**, 541–546 (1982).

P. J. Boland and F. Proschan, "Periodical Replacement with Increasing Minimal Repair Costs at Failure," *Operations Research*, **30**, 1183–1189 (1982).

R. Barlow, and L. C. Hunter, "Optimum Preventive Maintenance Policies," *Operations Research*, **8**, 90–100 (1960).

D. I. Cho and M. Parlar, "A Survey of Maintenance Models for Multi-unit Systems," *European Journal of Operational Research*, **51**, 1–23 (1991).

R. Cleroux, S. Dubuc, and C. Tilquin, "The Age Replacement Problem with Minimal Repair Costs," *Operations Research*, **27**, 1158–1167 (1979).

S. Epstein and Y. Wilamowsky, "An Optimal Replacement Policy for Life Limited Parts," *Operations Research*, **23**, 152–163 (1986).

N. Fard and X. Zheng, "Approximate Method for Non-Repairable Systems based on Opportunistic Replacement Policy," *Reliability Engineering and System Safety*, **33**(2), 277–288 (1991).

J. Huang and O. G. Okogbaa, "A Heuristic Replacement Scheduling Approach for Multi-Unit Systems with Economic Dependency," *International Journal of Reliability Quality and Safety Engineering*, **3**(1), 1–10 (1996).

N. Jack, "Repair Replacement Modeling over Finite Time Horizons," *Journal of the Operations Research Society*, **42**(9), 759–766 (1991).

N. Jack, "Costing a Finite Minimal Repair Replacement Policy," *Journal of the Operations Research Society*, **43**(3), 271–275 (1992).

V. Jayabalan and D. Chaudhuri, "Cost Optimization of Maintenance Scheduling for a System with Assured Reliability," *IEEE Transactions on Reliability*, **41**(1), 21–25 (1992).

D. A. Kadi and R. Cleroux, "Replacement Strategies with Mixed Corrective Actions at Failure," *Computers Operations Research*, **18**(2), 141–149 (1991).

K. C. Kapur and L.R. Lamberson, *Reliability in Engineering Design*, John Wiley and Sons, New York, 1977.

P. L'Ecuyer and A. Haurie, "Preventive Replacement for Multi-component Systems: An Opportunistic Discrete-time Dynamic Programming Model," *IEEE Transactions on Reliability*, **32**, 117–118 (1983).

L. M. Leemis, *Reliability Probabilistic Models and Statistical Methods*, Prentice-Hall, New Jersey, 1995.

V. Makis and A. K. S. Jardine, "Optimal Replacement of a System with Imperfect Repair," *Microelectronics and Reliability*, **31**(2–3), 381–388 (1991).

V. Makis and A. K. S. Jardine, "Note on Optimal Replacement Policy under General Repair," *European Journal of Operational Research*, **69**(1), 75–82 (1993).

T. Nakagawa, "Optimal Maintenance Policies for a Computer System with Restart," *IEEE Transactions on Reliability*, **33**, 272–276 (1994).

P. D. T. O'Conner, *Practical Reliability Engineering*, 2nd ed., John Wiley and Sons, New York, 1985.

O. G. Okogbaa, and X. Peng, "Loss Function of Age Replacement Policy for IFR unit under Transient Response," The NSF Design and Manufacturing Grantee Conference, Seattle, WA, January 1977, 7–10, 1997.

O. G. Okogbaa and X. Peng, "Time Series Intervention Analysis for Preventive/Predictive Maintenance Management of Multiunit Systems," *IEEE International Conference on Systems, Man, and Cybernetics*, **5,** 4659–4664 (1998).

K. Okumoto and E. A. Elsayed, "An Optimal Group Maintenance Policy," *Naval Research Logistics Quarterly*, **30,** 667–674 (1983).

S. Ozekici, "Optimal Periodic Replacement of Multi-component Reliability Systems," *Operations Research,* **36,** 542–552 (1988).

K. S. Park, "Optimal Number of Minimal Repairs before Replacement," *IEEE Transactions on Reliability*, **28,** 137–140 (1979).

W. P. Pierskalla and J. A. Voelker, "A Survey of Maintenance Models: The Control and Surveillance of Deteriorating Systems," *Naval Research Logistics Quarterly*, **23,** 353–388 (1976).

K. Pullen and M. Thomas, "Evaluation of an Opportunistic Replacement Policy for A Two-unit System," *IEEE Transactions on Reliability*, **35,** 320–324 (August 1986).

S. H. Sheu, "General Age Replacement Model with Minimal Repair and General Random Cost," *Microelectronics and Reliability*, **31**(5), 1009–1017 (1991).

S. H. Sheu, "Generalized Model for Determining Optimal Number of Minimal Repairs before Replacement," *European Journal of Operational Research,* **69**(1), 38–49 (1993).

B. D. Sivazlian and J. F. Mahoney, "Group Replacement of a Multi-component System Which Is Subject to Deterioration Only," *Advances in Applied Probability*, **10**, 867–885 (1978).

C. Tilquin and R. Cleroux, "Block Replacement Policies with General Cost Structures," *Technometrics*, **17**, 291–298 (1975).

C. Valdez-Flore, and R. M. Feldman, "A Survey of Preventive Maintenance Models for Stochastically Deteriorating Single-Unit Systems," *Naval Research Logistics*, **36**, 419–446 (1989).

H. C. Young and S. L. Chang, "Optimal Replacement Policy for a Warranted System with Imperfect Preventive Maintenance Operations," *Microelectronics and Reliability*, **32,** (6), 839–843 (1992).

CHAPTER 7

REUSE AND RECYCLING TECHNOLOGIES

Hartmut Kaebernick, Sami Kara
School of Mechanical and Manufacturing Engineering
The University of New South Wales Sydney, Australia

1 INTRODUCTION

Reuse and recycling technologies have become increasingly important over the last decade because they form part of an overall strategy toward sustainable development. Sustainability in product development and manufacturing is a real challenge if taken seriously, and it requires consideration of the entire product life cycle from material production through manufacturing, usage, and disposal. Life-Cycle Assessment (LCA) has already been widely applied with the aim to identify the main sources of environmental impacts during a product's life cycle. As a result, it is well known today that for the vast majority of industrial products there are two main sources of environmental impacts—namely, the material phase and the usage phase. The usage phase is being addressed by redesigning products toward better energy efficiency. The material phase needs to be addressed by using less materials or materials with lower impact factors. The reduction of material usage requires serious consideration of reuse and recycling technologies, which are aimed at reducing the production of new materials by bringing used materials or even entire used components back into the life cycle of new products.

2 RECYCLING TECHNOLOGIES

Recycling involves the separation of waste materials from the waste stream and then processing them as raw material for products, which may or may not be

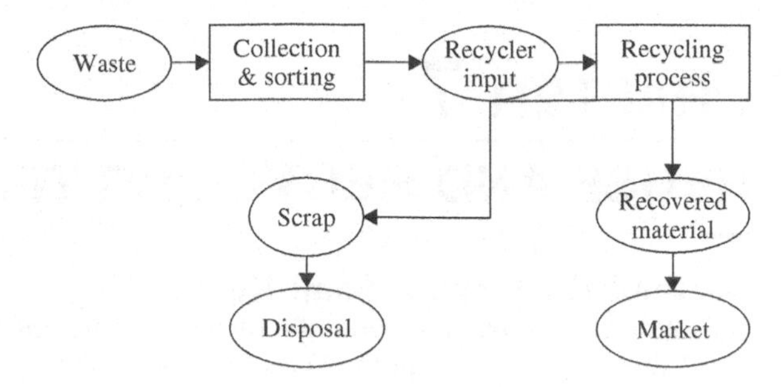

Figure 1 Recycling waste flow.

similar to the original. Nowadays, recycling companies adopt many different recycling processes as part of the overall waste flow. It starts with the generation of industrial or household waste. After collection and sorting, the waste is sent to a recycler as input material. Recycling processes are then used to generate recovered materials for the market or scrap, which is sent to disposal. The above Figure 1 shows the waste flow in recycling.

In order to improve the efficiency of the recycling process, the following economical aspects and product design guidelines should be considered.[1,2]

Economical Aspects:

- Try to recover and use recyclable materials for which a market already exists.
- If toxic materials have to be used in the product, they should be concentrated in adjacent areas so they can be easily detached.
- If nondestructive disassembly is not possible, ensure that the different materials can easily be separated into groups of mutually compatible materials. This is important, for instance, in efficient metal recovery and recycling.

Product Design Guidelines:

- Integrate as many functions in one part as possible.
- Minimize the types of materials used in the whole product.
- Use recyclable materials.
- Avoid use of polluting elements.
- Mark any parts made of synthetic materials with a standardized material code.

2.1 Process Overview

There are many processes between the waste input and the recovered material during recycling. Typical processes used during recycling can be disassembly, sorting, shredding, separation, remelting, or milling. The main steps are shown in Figure 2.

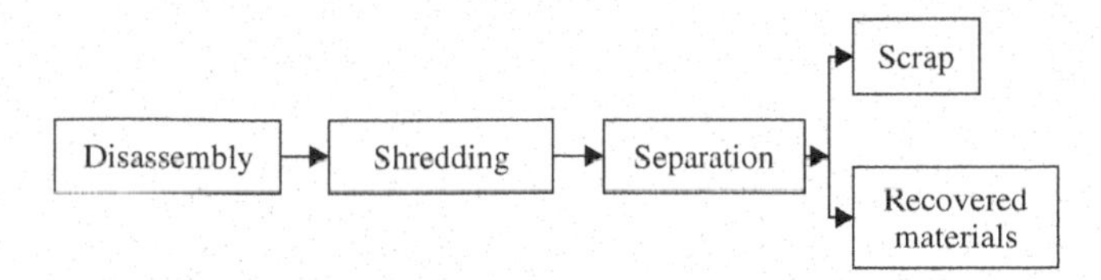

Figure 2 General steps in material recycling.

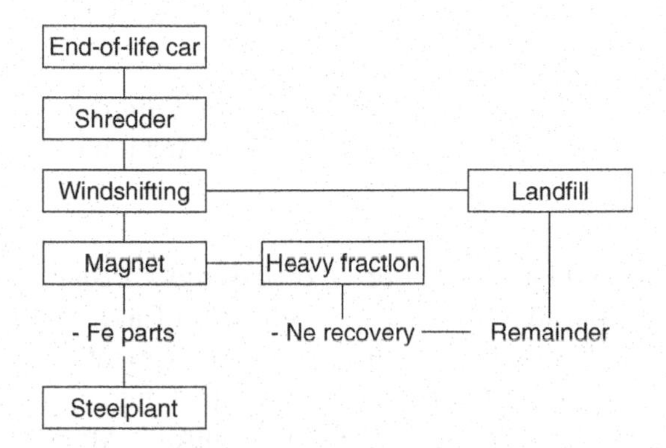

Figure 3 Flow of scrap cars at a shredder plant. (From Ref. 3.)

Disassembly is the first process in recycling before a discarded product goes into the shredding machine. Disassembly is necessary in order to maintain a required level of purity of the recovered material. After shredding, the waste is separated into various types of materials (e.g., ferrous metal, nonferrous metal, plastic). The separation method used for each type of material is based on the material's properties, such as particle size, density difference, magnetic property, electrical property, or differences in melting point. Figure 3 shows the recycling process for an end-of-life (EOL) car as an example.[3]

2.2 Disassembly

The disassembly process is covered in other chapters of this book, so it is only briefly discussed in the following section from a recycling perspective. The disassembly process can be classified into nondestructive or partially destructive disassembly. Although nondestructive disassembly is generally preferred over partially destructive disassembly, the demolition of cheap components of the product is favored.[4,5] Another way of classifying the disassembly process is based on the types of joints:

1. *Disassembly for threaded joints* is the most common group of processes. It allows unfastening by means of mechanized and/or manual processes (e.g., handheld single spindle compressed air tools, open-ended spanners, and screwdrivers).

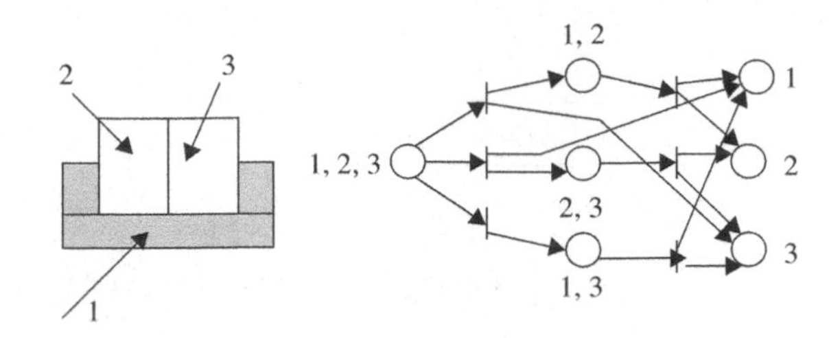

Figure 4 Example of a Petri Net (DPN) disassembly process.

2. *Disassembly for nonthreaded joints* is carried out by using special equipment for detaching nonthreaded joints such as rivets, bonds, or interference fits.

Conceptual Modeling of Disassembly Processes

It is possible to study the behavior of a disassembly system by using simulation models that lend themselves to experimentation. The methods used for modeling a disassembly process are grouped into two categories—namely, sequencing operations and planning and control.[5] Most of the methods are based on the theory of Petri Nets. An example of a sequencing operation is shown in Figure 4.

Design for Disassembly (DfD)

The technical and economical challenges of disassembly can be eased significantly if disassembly processes are considered during the design process—that is, if a product is designed for disassembly. The concept of design for disassembly (DfD) can be used to do the following:

- Facilitate maintenance and repair, thereby reducing costs
- Facilitate part/component re-use, thereby recovering materials and reducing costs
- Assist material recycling, thereby avoiding disposal and handling of waste
- Assist product testing and failure-mode/end-of-life analysis
- Facilitate product take-back and extended producer responsibility, thereby reducing liability and assisting in regulatory compliance

Two categories can be addressed during the design process, the disassembly embedded design and the active disassembly.[4,6] The first category, *disassembly embedded design*, groups products with special separation features, which are embedded in the product during manufacturing. These separation features can be activated during disassembly by means of a trigger (e.g., pressure and magnetic field). However, these techniques have several drawbacks. First, due to the product-specific nature of the separation features, considerable amounts of time and expertise are required from the designer to implement them. Second, the number of connections that can be linked and unfastened simultaneously with one disassembly action is constrained due to physical restrictions, which limit the opportunities for reducing the disassembly efforts.

The second category includes *active disassembly*. This concept enables separation of assemblies for recycling, through the use of smart materials or structures in the product design, by the application of a single or a combination of external triggers. Under this process, certain components are designed so they react to certain triggering conditions by separating.[4,7,8,9] Such assemblies may incorporate shape memory alloy (SMA) actuators in product housings. These actuators in the form of clips can undergo a shape change (shape memory effect) when exceeding a transitional temperature, and can then exert a force to affect disassembly.

Active disassembly techniques offer an answer to most of the limitations created by the first group (disassembly embedded design). This concept offers generic fastening solutions for diverse disassembly problems. Hereby, it is possible to apply one technique in different situations. Moreover, the number of connections that can be linked in order to be unfastened simultaneously is unlimited. Therefore, theoretically all connections can be unfastened by the application of a single trigger. Finally, the use of external triggers excludes the need for physical contact during the disassembly process. All these elements drastically increase the ease of disassembly.[4,6] For example, active disassembly has been used for the disassembly of mobile phones using shape memory polymers (SMP) for the screws.[4,6–9] The composition of the smart material (SMP) and polyurethane (PU) was employed during this process. Two different types of SMP fasteners were created for the experiments. These experiments proved it was possible for products to disassemble themselves at specific triggering temperatures using these smart material devices. The disassembly technique is termed *active disassembly using smart materials* (ADSM), and it has been successfully demonstrated on a variety of mobile phones.

Another example deals with waste products that contain battery power sources.[6,10] In this case, disassembly can be achieved using the residual electrical energy in waste batteries to heat the SMA actuator electrically. There is enough energy left in end-of-life batteries of mobile phones and vehicles to trigger at least four or six devices, respectively. Therefore, this technology would be utilized most efficiently in conjunction with the disassembly of a complete telephone, in a disassembly for recycling context.

2.3 Shredding

Process and Principle

Shredding is a process that transforms large items into small particles. Shredders are based on a hammermill principle. Wind-sifter and sorting belt also usually accompany such devices. Among various types of shredders, fragmentizers are the most widely used. They come in many different shapes and sizes but basically consist of a rotor with arms or hammers which impacts the refuse against fixed grates, bars, and anvils.[2]

Shredding works on the principle of reducing the desired input to smaller particles by using the following processes, which occur in various combinations:

- *Shearing* involves the actual cutting of the material. Similar to the action of scissors, shearing efficiency depends on the sharpness of the cutting edges working against each other and the tolerance of the space between them. Some companies have developed in-house technologies (such as hardened alloys) to maintain this tolerance and sharpness, ensuring a clean cut even after long operation times.
- *Tearing* involves pulling the material with such force that it comes apart. Some materials like fabrics, soft metals, plastics, and tyres, are more tearable than others.
- *Breaking* occurs for brittle materials such as glass, hard plastics, and certain metals, which tend to be broken or shattered in a shredder when the cutters are not sharp enough or loose.

Shredders can be used for disassembled components or for whole products. These days, it is possible to have a car, a washing machine, or a television as input of the shredder. After initial disassembly and sorting, shredders are also used for other products such as tires, hazardous wastes, medical wastes, municipal solid waste, industrial and commercial waste, and construction and demolition debris. In addition, the latest development in shredding technology allows computer and electronic hardware, plastics, papers, and solid metals to be shredded. Figures 5(a) and 5(b) show two special purpose shredders.[11] The STQ Series Four or "Quad" shaft shredder is designed to process bulk materials to a uniform particle size in a single pass, including wood, plastics, paper, tires, textiles, electronic equipment, and manufactured products. Shred-Tech's

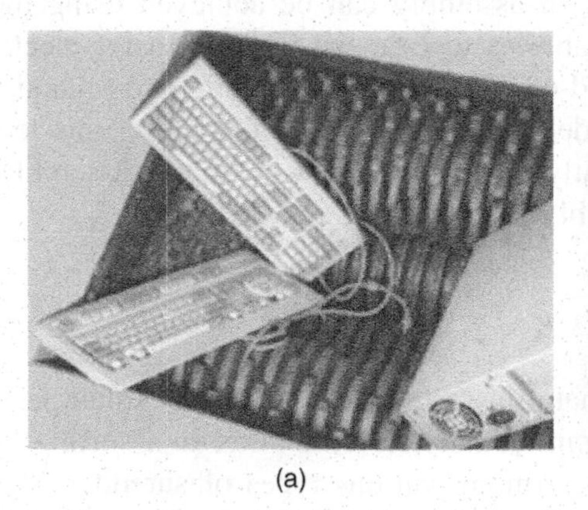

(a)

(b)

Figure 5 (a) The STQ Series Four or "quad" shaft shredder. (From Ref. 11.) (b) Shred-Tech's STG Metals Granulator/Shredder. (From Ref. 11.)

STG Metals Granulator is capable of reducing electronic hardware, including mainframes, computers, copper, zinc, and magnesium as well as steel-belted tires, to a 3/4-inch (19 mm) particle size at rates of up to 5 tons per hour.

There are quite a few companies, who develop shredding technologies, such as Mil Tek, Munson Machinery, Untha, Forus, Lindemann, Newell, Texas or Linx, SSI Shreddings System, and Shred-Tech. Among these, the last two are the leaders in design and manufacture of reduction systems and shredding machinery in a cost-effective way to solve the most difficult waste-reduction and recycling problems.

Grabbing

Grabbing is the process of seizing the material and pulling it down into the cutters. The function of grabbing depends on the size and shape of the hook on the cutter, as well as the weight and texture of the material coming into the shredder. For instance a large, light, smooth object, like a plastic panel, is relatively easy to cut but has a tendency to bounce or "float" on top of the rotating cutters. In this instance, it might be necessary to use a larger shredder or add a ram to push the material down into the shredder where it can be grabbed. Although, in theory, a shredder should only grab an optimum amount, some compressible materials, such as carpeting or paper, can be grabbed too easily and can choke the shredder if too much is grabbed at a time.

Cutter Geometry and Shaft Design

The shape and size of hooks on cutters vary according to the type of material they intend to grab (Figure 6).[12] Generally, the higher the position of the hook, the more material can be grabbed, which can increase the production rate for some materials. But it is important that the hooks do not grab more than the shredder's capacity; otherwise, frequent reversals may occur, resulting in process delays.

Figure 6 Hooks on shredder cutters. (From Ref. 12.)

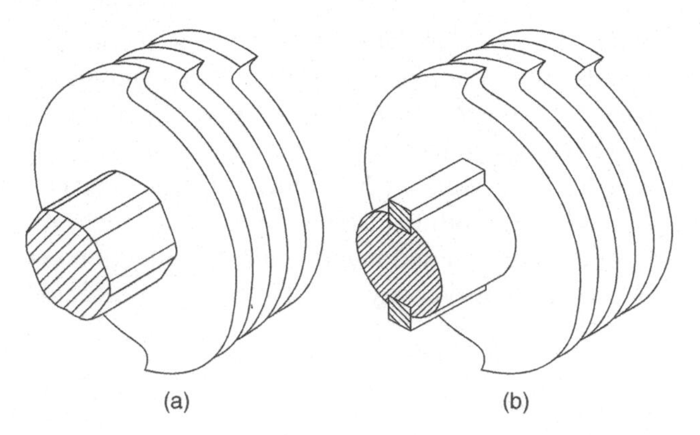

Figure 7 (a) Hex shaft. (From Ref. 2.) (b) Round/keyed shaft. (From Ref. 2.)

Traditionally, shredders were mounted with hex-shaped shafts (Figure 7(a)), which are preferred for high torque applications and are still considered to be stronger than round/keyed shafts (Figure 7(b)).[2] However, both hex shafts and cutters tend to wear down as the cutters become loose, and eventually the entire shaft and cutter must be replaced.

By contrast, a round/keyed shaft actually locks onto the cutters and performs very well under high loads. The economic and maintenance advantage of the round/keyed shaft is evident when it becomes worn, as only the key needs to be replaced.

Granulator

Another technology closely resembling the shredder is the *granulator*. Granulators are designed for universal applications of plastic recycling. It is necessary to have a disassembly process before granulation due to the way the cutters are designed. The companies that dominate the market are Tria, Getecha, Cumberland, Labotek, Rapidgranulator, and Zerma. Figure 8 shows a typical granulator.[13]

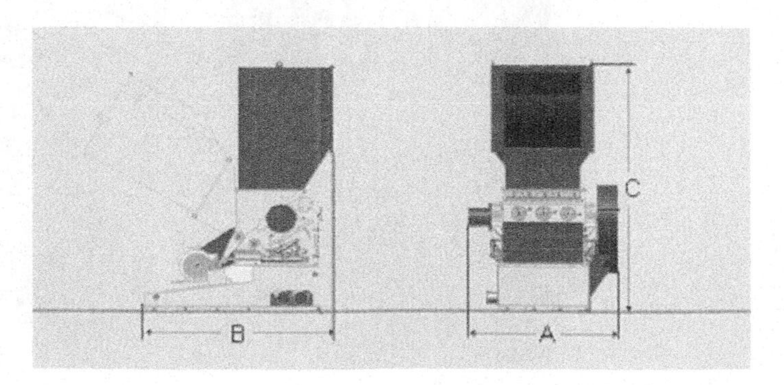

Figure 8 Zerma GSH granulator. (From Ref. 13.)

2.4 Separation Techniques

In order to recover valuable elements from the shredders, mainly ferrous and nonferrous metals, the shredded waste needs to be separated. At first, a separation process is carried out to separate the ferrous metal from the fluff of the shredder output. Then, nonferrous metals are recovered by using various separation techniques.

A number of separation techniques have been developed based on properties of the shredded materials, such as particle size, density difference, magnetic property, electrical property, or differences in melting point.

Magnetic Separation

Ferrous scrap is one of the most commonly recycled materials in the world today. More than 50 percent of the world's steel is produced from scrap. In 2002, in excess of 400 million tonnes of recycled ferrous scrap was used in the steelmaking process. The separation technologies developed for separating ferrous metal utilize the magnetic properties of the material.

Suspended Magnet Design. *Suspended magnets*, also known as *overband separators*, are designed to hang above conveyors for removing material on the conveyor as it passes under the magnet. Overband separators are self cleaning and are used where frequent contamination occurs in a product. Their self-cleaning function ensures continuing efficiency. Typical installations are over conveyors where tramp iron contamination must be removed to protect subsequent processing machinery or to provide a cleaner product. These separators can be fitted across the conveyor at right angles to the product flow, discharging to one side. Alternatively they can be fitted in line over the conveyor head pulley, discharging extracted iron forward of the product trajectory. Figure 9 shows an inline overband separator.[14]

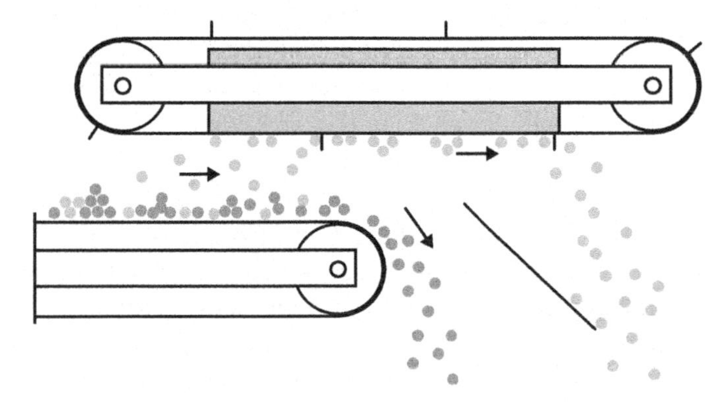

Figure 9 Inline overband magnetic separator. (From Ref. 14.)

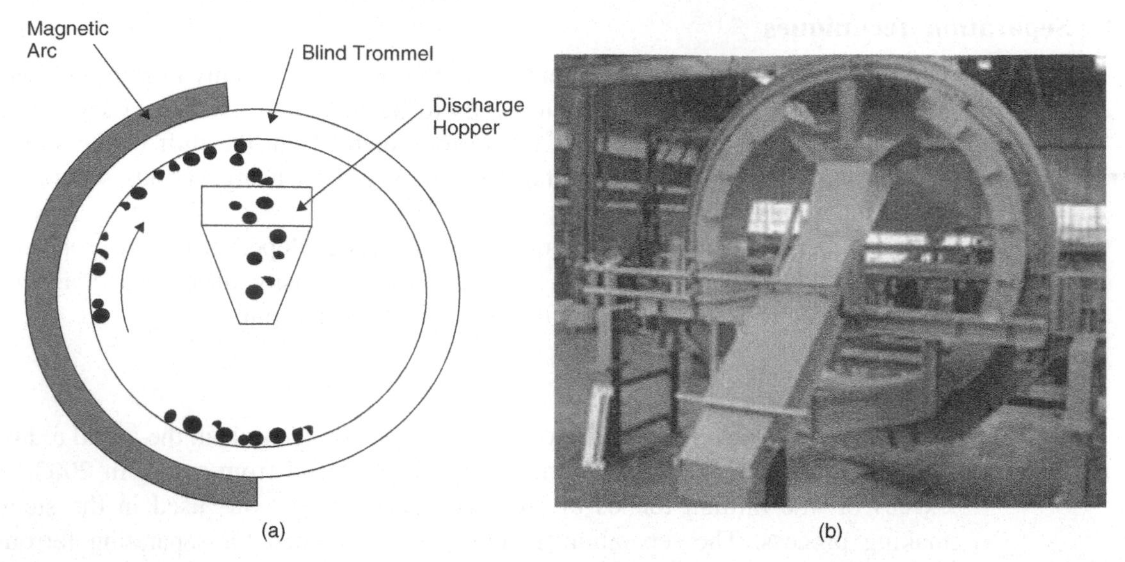

Figure 10 Trommel Magnetic Separator. (a) The working principle of the trommel magnetic separation. (From Ref. 13.) (b) The Eriez Trunnion Magnet System. (From Ref. 15.)

Trommel Magnetic Separator. A *trommel magnetic separator* produced by Eriez was developed to remove all ferrous metal from the stream discharged after shredding. This magnetic separator has proven to be effective for removing all sizes of particles and fragments.[13]

The system includes three basic components, namely the blind trommel, the magnetic sector, and the discharge hopper, as shown in Figure 10.[15] The function of the blind trommel is to transport the milled discharged material after the shredding process through the magnetic field, where it is picked up and transported to the discharge hopper.

Magnetic Pulley. Magnetic pulleys replace standard conveyor head pulleys and effectively convert the conveyer into a self-cleaning magnetic separator. As the conveyed material passes the head pulley and discharges in its natural trajectory, the magnets "scalp" large tramp metal from the waste stream, then discharge it as the belt pulls away from the backside of the pulley. Figure 11 shows a magnetic pulley system as designed by Eriez.[15]

Drum Magnetic and Rare Earth Roll Separator. The common magnetic drums use ceramic or Alnico magnet materials as a power source. A separator with such material provides good magnetic fields for removal of all sizes of iron contaminants in most applications. Ceramic and Alnico continue to be the magnets most frequently used to improve the product purity of dry bulk materials.

In addition, rare earth magnets are used today because of their high magnetic properties. The additional strength of the rare earth magnets helps in removing

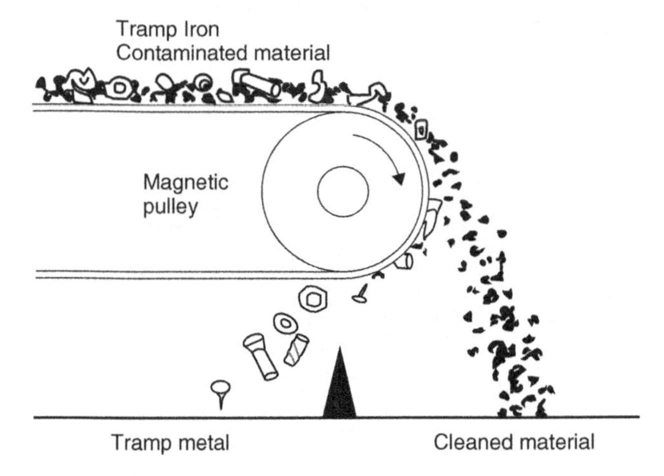

Figure 11 Magnetic pulley system designed by Eriez. (From Ref. 15.)

materials with low magnetic properties or very fine iron contaminants from a wide variety of powdery, dry-bulk materials, as well as slurries. Rare earth drums are effective in treating or purifying large quantities of bulk materials such as foods, plastics, abrasives, metal powders, ceramic material, paper, glass cullet, soda ash, kaolin clay, chemicals, gypsum, and quartz powder. They remove very fine, ferrous particles, locked particles, and even strongly paramagnetic particles.[15] Magnetic separators with high-intensity rare earth Nd–Fe–B (neodymium–iron–boron) magnets are used for removing magnetic contamination with low magnetic properties and fine iron particles.

However, the principle of this technique is the same as for other magnetic separation techniques. As the material reaches the drum, the magnetic field attracts and holds ferrous particles to the drum shell. As the drum revolves, it carries the material through the stationary magnetic field. The nonmagnetic material falls freely from the shell, while ferrous particles are held firmly until they are carried out of the magnetic field.

There are two systems, dry and wet drum separators. There are two distinct applications for wet drum magnetic separators. One application is the recovery of magnetite or ferrosilicon in a heavy media process. The other is the concentration and recovery of magnetite from iron ore. Figures 12(a) and 12(b) show wet and dry systems, respectively.[15]

The drum separators can also be used as a horizontal conveyor for the waste flow. Moreover, there are several types of drum configurations, such as a single-stage drum or a two-stage drum. The difference between them is the purity of the material that is recovered.

Metal Detector. Although magnetic devices are effective at eliminating contamination by ferrous material, nonferrous metals, such as brass, aluminum, and

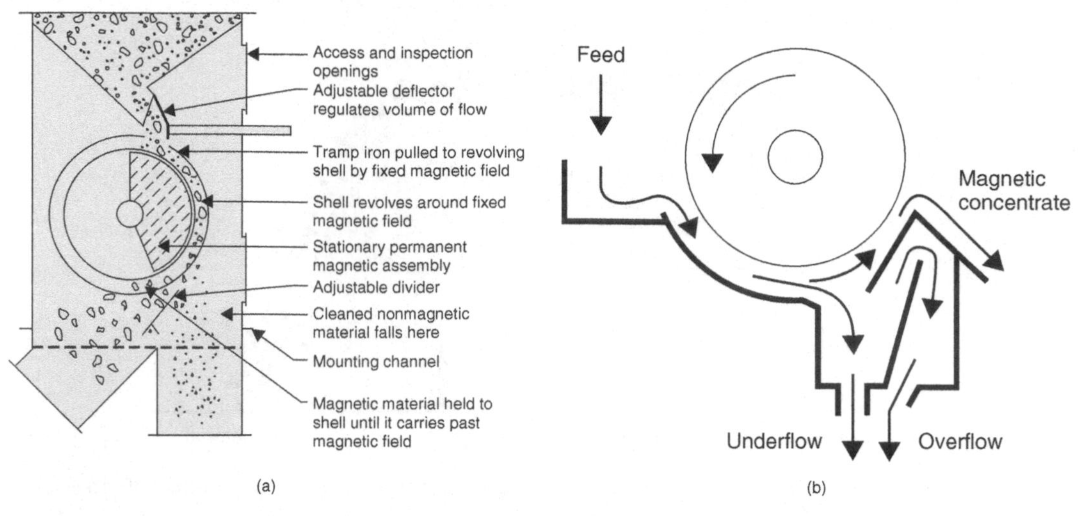

Figure 12 Magnetic drum separators. (a) Dry drum separator. (From Ref. 15.) (b) Wet drum separator. (From Ref. 15.)

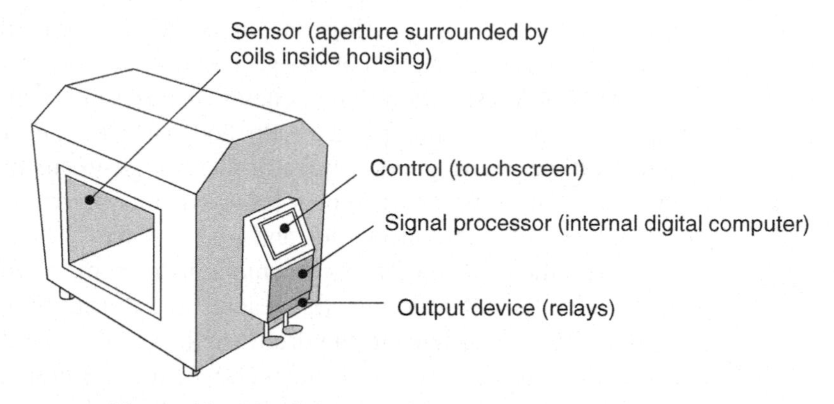

Figure 13 Metal detector components. (From Refs. 2, 16.)

stainless steel, are difficult to remove by using these techniques. Metal detectors similar to those used in airport detectors can be used to address this problem. The metal detector can be incorporated at several stages in the process, wherever there is a chance to detect metal particles.

An industrial metal detector consists of four main components: the output device, signal processor, control, and sensors, as illustrated in Figure 13.[2,16].

The reaction of the sensor is transmitted to the filter and the signal is interpreted by an electronic device. If a metal material is detected by the sensor, the output device is activated. The output can be an alarm or a reject device to remove the metal from the stream. An automatic reject device can be synchronized with the detection. The feed device is generally a conveyor, but chutes, pipelines, or

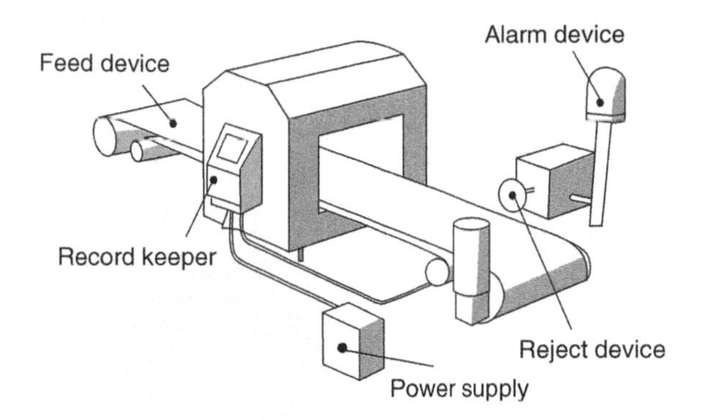

Figure 14 Metal detection system. (From Refs. 2, 16.)

vibrating trays may also be used. Figure 14 shows a complete metal detection system.[2,13,16]

Nonferrous Metal Separation

Eddy currents are usually used to separate the nonferrous metal (e.g., aluminum, brass, copper, lead). Eddy current separation is used to separate metals based on a higher conductivity-to-density ratio. This technology is able to separate and recover aluminum and other nonferrous metals from household, industrial, and incinerated waste, including inert plastics and other materials.

The working principle of the eddy current separator is a rotor comprised of magnet blocks, either standard ferrite ceramic or the more powerful rare earth magnets (depending on the application), which are spun at high revolutions (over 3000 rpm) to produce an *eddy current*. This eddy current reacts with different metals, according to their specific mass and resistivity, creating a repelling force on the charged particle. If a metal is light and conductive, such as aluminum, it is easily levitated and ejected from the normal flow of the product stream making separation possible. Separation of stainless steel is also possible depending on the grade of material. There is no question that eddy current separators have been a crucial development for the recycling industry.[1,2] Figure 15 shows an eddy current system. Some eddy current separators on the market offer the separation in three fractions of different combinations of materials:

1. A ferrous metal fraction, a nonferrous metal fraction, and a nonmetallic material fraction
2. An aluminum fraction, copper and lead fraction, and a nonmetallic material fraction
3. A composite packaging and soft aluminum fraction, a rigid aluminum fraction, and a nonmetallic fraction

The eddy current separator has a very important role, because of the high value of the nonferrous materials recovered.

Figure 15 Eddy current separator. (From Ref. 2.)

Other Separations

Froth Flotation. Froth flotation is often used to separate solids of similar densities and sizes.[15] It is especially useful for particle sizes below 100 mm, which are typically too small for gravity separation using jigging and tabling.

There are several different types of froth flotation systems in use today, including the mechanical type, of which there are many subtypes, and the flotation column. Figure 16 shows a mechanical froth flotation system.[17]

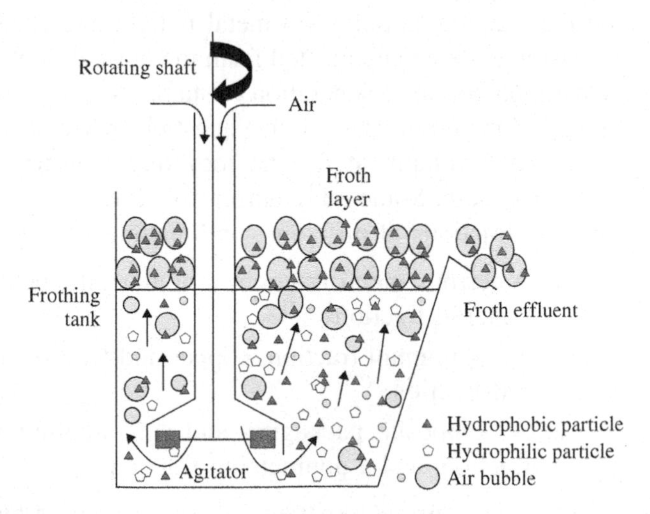

Figure 16 Sub-aerated mechanical froth flotation apparatus. (From Ref. 17.)

The froth flotation device consists of a frothing tank containing the solution and solid-particle mixture. An agitator provides continuous recirculatory mixing of the solids and liquids. Air is introduced through the shaft and enters the solution at the bottom of the agitator (sub-aerated). The introduction of air creates bubbles that are stabilized by adding a frothing agent. The hydrophobic particles, activated by an added collector agent, adhere to the air bubbles. The air bubbles rise to the surface of the solution, where they collect as a froth layer (the concentrate stream), which spills over into a froth-collection tray. The froth may also be mechanically skimmed off. The hydrophilic particles remain in solution, which is eventually decanted off as waste (the tailing stream).

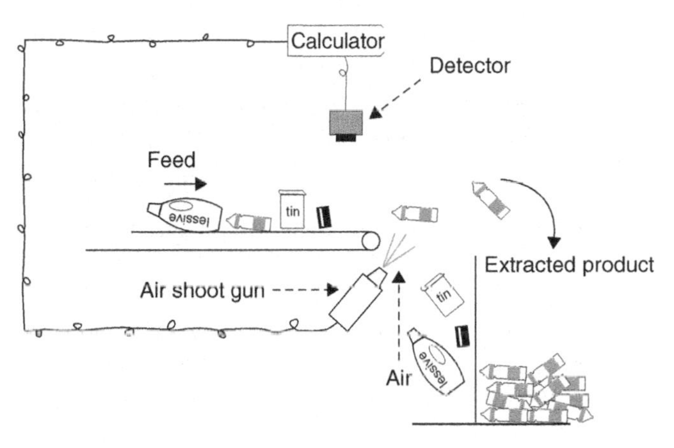

Figure 17 Optical sorting, IR. (From Refs. 1, 16.)

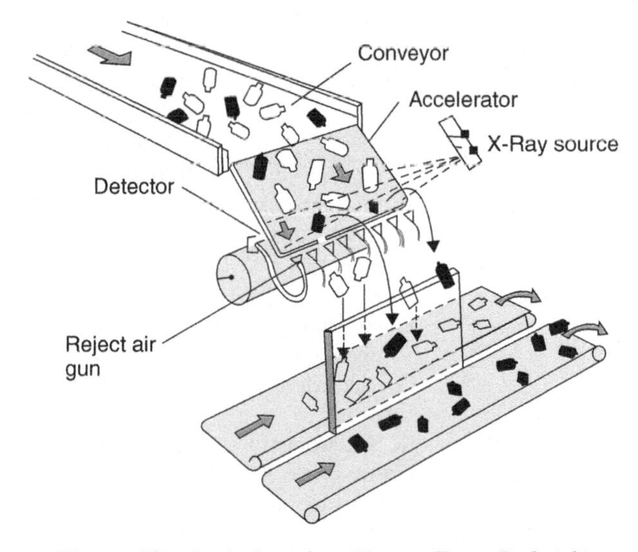

Figure 18 Optical sorting, X-ray. (From Ref. 16.)

Infrared Optical Sorting. This sorting technique identifies materials by their chemical nature using their spectrum of reflexion in the infrared region. This type of sorting is particularly useful for plastic packaging wastes (e.g., plastic bottles). Figure 17 shows an infrared optical sorting system.[1,16]

X-Ray Optical Sorting. This sorting technique identifies material that is excited by X-rays. This type of sorting is particularly useful for plastic scrap with halogen (e.g., chlorine, bromide, fluorine). Figure 18 shows an X-ray optical sorting system.[16]

Summary of the Different Techniques

There are other separation techniques, such as gravity separation (Wilfrey Table) or thermic separation (using the thermal behavior). This section illustrated some of the most widely used separation techniques for recycling. Table 1 is a summary of separation methods.

Table 1 Summary of Separation Techniques Used for Recycling.

Type of Process	Methods	Devices	Separation Principles
Mechanical sorting	Sieving	Horizontal vibrating sifting screen, rotary trommel, rotary disc screen	Size, rigidity
	Table	Wilfery Table	Size, weight, shape
Acrolithic sorting	Acrolithic separation	Acrolithic separator	Air, density, size, shape, bearing pressure
Hydraulic sorting	Hydraulic separation	Hydraulic classifier (forth flotation)	Floating, Archimedean pressure
Magnetic sorting	Magnetic separation	Drum magnetic separator, metal detector, magnetic pulley, suspended magnet separator, trommel magnetic separator	Magnetic properties, magnetic force and repulsion (ferrous metal)
Electrical conductivity sorting	Electromagnetic separation	Eddy current separator	Repulsion (nonferrous metal)
	Electrostatic separation	Electrostatic separator	Electric potential field
	Conductivity sorting	Conductivity sorter	Conductivity
Optical sorting	IR Spectrometry	IR detector	IR spectrum of reflexion
	Fluorescence X Spectrometry	X-ray excitator	Fluorescent spectrum (X-ray)
Thermal sorting	Melting	Thermal separator	Heat behavior

3 REUSE

3.1 Reuse Strategy

It has been admitted that reuse is the ultimate way of increasing sustainability, but it is not easy to be applied in reality. The concept of *reuse* stems from the fact that a certain number of parts and/or subassemblies have a design life that exceeds the life of the product itself. This makes the idea of reuse practical. It has already been argued that product take-back can pay itself. It has already been proven that reuse of electric motors offers potential for profitable product takeback.[18,19,20] The literature further reveals that reusing small electric motors of consumer products is technologically feasible, is economically attractive, and does not compromise product quality. Reuse, the highest level of product recovery, if made possible will promote an economically competitive end-of-life treatment and will reduce the negative environmental impacts of discarded products. However, this strategy requires sound methods to maximize the use of products by reusing the used parts. Some manufacturers have already been reusing the used parts of products at the end of their first life. For example,[19,20] Xerox Corporation promotes reuse of parts of its photocopiers. Another example is Kodak's one-use camera that employs a number of used parts such as flash and internal mechanism. There is often confusion between the terms that are applied in the reuse strategy such as reuse, repair, refurbishing. The following figure 19 shows in detail the difference between these terms.[21,22]

Direct reuse is a basic solution for the reuse strategy. Furthermore, direct reuse does not generally require any disassembly as sufficient economic value still exists in the product, allowing it to be passed on to another user without any value-adding operations. Alternatively, the used product may be repaired. *Repair* refers exclusively to the process of restoring any damage such as broken parts. It does not generally include a major disassembly solution.

The refurbishing process is very similar to that of remanufacturing. The difference lies in the levels of disassembly and rework.[22] Terms such as *rebuilding*, *refurbishing*, *reconditioning*, or *overhauling* are also frequently used. However,

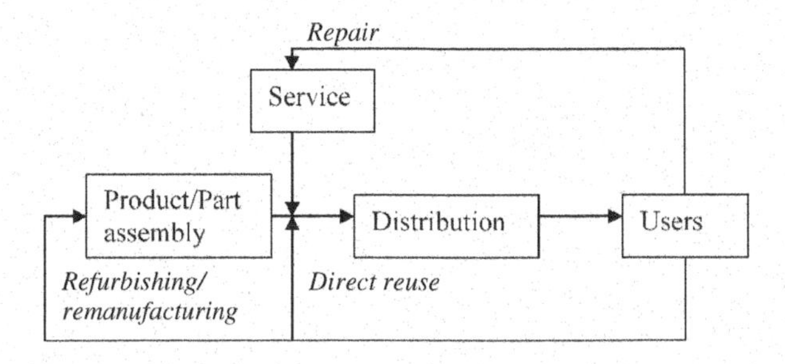

Figure 19 Processes involved in reuse.

remanufacturing is the standard term for the process of restoring used durable products to a "like new" condition. Remanufacturing is also referred to by different names across different nations and industries. For instance, the Japanese literature refers to *inverse manufacturing* and in Sweden it is sometimes called *backtracking manufacturing*.[21] There are also different names depending on the area of application. For example the car industry often refers to *rebuilding,* while the tire remanufacturing industry refers to *retreading*.

Remanufacturing, which involves processes such as disassembly, cleaning, inspection testing, and reassembly, is the process of rebuilding products that have reached their end of life (EOL). A remanufactured product could either be restored to original specifications or be modernized and upgraded with new features.[23] This definition introduces the concept of an upgraded part. Indeed, when a component of a used product is still good, it may be integrated into a new product that is more complex than the original product. Although there are a number definitions for remanufacturing,[21,22] there is one common point in all of these definitions, which is the restoration of used products to a like-new condition, providing them with performance characteristics and durability at least as good as those of the original product. Restoring the EOL product to a like-new quality and performance is the central criterion for remanufacturing and also distinguishes it from other product recovery strategies. Furthermore, remanufacturing is one of the important stages in the implementation of the reuse strategy.

Although reuse has great potential to increase sustainability, it is very difficult to apply. Reuse is confronted with several uncertainties, the most common of which is the uncertainty of the product's quality after use. One of the major barriers in reusing the used parts is the unavailability of reliable methods to assess the reliability of used parts.[18,24] There are a number of other areas—for example, maintenance, accelerated life testing and machine condition monitoring—which can be explored to find out the path to determine the reuse potential. Reliability assessment of parts, machines, and plants is an ongoing activity in the area of design, development, and maintenance. The design and development phases mainly focus on the assessment of the design life of new parts or products. Accelerated life testing is one of the commonly used methodologies in these areas. However, maintenance procedures address the life assessment under a different perspective. Machine condition monitoring, analyzing the machine operating data and statistical pattern analysis, cover the broader areas of remaining life assessment techniques. Vibration analysis, oil/lubricant analysis, Weibull analysis, and survival analysis are among the most pertinent techniques employed to determine mean time to failure (MTTF) or mean time between failures (MTBF).

Therefore, finding a suitable methodology to assess the remaining life of used parts with reasonable certainty will play a pivotal role for subsequent decisions regarding reuse. There are currently three options used to address this issue: (1) maintenance data analysis, (2) testing the product at the end of its first life cycle, and (3) recording and analyzing life history (operating data) of the product to make a reuse decision.[18]

3.2 Maintenance Data and Analysis

The objective of maintenance is to preserve the condition of the products so as to fulfill their required functions throughout their life cycle. Maintenance is important to control the conditions of products.[25] Figure 20 presents the different maintenance activities.

The condition base maintenance (CBM) is a concept to recognize the failures through monitoring or diagnostics, and then to take preventive actions. Within many manufacturing facilities, data collection has become a cornerstone maintenance practice. As a result, CBM has emerged as an alternative to traditional planned maintenance or run-to-failure operation. CBM, based on sensing and assessing the current state of the system, emerges as an appropriate and efficient tool for achieving near-zero breakdown time through a significant reduction and elimination of downtime due to process or machine failure.[25]

Data are mainly collected on machine vibrations, temperatures, oil, and running speed. Then the condition of the machine under consideration can be assessed and protected by using trend analysis softwares. However, the data collected during the first life of a product can also be used to assess the remaining useful life, which is essential for the implementation of the reuse strategy.

Weibull and Statistical Pattern Analysis

A number of life prediction models have been developed, which utilize reliability and maintenance data.[18,25,26] This philosophy has been based on the fact that the failure pattern can be divided into two distinct phases, namely a stable and an unstable phase. These two phases are distinguished from each other by using statistical process control methods. The selection of a particular method to predict the remaining life is dependent on the way in which the machinery

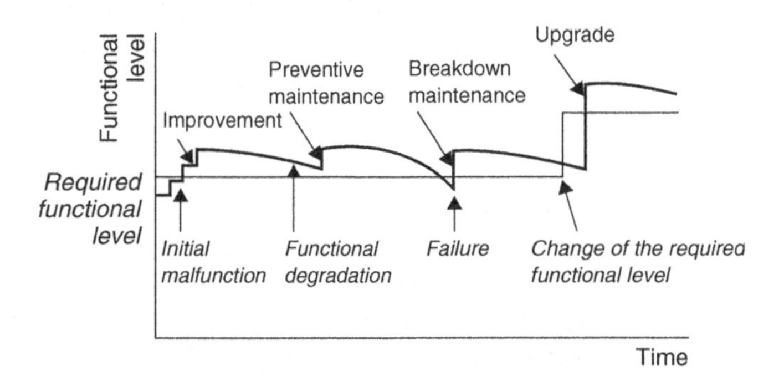

Figure 20 Maintenance activities. (From Ref. 25.)

progresses to failure. The first method relies entirely on a reliability model, while the second method employs a combination of reliability and condition monitoring measurements.

There are other methodologies that combine Weibull analysis and statistical methods to predict failures by using computer algorithms. Weibull analysis uses failure reference and mean time to failure (MTTF) to forecast failures, whereas statistical pattern analysis uses test data to identify a statistical pattern such as trend lines. Performance information is collected during the system's or component's operating life. The proposed combined method generates two possible dates of failure: one based on MTTF and another one using trend-line extrapolation. However, the experimental results of the proposed method show a lack of accuracy in predicting the next failure. The literature states the following factors that affect the accuracy:[27]

1. Factors affecting the accuracy of Weibull analysis:
   a. uncertainty in the failure datum
   b. uncertainty in the failure mode
   c. uncertainty in the date of manufacture
   d. lack of knowledge of the actual operating time
   e. lack of knowledge of the stress levels applied to the item
2. Factors affecting the accuracy of the statistical pattern method:
   a. accuracy of measurements
   b. lack of knowledge of actual operating time

The results suggest that the accuracy of the forecast strongly depends on the knowledge of the operating information such as time and stress. Therefore, existing maintenance processes are not appropriate to provide accurate operating times and operating stress data.

The fundamentals of prognostics have been widely discussed in the literature with an emphasis on the estimation of remaining life and interrelationships between accuracy, precision, and confidence. They highlight the distinction between static and dynamic views of failure distributions. Based on this, a number of methodologies have been suggested that measure both the accuracy and the uncertainty of remaining life estimates.

Vibration Analysis

Vibration analysis has always been one of the main methodologies in the area of condition based maintenance. Although the majority of the work in this area is related to fault detection, recent literature in the area of CBM emphasizes the importance of remaining life prediction. For instance, Li et al. and Al-Najjar emphasize the need to fully utilize the useful lives of rolling element bearings.[28,29] These works relate bearing failure modes to the observed vibration spectra and their development patterns over the bearings' lives. They argue that the accurate prediction of remaining life requires: (1) enough vibration measurements, (2) numerate records of operating conditions, (3) better discrimination between frequencies in the spectrum and (4) correlation of (2) and (3). This is because life prediction depends on accurate knowledge of primary, harmonic and side-band frequency amplitudes, and their development over time. Their findings further reveal that more of the potential life of bearings can be used if more accurate data and better records are available.

In another model, random vibration theory is used to estimate the fatigue damage or fatigue life of a component.[30] The proposed model provides a quick prediction of the fatigue damage or fatigue life when a component is subjected to variable-amplitude loading that has a certain random nature. There is also a stiffness-based prognostic model for bearing systems based on vibration response analysis and damage mechanics.[31] The proposed model is based on the relationship between the vibration spectral features (natural frequency and the acceleration amplitude at the natural frequency), bearing running time, and bearing failure lifetime, based on a dynamic stiffness analysis and accumulative damage mechanics.

Fatigue Analysis

Fatigue analysis is another methodology used to predict remaining life by using maintenance data. Stochastic models are used for the fatigue-induced crack propagation in metallic materials.[32] They use this crack propagation model for damage monitoring and remaining life prediction of stressed structures. Nonlinear stochastic models are also used for prediction of fatigue crack damage in mechanical structures.[33] These models are based on the principle of Gauss-Markov processes and are formulated under the assumption that crack length is lognormal distributed instead of the more common assumption that crack growth rate is lognormal distributed. The model structure allows estimation of the current damage state and prediction of the remaining service life.

Fatigue analysis is also used for calculating the reliability of an assembly of rotating parts subjected to fatigue failure.[34] The reliability of the entire assembly can be estimated for both identical and nonidentical components. Existing literature compares several approaches to fatigue life prediction using a real automotive engineering case study, and taking into account that optimization, based on fatigue life, requires accurate relative distribution rather than exact values.[35] The paper concludes that although both the quasistatic and frequency domain approaches are potentially more efficient than transient dynamic analysis, parameter sensitivity of the frequency domain approach may preclude its eventual use.

The fatigue life prediction of components and structures subjected to random fatigue—to cyclic loading the amplitude of which varies in an essentially random manner—was also investigated.[36] These studies particularly concentrate on the general problem of directly relating fatigue cycle distribution to the power spectral density (PSD) by means of closed-form expressions that avoid expensive digital simulations of the stress process. Numerical simulations and theoretical considerations, carried out in the paper, show that the statistical distribution of fatigue cycles depends on four parameters of the PSD, and the methods proposed in the literature provide reliable results only in particular cases. The suggested methodology lays the foundation for a more precise evaluation of the fatigue cycle distribution.

MTTF, MTBF, and MFOP

The concept of a maintenance free operating period (MFOP) can be used as an alternative to mean time between failures (MTBF). MFOP is favored over MTBF due to several drawbacks associated with the traditional reliability requirement for MTBF. The major advantage of MFOP is that it tracks behavior of the system throughout the life of the system. MFOP will force the suppliers to analyze and understand various mechanisms of failure that will further help to improve the reliability of the overall system. Two mathematical models have been presented to predict the maintenance free operating period survivability (MFOPS), one using the mission reliability approach and the other using the alternating renewal theory.

The availability analysis based on the concept of MFOB is also discussed in the literature, and is applied to a steam generation system, consisting of three subsystems, and a power generation system, consisting of four subsystems arranged in series, with three states, namely good, reduced, and failed.[37] The analysis is based on the assumption of constant failure and repair rates for each working unit. The mathematical model is developed on the birth–death process. Steady-state availability and the MTBF expressions have been proposed.

3.3 Lifetime Monitoring

Lifetime Data Recording and Collection

The collection of life cycle data is an important process in lifetime monitoring. The McLean paper describes the development of a wireless surveillance system for practical application in the factory environment.[38] Figure 21 shows a current embodiment of a wireless system.

A fully wireless condition-based monitoring (CBM) data collection has several advantages. With such a system, operators can monitor sensor data or be informed of alarm conditions. McLean claims that a wireless data-gathering network must be designed for easy installation and easy use.[38] As a result, latest research and development in this field have lead to the concept of the smart product, a product having built-in capabilities of recording the operating data, from which the lifetime monitoring data can be downloaded by using either a plug-in device or an aforementioned wireless system. Two examples from this group of data collection systems, the Watchdog and the Life Cycle Unit, will be described in more detail in the following section.

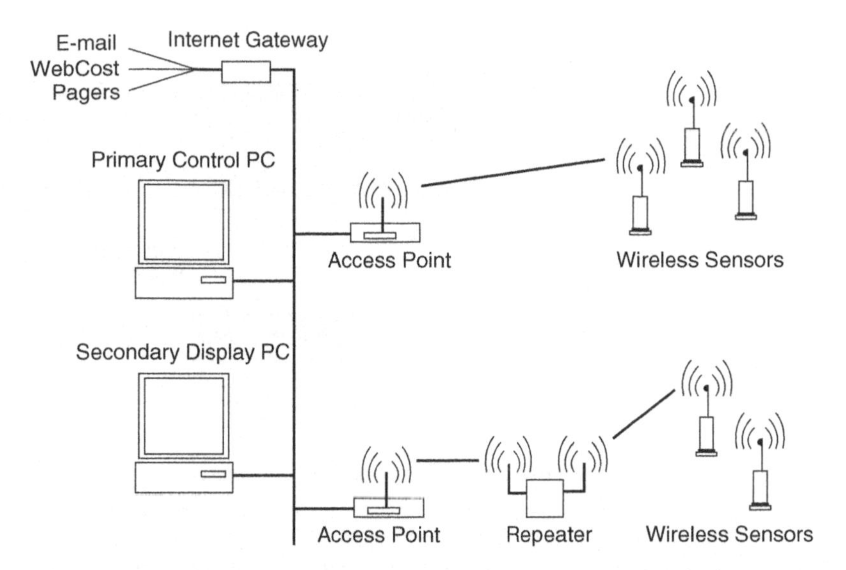

Figure 21 Embodiment of a wireless CBM data-gathering system. (From Ref. 38.)

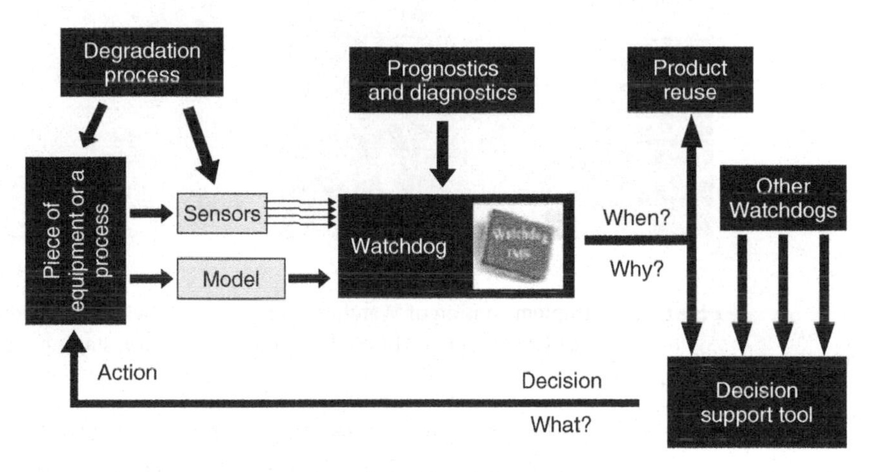

Figure 22 Schematic representation of the Watchdog maintenance system. (From Ref. 25.)

Watchdog Agent™. The Watchdog Agent™ bases its degradation assessment on the reading from multiple sensors.[25,39] These sensors measure the critical properties of a process or machinery. The Watchdog Agent™ detects and quantifies the degradation by quantitatively describing the corresponding change of sensor signature. Statistical models are then used to predict the future behavior and thus forecast the behavior of the process. Figure 22 shows a schematic intelligent system.[25]

The goal of this type of watchdog is to answer the following questions:

- When is the process going to fail or degrade?

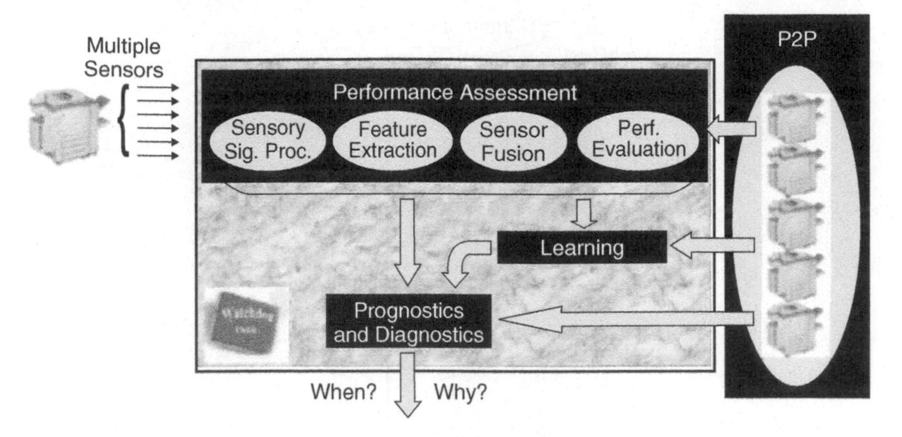

Figure 23 Functionality of the Watchdog Agent™. (From Ref. 39.)

Figure 24 Implementation of Watches Agent™. (a) Elevator door equipped with a Watchdog Agent™. (From Ref. 15.) (b) Material handling device for staging. (From Ref. 15.)

- Why is the performance degrading, and what are the causes?
- What is the most critical object in the system maintenance and repair?

The functionalities of the watchdog system are multisensor assessment, forecasting, and diagnosing the reasons of performance degradation, as shown in Figure 23.[39]

A number of Watchdog Agents™ have already been implemented in industrial and service facilities. For example, a commercial elevator door has a Watchdog Agent™ based on the logistic regression to evaluate the performance of the elevator door (Figure 24(a)). Another Watchdog Agent™, based on the frequency signal analysis, has been implemented in a material-handling device in order to assess the gearbox performance (Figure 24(b)).[15]

Finally, Watchdog Agent™ is a tool for multisensor assessment and prediction of machine or process performance. The link to the reuse technology is that this

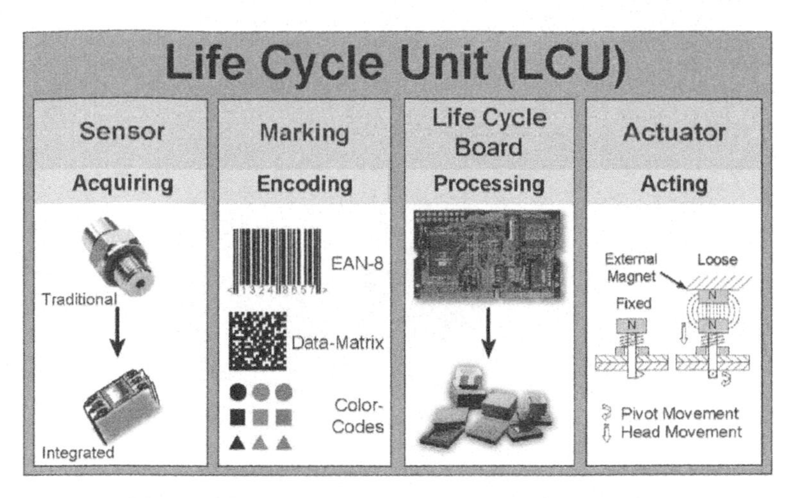

Figure 25 Life-cycle unit (LCU). (From Ref. 41.)

kind of tool can identify a component with significant remaining useful life, which could be cost-efficiently disassembled and reused.

Life Cycle Unit. The life cycle unit (LCU) is another tool developed for collecting operating data during the lifetime of a product.[40,41] The LCU enables the acquisition of operational parameters and the stress acting on wear components. An LCU is a product accompanying information system that acquires relevant product and process data during the entire product life span. It processes, stores, and transfers the data for evaluation. Figure 25 shows the functions of the LCU in details.

The unit composed of a sensor for acquiring data, a processor for processing the data, and an actuator for utilizing the collected data. In this example, the data are utilized to activate a disassembly process by using actuators. The data are also helpful to expand existing business areas and to open up new areas in product-related services such as tele-diagnosis, guaranteed availability and reliability, optimized spare part-supply, condition based maintenance (CBM), selling use instead of selling products, or reuse of products and components. Figure 26 shows four distinctive products, namely a shock absorber, a car engine, a bogie for fright wagons and a hose line with an embedded LCU to collect lifetime monitoring data. For instance, the bogie for fright wagons was developed under the framework of the research project "Light and Low Noise Freight Wagon Bogie LEILA DG."[40,41]

The main physical values analyzed are vibration, force, temperature, current, and magnetic field. The systems mainly rely on threshold analysis, vibration, analysis by using fuzzy logic, or neural networks. In order to determine the wear condition and to forecast the failure behavior of the components, stress or operational parameters that affect the condition (causes of wear) and parameters

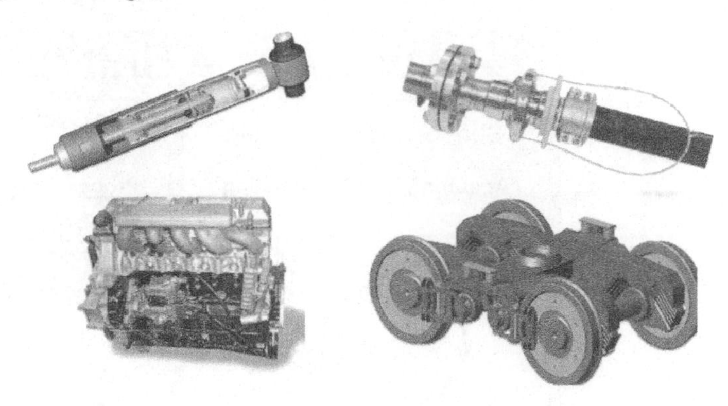

Figure 26 Products with embedded LCUs. (From Ref. 42.)

(e.g., vibrations), that are directly caused by the condition (wear itself) were collected. Then the test of the life span for a selected component, a cylinder roller bearing was realized. The approach was verified with different stress cases and implemented in a testbed, to determine the individual life span of cylinder roller bearings. This project is currently at the implementation stage in practice.

Lifetime Data Analysis

Remaining life assessment by life-cycle data analysis considers statistical as well as condition-monitoring data analysis for decision making on reuse. The methodology addresses the problem of reliability assessment of used parts by considering two important aspects. First, it assesses the overall reuse potential of components with a clear understanding of the failure mechanism. Second, it determines the actual (used) life of the components by analyzing the operating history of components.[24,26] The remaining useful life is a function of the component's overall mean life and the actual (used) life under the operating conditions of use. The mean life (L_{M}) and the actual life (L_{A}) of components under given conditions of use represent two distinct perspectives—static and dynamic—and therefore, they need to be addressed differently. L_{M} basically represents the component's total functional life under stated conditions of use, and it is estimated by analyzing time-to-failure data of a family of components operated under the same conditions of use. The accuracy and authenticity of the L_{M} estimation becomes better with increasing amounts of available statistical data. The aforementioned maintenance data analysis techniques, in particular Weibull analysis, are useful for this purpose. For instance the concept of MTTF can be used to estimate the mean life of a component. By contrast, L_{A} is dynamic in the sense that it mainly depends on the real conditions of use, and its assessment is based on the actual conditions of use, which requires lifetime analysis.

One of the major obstacles for the development of a methodology based on life-cycle data analysis is the unavailability of operating information, particularly

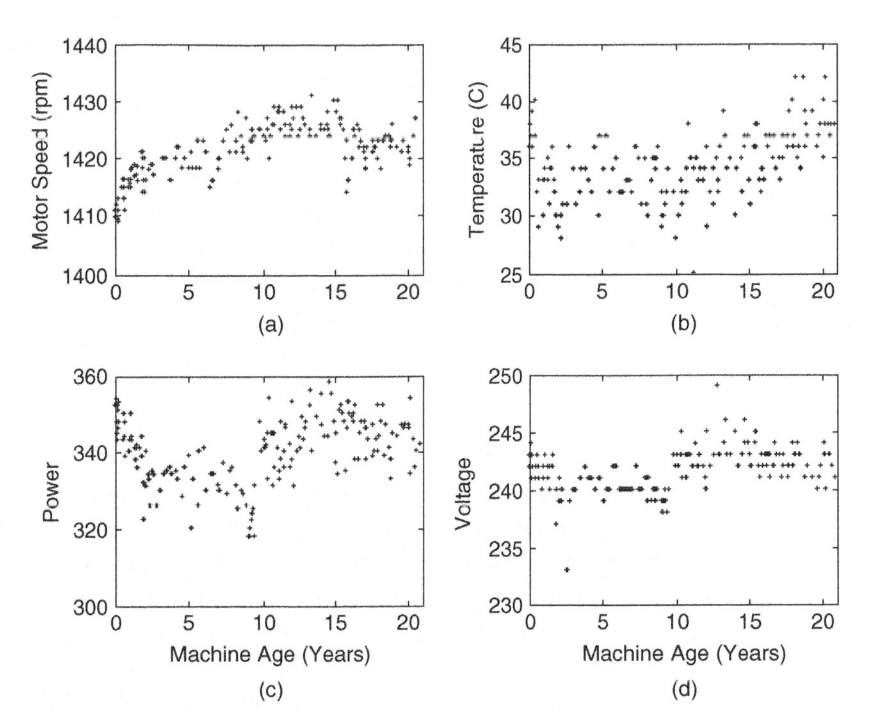

Figure 27 Pattern analysis of the accelerated life testing data. (From Ref. 42.)

in the case of consumer products such as washing machines or refrigerators. Accelerated lifetime testing is a short-term solution until the real-time operating data are available. The acceleration is basically usage rate acceleration, in which the machine operates continuously. Therefore, the tools of accelerated life testing that are extensively used in manufacturing practice offer the attractive benefit of requiring relatively small investment in terms of time and resources. Figure 27 shows the lifetime data collected during the accelerated life testing of a washing machine motor.[42] Extrapolation of data from high stress levels (e.g., pressure, voltage, temperature, and corrosive media) through a physically reasonable statistical model can provide life estimates at lower, normal stress levels. The analysis tools used in this context are based on pattern analysis and recognition by using various methods such as statistical pattern analysis and artificial intelligence techniques (e.g., neural networks and expert systems).[26]

The first type of methodology is based on statistical techniques. Their objective is to devise a monitoring system that must be able to distinguish between the variations of the monitored variable that are caused by the variations of the operation conditions and those that are due to arising and progressing of failures and misoperations.

One of the approaches is based on multiple linear regression, which is one of the most commonly used statistical techniques to represent a quantity through a model describing this quantity as a function of other variables independent from it. These techniques are particularly useful in the case of a strong trend of data. In addition, regression analysis is a useful technique, particularly because of its ability to provide an answer on how each of the explanatory variables contributes toward the estimation of the response variable.

Another is based on the mathematical theory of *kriging*. Kriging techniques are modified forms of multiple linear regression. They are often associated with the acronym BLUE (best linear unbiased estimator).[26] Ordinary kriging, also called *punctual kriging*, is the simplest and most common form of kriging. It is suitable for data that show weak trends. Universal kriging is similar to ordinary kriging but is used when a strong trend is present in the data samples. *Co-kriging* is a multivariable extension of kriging. It can be used to estimate one variable from several variables by utilizing the concept of estimating primary and secondary variables at the same time. This method is used to estimate the value of a variable in a point by supposing that it is related to the known values of the variable within a zone surrounding the point. This estimation of the value of the variable in a point is performed as a linear combination of the weighted values of the variable in the range. The weights are selected in such a way as to minimize the variance of the error of the estimate and to not introduce artifacts in the estimated values. The kriging techniques have the following advantages:

- No reference baseline is required to individuate the model.
- The model is immediately operational.
- A limited number of samples suffice to meet the requirements for estimation.
- The model provides accurate estimates.
- Measurement error can be taken into account.

Kriging techniques have a major problem in having singular matrices when there are a lot of data sets with the same data value. This scenario results in a large number of zero entries in the distance matrix that makes the determinant of the matrix equal to zero. In other words, kriging procedures are more suitable for data sets with distinct data points.[26]

Weaknesses in the traditional techniques have been found in certain areas if there are parametric trend fluctuations and a high number of inputs. Research indicates that a neural network is a powerful data-modeling tool that is able to manipulate complex input/output relationships in order to cope with the shortcomings of the previous techniques.[26] As a result, the use of artificial intelligence, in particular neural networks, has become an attractive alternative to meet the requirements of the predictive analysis because of the following advantages:

- It can detect the nonlinear relations between variables.
- There is no restriction to the number of input variables.
- Variables do not need to be independent.

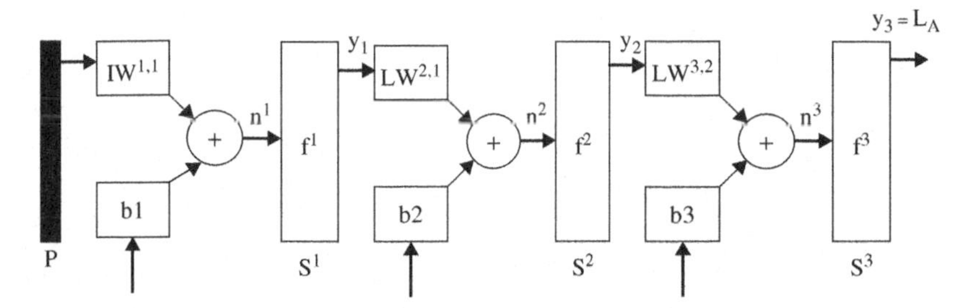

Figure 28 The structure of the neural network. (From Ref. 26.)

- There are higher processing speeds after learning.

Figure 28 shows a neural network model developed for lifetime prediction of household appliances. The architecture of the proposed network has three layers. This includes two hidden layers of sigmoid (tansig) neurons followed by an output layer of a linear neuron (purelin). The purpose of hidden layers with nonlinear transfer functions (f) is to allow the network to learn nonlinear and linear relationships between input variable (P) and output variables (y). The linear transfer function in the output layer lets the network produce outputs outside. (L_A, the output of the model, is the actual life of a component at given conditions.)

However, one of the major problems with neural network models is the requirement of huge amounts of data before a network can be trained to produce acceptable results. At the same time it is very difficult to gather a sufficient amount of condition monitoring information, particularly in consumer products. Furthermore, this technique needs a laborious phase of training and learning and the availability of a statistically significant database that must cover all the envisaged operating conditions.

The literature suggests that the monitored system based on the kriging technique is not affected by some problems common to the other two models. These problems are the requirement of a large amount of data for their tuning, both for training the neural network and defining the optimum plane for the multiple regression. This is valid not only in the system starting phase but also after a trivial operation of maintenance involving the substitution of machinery components having a direct impact on the observed variable.

3.4 Future Strategy: Selling Services, Not Products?

The main aim of the "selling use or services" approach is to sell the use of a product as a service rather than the product itself. Manufacturers or third-party companies can act as a service provider. The service provider offers functionality of the product to the customer without passing the ownership of the product. As a result, the service provider manages the cost of investment, logistics, operations, maintenance, and disposal. Customers pay only for the use of the services, not for the product itself.[43,44] Selling service becomes more competitive if the

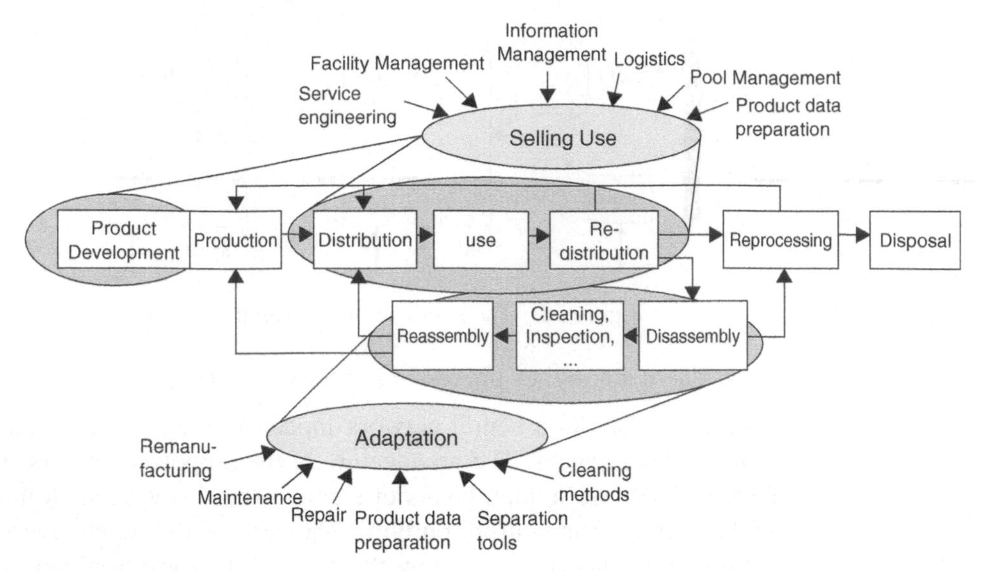

Figure 29 Selling use strategy within the context of closed-loop manufacturing. (From Ref. 43.)

product can be transferred to multiple usage phases of adaptation, as shown in Figure 29.[43]

Adaptation processes are necessary due to the change of technical necessities and user needs. The types of adaptations are maintenance, repair, remanufacturing, up and downgrading, enlargement, and reduction, as well as rearrangement and modularization. Adaptation often requires disassembly, cleaning, treatment, component supply and removal, inspection, sorting, and reassembly. A balanced strategy of preventive maintenance and repair preserves or increases the residual value of the product. The product that will be used for the selling services purpose needs to be designed with modularity, integratability, customization, convertibility, and diagnosability in mind in order to support customer-driven adaptability. The consideration of these properties during the early design phase increases the applicability and the availability of a product in multiple usage phases.[43]

In the selling-service approach, the service provider is responsible for the availability of the service at the right time, at the right place, and with an acceptable quality. Therefore, the service provider needs quality, information, and communication systems to guarantee product pursuit and access. Leasing, renting, and service contracts regulate the responsibilities between the customer and the service provider.

Selling service becomes competitive to selling products, once the idle capacity cost is higher than the extra costs required for logistics and information management (Figure 30).[43] Logistics include all necessary processes to provide the

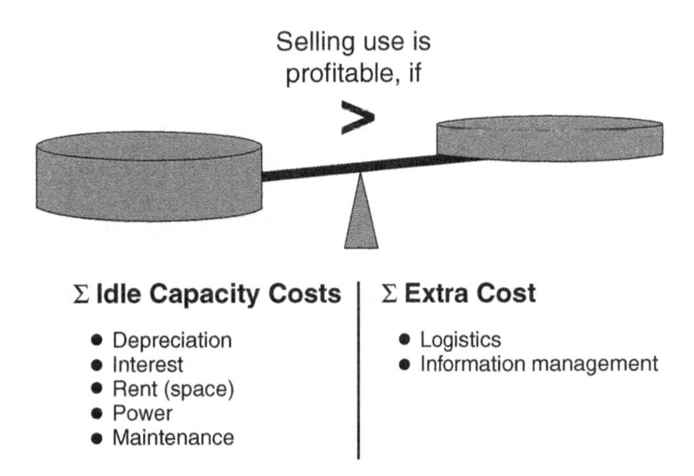

Figure 30 Justification of selling service instead of product. (From Ref. 43.)

service. Thus, the cost of adaptation between usage phases and the transport of the product are included as well.

The challenges to the implementation of the selling service strategy are the rapid development of technology and the change in customer taste and market requirements. In order to meet these challenges, there must be collaboration between governments, companies, and user groups.

4 CONCLUDING REMARKS

The two activities described in this chapter, namely the recycling of materials and the reuse of products and components, are at different stages of development and implementation. Recycling of materials, being the first step toward reducing the production of new materials, is already a very advanced technology. Shredding and separation techniques provide reasonably high purity of recycled materials, although there are still intensive developments going on to improve the purity further so that the differences between recycled and new materials become smaller or might even disappear altogether. The other activity of reusing old products and their components for the production of new products is still at the beginning of its development. The main hurdles are reliable lifetime prediction and negative consumer perception, which have to be overcome before a major breakthrough toward implementation can be achieved. However, in view of the ever-increasing demand for sustainability, it is only a matter of time before these strategies will be adopted more widely.

REFERENCES

1. B. A. Hegberg, G. A. Brenniman, and W. H. Hallenbeck, *Mixed Plastic Recycling Technology*, Noyes Data Corp, Park Ridge, NJ, 1992.

2. T. J. Veasy, R. J. Wilson, and D. M. Squires, *The Physical Separation and Recovery of Metals from Wastes*, Gordon & Breach Science Publishers, Amsterdam, 1993.
3. H. Yamamoto and S. Shibata, "Inverse Factory for End-of-Life Cars—complete Dismantling System," 1st International Symposium on Environmentally Conscious Design and Inverse Manufacturing, EcoDesign99, 1999.
4. M. Santochi, G. Dini, and F. Failli, "Computer-aided Disassembly Planning: State-of-the-art," *Annals of the CIRP*, **51** (2), 1–23 (2002).
5. S. Kara, P. Pornprasitpol, and H. Kaebernick, "A Selective Disassembly Methodology for End-of-Life Products," *Assembly Automation*, **25** (2), 124–134 (2005).
6. B. Willems, W. Dewulf, and J. Duflou, "End-of-Life Strategy Selection: A Linear Programming Approach to Manage Innovations in Product Design," Proceedings of the 11th International CIRP Life Cycle Engineering Seminar, Belgrade, 2004.
7. J. D. Chiodo, N. Jones, E. Billet, and D. J. Harrison, "Shape Memory Alloy Actuators for Active Disassembly Using 'Smart' Materials of Consumer Electronic Products," *Materials and Design*, (23), 471–478, (2002).
8. T. Suga and H. Hosada, "Active Disassembly and Reversible Interconnection," Proceedings of the IEEE International Symposium on Electronics and the Environment, pp. 330–334, 2000.
9. S. Koyu, O. Hideo, T. Masnobu, and Y. Takeo, "Study of Auto-Disassembly System Using Shape Memory Materials," Proceedings of the 3rd International Symposium on Environmentally Conscious Design and Inverse Manufacturing, Tokyo, Japan, pp. 504–509, December 2003.
10. N. Jones, D. Harrison, H. Hussein, E. Billett, and J. Chiodo, "Electrically Self-powered Active Disassembly," *Journal of Engineering Manufacture*, **218** (7), 689–697, (2004).
11. Shred-Tech (www.shred-tech.com).
12. SSI Shredding Systems, Inc. (www.ssiworld.com).
13. Zerma Machinery and Recycling Technologies, (www.zerma.com).
14. Jaykrishna Magnetics (www.jkmagnetics.com).
15. Eriez, Magnetic Drum Separator (www.eriez.com).
16. Global Magnetics, (www.globalmagnetics.com).
17. Argonne National Laboratory, (http://www.es.anl.gov/Energy_Systems).
18. S. Kara, I. M. Mazhar, and H. Kaebernick, "Lifetime Prediction of Components for Reuse: An Overview," *International Journal of Environment and Technology*, **4** (4), (2004).
19. H. Kaebernick, S. Kara, and M. Sun, "Sustainable Product Development and Manufacturing by Considering Environmental Requirements," Robotics and Computer—*Integrated Manufacturing Journal*, **19** (6), (December 2003).
20. H. Kaebernick, M. Anityasari, and S. Kara, "A Technical and Economic Model for End-of-Life (EOL) Options of Industrial Products," *International Journal of Environment and Sustainable Development*, **1** (2), 171–183, (2002).
21. N. Jackobsson, "Product Service Systems: Panacea or Myth?" Ph.D. dissertation, Lund University, Sweden, 2004.
22. R. Steinhilper, *Remanufacturing: The Ultimate Form of Recycling*, Fraunhofer IRB Verlag, 1998.

23. E. Sundin, "Product Properties Essential for Remanufacturing," 8th CIRP International Seminar on Life Cycle Engineering, Varna, Bulgaria, June 2001.
24. S. Kara, M. I. Mazhar, H. Kaebernick, and H. Ahmed, "Determining the Reuse Potential of Components Based on Life Cycle Data," *Annals of the CIRP*, **54** (1) (2005).
25. S. Takata, F. Kimura, F. J. A. M. van Houten, and E. Wetkamper, "Maintanence: Changing Role in Life Cycle Management," *Annals of the CIRP*, **53** (2), 1–13, (2004).
26. I. M. Mazhar, S. Kara, and H. Kaebernick, "Reuse Potential of Used Parts in Consumer Products: Assessment with Weibull Analysis," Proceedings of the CIRP Seminar on Life Cycle Engineering, June 21–22, Belgrade, Serbia & Montenegro, 2004.
27. K. Fitzgibbon, R. Barker, T. Clayton, and N. Wilson, "A Failure-Forecast Method Based on Weibull and Statistical-pattern Analysis," Reliability and Maintainability Symposium IEEE, pp. 516–521, 2002.
28. Y. Li, T. R. Kurfess, and S. Y. Liang, "Improved Effectiveness of Vibration Monitoring of Rolling Bearings in Paper Mills," Proceedings of the IMechE Part J, *Journal of Engineering Tribology* (Professional Engineering Publishing) **212** (2), 111–120, (1998).
29. Al-Najjar, "Accuracy, Effectiveness and Improvement of Vibration-based Maintenance in Paper Mills: Case Studies," *Journal of Sound and Vibration*, **229** (2), 389–410, (2000).
30. H. Y. Liou, and C. S. Shin Wu, "A Modified Model for the Estimation of Fatigue Life Derived from Random Vibration Theory," *Probabilistic Engineering Mechanics*, **14** (3), 281–288, (1999).
31. Q. Jing, B. S. Brij, S. Y. Liang, and C. Zhang, "Damage Mechanics Approach for Bearing Lifetime Prognostics," *Mechanical Systems and Signal Processing*, **16** (5), 819–829, (2002).
32. A. Ray, and S. Tangirala, "Stochastic Modeling of Fatigue Crack Propagation," American Control Conference, IEEE, vol. 1, pp. 425–429, 1997.
33. A. Ray, and S. Tangirala, A Nonlinear Stochastic Model of Fatigue Crack Dynamics," *Probabilistic Engineering Mechanics*, **12** (1), 33–40, (1997).
34. A. K. Sheikh, M. Ahmed, and M. A. Badar, "Fatigue Life Prediction of Assemblies of Rotating Parts," *International Journal of Fatigue*, **17** (1), 35–41, (1995).
35. M. Haiba, D.C. Barton, P. C. Brooks, and M. C. Levesley, "Review of Life Assessment Techniques Applied to Dynamically Loaded Automotive Components," *Computers & Structures*, **80** (5–6), 481–494, (2002).
36. G. Petrucci, and B. Zuccarello, "On the Estimation of the Fatigue Cycle Distribution from Spectral Density Data," Proceedings of the IMechE Part C, *Journal of Mechanical Engineering Science* (Professional Engineering Publishing), **213** (8), 819–831, (1999).
37. N. Arora, and D. Kumar, "Availability Analysis of Steam and Power Generation Systems in the Thermal Power Plant," *Microelectronics and Reliability*, **37** (5), 795–799, (1997).
38. C. McLean, Wireless data gathering system for condition based maintenance, www.techkor.com/techset.htm.
39. J. Ni, J. Lee, and D. Djurdjanovic, "Watchdog—Information Technology for Proactive Product Maintenance and Ecological Product Reuse," Proceedings on Colloquium on E-Ecological Manufacturing, Technical University of Berlin, Germany, March 2003.
40. G. Seliger, A. Buchholz, and U. Kross, "Enhanced Product Functionality with Life Cycle Units," Proceedings of the Institutions of Mechanical Engineers, Part B, *Journal of Engineering Manufacture*, **217**, (B9), 1197–1202, (2003).

41. G. Seliger, U. Kross, and A. Buchholz, "Efficient Maintenance Approach by On-board Monitoring of Innovative Wagon Bogie," Proceedings of the IEEE ASME International Conference on Advanced Intelligent Mechatronics, Kobe, Japan, pp. 407–411, 2003.
42. I. M. Mazhar, S. Kara, and H. Kaebernick, "Reliability Assessment for Reuse of Components in Consumer Products—A Statistical and Condition Monitoring Data Analysis Strategy," Proceedings of the 4th Australian Life Cycle Assessment Conference, Sydney, Australia, February 23–25, 2005.
43. G. Seliger, "Global Sustainability—A Future Scenario," Proceedings of the Global Conference on Sustainable Product Development and Life Cycle Engineering, Berlin, Germany, September, 2004.

CHAPTER 8

DESIGN FOR REMANUFACTURING PROCESSES

Bert Bras
George W. Woodruff, School of Mechanical Engineering
Georgia Institute of Technology, Atlanta, Georgia

This chapter discusses qualitative design for remanufacturing guidelines. It provides an overview of the industry, including typical facility level processes, so we can better understand the rationale for the given design guidelines. A distinction will be made between overarching guidelines and specific component hardware-oriented design guidelines.

1 INTRODUCTION TO REMANUFACTURING

Many are unfamiliar with the basics of remanufacturing. In this section, we will focus on what these are and why remanufacturing is done.

1.1 Definitions

Remanufacturing is both a new and old phenomenon. It is receiving a lot of attention nowadays from an environmental point of view (as evident from this book), but has been practiced for a very long time. In remanufacturing, a nonfunctional or retired product is made like-new through a series of industrial operations.[1] A minimum definition is "bringing a product back to sound working order." Remanufacturing basically is the process of disassembly of products, during which time parts are cleaned, repaired, or replaced, and then reassembled to sound working condition. In that context, the Remanufacturing Institute (www.reman.org) considers a product remanufactured if the following conditions are met:

- Its primary components come from a used product.
- The used product is dismantled to the extent necessary to determine the condition of its components.
- The used product's components are thoroughly cleaned and made free from rust and corrosion.
- All missing, defective, broken, or substantially worn parts are either restored to sound, functionally good condition, or they are replaced with new, remanufactured, or sound, functionally good used parts.
- To put the product in sound working condition, such machining, rewinding, refinishing, or other operations are performed as necessary.
- The product is reassembled, and a determination is made that it will operate like a similar new product.

Remanufacturing is viewed differently from recycling in that the geometry of the product is maintained, whereas in recycling the product's materials are separated, ground, shredded, and molten for use in new product manufacture. Remanufacturing is viewed by many as a special form of recycling. The U.S. Code of Federal Regulations, for example, allows remanufactured products to be claimed as recyclable (see 16 CFR 260.7), provided that certain conditions for such claims are met. The German Engineering Standard VDI 2243 uses the phrase "product recycling" to denote product remanufacture in contrast to *material recycling*.[2] And the European End of Life Vehicle (ELV) Directive allows reuse to count as a form of recycling.[3]

More complicated are the differences between *reconditioned*, *repaired*, *refurbished*, and remanufactured. These terms are often used synonymously, but they convey different meanings, dependent on the audience. The U.S. Code of Federal Regulation, Title 16—Commercial Practices, Part 20, "Guides for the Rebuilt, Reconditioned and Other Used Automobile Parts Industry," Paragraph 3—states:

> It is unfair or deceptive to use the words "Rebuilt," "Remanufactured," or words of similar import, to describe an industry product which, since it was last subjected to any use, has not been dismantled and reconstructed as necessary, all of its internal and external parts cleaned and made rust and corrosion free, all impaired, defective or substantially worn parts restored to a sound condition or replaced with new, rebuilt (in accord with the provisions of this paragraph) or unimpaired used parts, all missing parts replaced with new, rebuilt or unimpaired used parts, and such rewinding or machining and other operations performed as are necessary to put the industry product in sound working condition.

Similarly, the word *repair* is often considered as bringing a product back to a basic functional condition by removal a single fault condition, whereas remanufacturing goes much deeper and brings the product back to almost or better than new condition.

1.2 Potential Benefits

The primary benefits of remanufacturing arise from a reuse of resources. In contrast to recycling, remanufacturing tries to retain the geometrical shape of parts. Hence, the need for material forming process, and associated energy expenditures, is reduced. This is why remanufacturing is often seen as a key strategy for long-term sustainable development, and proponents tout huge energy savings compared to traditional manufacturing processes.[4] Benefits to society include:

- Cheaper goods and hence a higher standard of living
- The creation of new jobs, since remanufacture is heavily labor intensive[5]

A more technical potential benefit is that remanufacturing has the capability of bringing used products back to equal or *better* than new condition. Many argue that it is impossible to substantiate this claim in general, due to the fact that wear will degrade products. Nevertheless, remanufactured products often contain (new) components that are better than the original components the product came with from the original assembly line. Bearings, bushings, and motors are just some examples of parts that have increased in performance, quality, or life expectancy due to better materials and manufacturing processes. Engines may receive better emission control or fuel injectors for increased efficiency. Even large U.S. Navy ships that have undergone refits are arguably better than new when we consider their weapon systems and engine upgrades (consider the battleship *U.S.S. Missouri*, which served in both World War II and the first Gulf War). In such cases, the phrase *upward remanufacturing* is used to denote the better-than-new state of the product after remanufacturing.

1.3 Size of Industry

In the late 1990s, approximately 73,000 U.S. firms sold an estimated $53 billion worth of remanufactured products.[6] This number does not include remanufacturing operations within the U.S. Department of Defense, arguably the largest remanufacturer, which constantly remanufactures military equipment (from combat equipment to radar systems) to extend the service life of these products.[7] Remanufactured automobile parts such as engines, transmissions, alternators, and starter engines accounted already for almost 10 percent of the total production in Germany.[2] In the 1980s, replacement parts for automobiles were the largest application of remanufacturing in the United States.[5] These reports did not include the now-thriving business of toner cartridge remanufacturing. Even household appliances and industrial hand tools are currently remanufactured by third parties and (in some cases) Original Equipment Manufacturers (OEMs). Wherever a product can be cost effectively remanufactured and can be priced significantly lower than a new product, there is potential for remanufacturing. Even the electronics industry is considering reuse and remanufacture more than in the past.

The market and business potential for remanufacturing is even larger if leasing is considered.[8] Of the $668 billion spent by business on productive assets by U.S. businesses in 2003, $208 billion was acquired through leasing.[9] For General Electric alone, leasing is a $10 billion business. Several manufacturers have realized the benefits of remanufacturing their own products after a (first) lease. The Xerox Corporation had already saved around $200 million in 1991 by remanufacturing copiers returned at the expiration of their lease contracts. Fuji-Xerox integrates remanufactured components into "new" copiers wherever appropriate.[10,11] In addition to selling its products, Caterpillar is also leasing them ("selling miles"), which allows penetration into markets that before could not afford a Caterpillar product.[10,11] Caterpillar now has a remanufacturing division that also provides its services to other companies. Some aircraft engine manufacturers can sell engines below manufacturing cost because they have integrated maintenance and financing contracts as part of the engine sales. All these examples are products that are relatively expensive and are of high capital value. Kodak, however, is a classic example of a company that has created a fully integrated reuse and remanufacture strategy around its (inexpensive) reusable Funsaver camera line. A key aspect is that users want to return the camera because they need the film to be developed, allowing for recovery of the product. Toner cartridges are another example of low-cost products/components that are being remanufactured.

Remanufacturing is most viable for products that have a high replacement cost or have valuable components that can cost-effectively be reused or reconditioned. The practice of remanufacturing is most predominant in the U.S. Department of Defense and most visible in the automotive industry. Remanufacturing is (still) mostly performed by third-party remanufacturers because original equipment manufacturers (OEMs) do not consider it a core business. More and more OEMs, however, recognize the value of remanufacturing, especially if business strategies move from selling products to leasing products and/or selling services where the OEM retains ownership of the product.

1.4 Consumer Demand

Although prices vary for remanufactured parts and products, they are typically significantly lower than equivalent new products. The lower price is the primary reason for consumer demand for remanufactured parts. Early studies showed that customers preferred remanufactured parts over new if remanufactured parts were offered at about 57 percent of the new part price.[12] To offer remanufactured parts at such a discount price compare to new parts, there must be a significant price margin between the sales price of the remanufactured product and the price that the remanufacturer has to pay for the used product/core. Some remanufacturers prefer to remanufacture foreign parts, because they usually sell for more in the local market than domestic components.

Despite the price advantage, some customers are still wary of remanufactured parts because they feel that these "used" parts are inferior compared to new parts. Many remanufacturers, therefore, back their work with a warranty, although it often covers only the cost of replacement parts and not labor.

Many times, however, the consumer does not have a choice because new parts may simply not be available anymore, and there is no other option than remanufactured parts. Original equipment manufacturers also rely on remanufacturing for parts supply, allowing them to stock fewer new parts and allowing their suppliers to free up tooling in production facilities for production of parts for newer products. Especially in the automotive industry, the OEM parts supply relies heavily on OEM-sanctioned remanufacturing efforts.

2 BASIC REMANUFACTURING BUSINESS PRACTICE

Remanufacturing is wide spread and covers many industries, but some general observations can be made about the industry that will help us understand how and where changes in product design can be beneficial.

2.1 Basic Business Scenarios

Different business practices exist with different combinations of actors. In Figure 1, a schematic of possible product flows is shown between different actors in the remanufacturing business practice. Two basic scenarios exist:

1. OEM manufacturers, sells (or leases), recovers, remanufactures, and resells products and parts.
2. OEM manufacturers and sells products, but third-party actors independently capture, remanufacture, and resell the used products and parts.

The second scenario is the predominant business scenario, but the first scenario is gaining momentum in certain industries. A hybrid scenario—that is, direct collaboration between OEMs and third-party remanufacturers, is also possible and frequently seen in automotive parts remanufacturing.

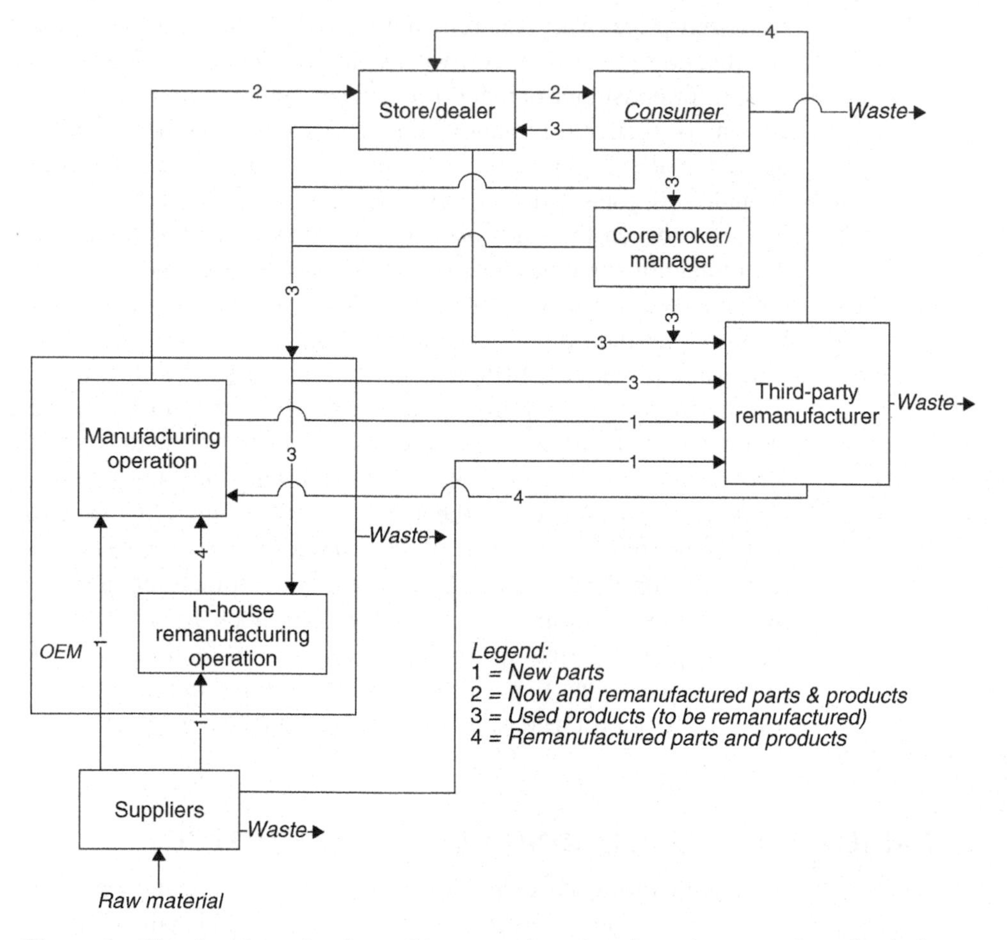

Figure 1 Simple schematic of possible part and product flows in remanufacturing industry.

2.2 The Industry's Raw Material: Cores from Consumers

Product and part cores are the lifeblood of remanufacturing. Without cores to remanufacture, there is no business. These cores have to be collected from users/consumers. As can be seen in Figure 1, the primary interface between a consumer and the OEM is the OEM's dealer or a store. Stores often have the option of returning used products directly to the OEM (e.g., in the case of single-use cameras), or directly to a remanufacturer (e.g., automotive parts). Consumers also send or give their discarded products to core brokers (e.g., automotive scrap-yards) and managers who supply these cores to remanufacturers. In the inkjet and laser toner industry, the OEM frequently uses direct-mail returns from the consumer. In the latter cases, the OEMs pay for shipping, but that may not be enough to compensate consumers or stores for the inconvenience so additional money is often paid for returned cores as an added incentive. In the automotive sectors, *core charges* sometimes up to 10 to 20 percent of the remanufactured product sales price are used to ensure that consumers return a core. Often, an extra stipulation is that it has to be a "rebuildable" core. Despite these strategies, obtaining cores at a reasonable cost (and at the right time) can be a major difficulty and hurdle for many companies.

2.3 Third-Party Remanufacturing

The vast majority of the firms that make up the remanufacturing industry are third-party companies who do not manufacture the original product. They are often referred to as *independent remanufacturers*. Third-party remanufacturing is very common in the remanufacturing industry in the United States and are most visible in the automotive aftermarket. Remanufacturing in the automotive industry has been around for a long time. In recent years, however, there has been a growth trend in this market, as more businesses are tapping into this profitable field. Automotive part retailers have seen a recent growth in the volume of remanufactured parts. Some of the components currently being remanufactured include clutches, brake shoes, engine blocks, starters, alternators, water pumps, and carburetors. There are national organizations of automotive remanufacturers, such as The Automotive Parts Rebuilder's Association, and publications, such as *Automotive Rebuilder* magazine, which support facets of the automotive remanufacturing industry.

Automotive part remanufacturers often obtain the cores that they remanufacture from core suppliers/brokers, from salvage yards, or by one-for-one core exchanges from stores and dealers. Parts are often purchased from (specialized) suppliers, or even directly from the OEM (see Figure 1).

Larger organizations frequently disassemble and rebuild the parts in large batches, in an assembly line manner, in order to obtain economies of scale. Smaller businesses often remanufacture parts on an individual basis in order to maintain a lean inventory of diverse models. For example, several years ago at a local Atlanta alternator and starter remanufacturer, alternators and starters were being remanufactured in batches—several of the same alternators on a given day—in an assembly line style. However, the business practice was changed such that each employee would remanufacture one product at a time, from start to finish. This resulted in a much leaner inventory, as well as employees who could perform all facets of the remanufacturing process.

Most remanufacturers will disassemble a unit, clean all functional parts, add grease, sealants, or paint to protect them, replace all worn parts, refurbish the exterior, reassemble it, and test the reassembled unit. However, there is a wide range of quality levels when it comes to remanufactured components. Remanufacturers who are conscientious about quality will frequently use higher-quality replacement parts than the original (and called for by specifications), repaint assemblies (prevents corrosion and grease build-up), and monitor the quality of replacement parts, in order to prevent warranty returns, as well as to build their reputation as a quality remanufacturer.

Third-party remanufacturers typically sell their products directly to replacement parts stores. In some cases, they are contracted by OEMs to remanufacture replacement parts and in that case, they tend to act as suppliers for OEMs, who sell the remanufactured parts through their existing dealer networks.

2.4 Original Equipment Manufacturers (OEM) Remanufacturing

Despite a growing interest in remanufacturing, most products are being remanufactured more out of serendipity than by design. Entrepreneurs tend to take advantage of the intrinsic value in (used) products by collecting and remanufacturing them profitably. Hence, unless an OEM is doing the remanufacture, there is little incentive to design products for remanufacture because that would only make it easier for entrepreneurs to start remanufacturing these products and potentially take away market share from new(er) products or OEM replacement parts.

A crucial issue for any organization considering remanufacture is the economic incentive. OEMs entering the remanufacturing market can gain a competitive edge by using their design capacity to make their products easy to remanufacture. OEMs can also offer a remanufactured warranty equivalent to the original warranty. And, they can use existing store/dealer networks and distribution networks for marketing and sales. The combination of these factors would give OEMs an edge. Most importantly, in order for OEMs to enter into remanufacturing ventures, a paradigm shift must occur whereby remanufacturing is an important design and business driver.[13] This shift has already occurred in certain industries. For example, Pratt & Whitney is involved in remanufacture of the aircraft engines they manufacture.[5] General Electric remanufactures transformers, control systems, and other components. Caterpillar recently offered its internal remanufacturing services to external parties and formed a new remanufacturing division for this purpose. One of the most cited examples is the Xerox

Corporation, which has been involved in the practice of remanufacturing for many years. Xerox (and its global partners) developed a program whereby it reclaims its copiers at the end of their service life (or lease). Xerox then remanufactures its copiers by reusing parts and replacing worn-out parts, and returning them to service.[14] They assemble remanufactured and new copying machines on identical assembly lines with identical quality standards. Xerox has worked extensively in the design of their products to make their copiers easy to disassemble, modify, and reassemble. They can do this by using modular attachment methods, and by standardizing the parts used in the copiers. This practice allows Xerox to develop a broad variety of products using a large base of modular parts. One hurdle that OEMs face when promoting remanufactured products is that a number of states in the United States have policies blocking or limiting the use of remanufactured parts in products that are sold or leased as new.

In general terms, to successfully integrate remanufacturing into their business, OEMs must focus on providing value to consumers at different levels:[15] (1) initial sale/lease (compete based on features, performance and price), (2) performance-sensitive (early) reuse (technology is still relatively current, higher price, testing and refurbishment required), (3) price-sensitive (later) reuse (older technology, lower price, not necessarily refurbished), (4) service and support (replacement parts), (5) second market reuse (other industries find another use for goods), (6) recycle materials (lowest economic value but landfill is avoided). Benefits to the producer include:

- If price elasticity is significant, their market share grows.
- The trade-in value encourages customer loyalty and repeat business.
- Data on product failures can be used to create product improvements.
- The existing product distribution system used by the OEM can be converted into a two-way street in which products' used cores are returned.[5]

Good discussions on the opportunities, barriers, and steps needed for increasing the roles of OEMs in remanufacture are given in References 5, 13, 16–18.

2.5 Customer Returns—A Driver for OEM Remanufacturing

Customer returns can be an important initial driver for OEMs to consider remanufacturing as a business strategy option. When a customer returns a defective product to a retail or online store, it is invariably sent back to the original supplier, who can then simply replace it with a new one and discard the defective product, or repair or remanufacture the defective product. In many cases, however, customers return products for other reasons than defects. Mostly, they can be restocked immediately in the store, but they may also be sent back to the original supplier/manufacturer if, for example, the packaging is damaged or a suspicion of damage exists. Such returns are inspected, repackaged, and often resold as "factory refurbished" by the original manufacturer or its dealers. Consumer laws prevent such items from being labeled *new*, even though they may never have been used at all. It is estimated that about 4 percent of total logistics costs are spent on returns,[19] which equated to over $35 billion per year in the late 1990s in the United States alone. This has led to an emergence of specialized firms that focus on *reverse* logistics. Large consumer product manufacturers (with correspondingly large return volumes), such as Dell and Black & Decker, have dedicated refurbish, recondition, and/or remanufacturing centers handling their returns. Because of the reliance on manual labor, many of these operations have been moved to low-cost countries such as Mexico and China.

3 REMANUFACTURING FACILITY PROCESSES

The preceding sections offered a high-level overview of the remanufacturing industry and its actors. In this section, we descend to the facility level to illustrate typical remanufacturing process. The discussion is primarily focused on mechanical products (e.g., automotive and manufacturing equipment) as the main example due to their widespread remanufacture.

3.1 Typical Facility-Level Remanufacturing Process

In order to understand how to *design* for remanufacturing, one needs to know the basic processes. Remanufacturing spans many industry sectors. As in manufacturing, no single uniform process exists. The following twelve processes, however, can be found in any remanufacturing facility:

1. Warehousing of incoming cores, parts, and outgoing products
2. Sorting of incoming cores
3. Cleaning of cores
4. Disassembly of cores and subassemblies
5. Inspection of cores, subassemblies, and parts
6. Cleaning of specific parts and subassemblies
7. Parts repair or renewal
8. Testing of parts and subassemblies
9. Reassembly of parts, subassemblies, and products
10. Testing of subassemblies and finished products
11. Packaging
12. Shipping

Design for remanufacturing processes should be focused on reducing the cost, effort, and overall resource expenditure of these processes by proper product design. In general, deeper insight is needed, however, and a detailed study of actual processes may need to be performed similar to design for manufacturing efforts. To give a flavor of what are critical-process issues, Hammond, et al., 1998[20] conducted a survey in which a number of automotive remanufacturers

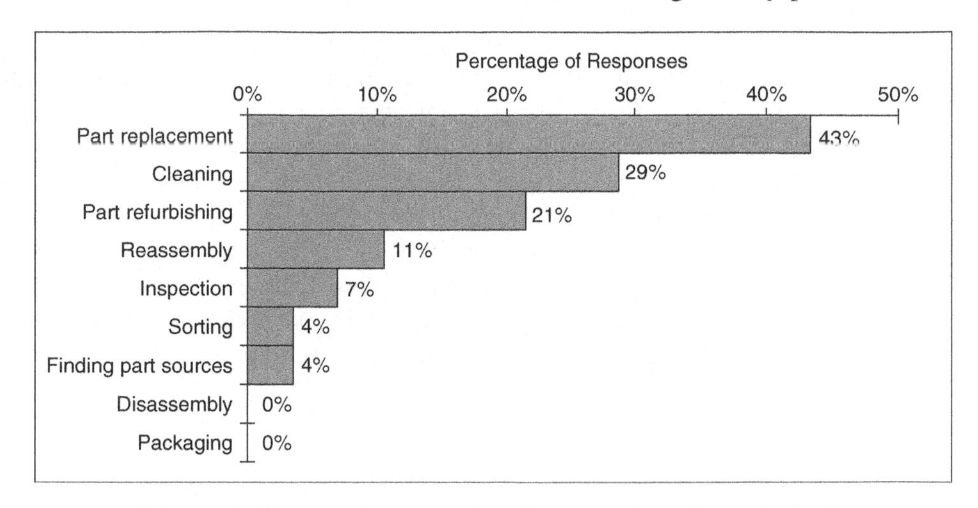

Figure 2 Most costly remanufacturing operations. (From Ref. 20)

provided insight into their most costly remanufacturing operations. As shown in Figure 2, part replacement tops the list, followed by cleaning and refurbishing (see Ref. 20 for more survey results). Although not exhaustive, such survey results are indicative of inherent product design problems. These processes, and associated issues, are discussed in more detail in the following sections.

3.2 Inspection and Testing Processes

Based on experience, technicians typically know what kind of failures are common to particular products, and can estimate what it will take to remanufacture them. Most unforeseen variations costs are related to such items as hidden failures that have damaged major parts to the point that replacement of these parts would add too much cost to the component. Parts that have a damaged or broken core frequently cannot be remanufactured at all. In the case of starter and alternator remanufacture, there are usually one or two relatively inexpensive parts that cause the starter or alternator to fail. For example, certain alternators are installed very close to the exhaust manifold, such that the bearing routinely fails because of the heat. Another type of alternator has a rectifier that consistently fails. Many remanufacturers have learned to recognize these defects and replace these parts every time, and by doing so, they can assure a quality product.

Material failures, if not at the surface, are often more difficult to inspect and require nondestructive inspection and testing methods. The American Society for Nondestructive Testing (ASNT) provides a good classification, overview, and primer of available nondestructive testing (NDT) methods (www.asnt.org). Basically, NDT methods fall into six major categories: visual, penetrating radiation, magnetic-electrical, mechanical vibration, thermal and chemical-electrochemical. Each has its specific purpose, advantages, and disadvantages. Depending on the product being remanufactured and its value, NDT may be required (e.g., aircraft parts). Sometimes, especially for low-value components, simple replacement (e.g., of bearings) may be preferred over the expense of testing.

One option is to include a service life counter or condition-monitoring device into the product design. The purpose of these devices is to give insight into the actual service that the product has seen over a given time period. Devices can range from simple page counters on copiers and printers to sophisticated systems that include sensing of loading and environmental conditions. This latter is used in many high-value product systems (like aircraft and military equipment) and is especially valuable if the devices can be tied to condition-based maintenance systems.

3.3 Cleaning Processes

Various types of cleaning processes are available for industrial manufacturing and remanufacturing operations. Table 1 lists a general classification of cleaning processes reported in various literature sources.

The cleaning processes presented in Table 1 can also have variations in application and material handling methods used. For example, aqueous cleaning can be done in spray and/or immersion and can use belt conveyors, rotary drums, rotary tables, and other methods.[21]

Thermal cleaning uses heat to loosen oil, grease, dirt, paint, adhesives, rust, and other contaminants from metal surfaces. Organic contaminants, such as oils and greases, oxidize and burn away. This is perhaps, the most important benefit of thermal cleaning, because it eliminates the costs of disposal of these organic

Table 1 Classification of Cleaning Processes

Mechanical		Chemical	
Brushing	Wire Fiber	Aqueous	Acidic Neutral
Steam-Jet Cleaning			Alkaline
Abrasive blasting	Sand Metal shot Silicon carbide	Solvent	Spray Immersion Vapor degreasing
	Plastic	Ultrasonic	
	Corncobs	Biological	
	Nut shells CO_2 pellets	CO_2	Liquid Supercritical
	Ice pellets	Pickling	
Tumbling		Salt-bath	
Vacuum de-oiling		Thermal	Burn off
Scrubbing/wiping			Plasma

contaminants in liquid form that is typical of water-based washing operations. Blasting must be subsequently used, because the heat does not actually bake the foreign material off, but dries it out, leaving ashes and surface oxides that require blasting to remove. Ovens to perform thermal cleaning have long operation cycle times, so batching is necessary. Thermal cleaning (using ovens), over the last decade of stricter environmental regulations, has become the cleaning method of choice for many remanufacturers of all sizes and product lines.[22] Plasma cleaning is an alternative that cleans surfaces by chemical reactions of activated gas species with surface contaminants, creating volatile products that will be pumped away.[23] The gases used for plasma (e.g., oxygen) are fairly inexpensive, accessible, and easy to handle. In oxygen plasma cleaning, contaminants are fully oxidized into CO, CO_2 and H_2O.[24] It leaves parts dry after treatment, is suitable for cleaning parts with complex shapes, and does not require any special safety measures. It also produces little or no waste.[25] However, it cannot clean inorganic materials or clean thick organic films effectively or efficiently. For the latter reason, plasma cleaning is usually accompanied by wet cleaning to remove heavy soils from parts.[21] Clearly, no thermal cleaning process is possible when cleaning plastic components, lighter metal components, or heat-treated components. Figure 3 samples of (small) thermal cleaning ovens are shown with cores that have been thermally cleaned as well.

Solvent-based chemical cleaning has become probably the least desirable technology for cleaning, especially if the solvents used are cataloged by the Environmental Protection Agency (EPA) as air, water, or land hazardous contaminants. For example, chemical degreasing operation using solvent-based solutions like Perchloroethylene release volatile organic compounds (VOCs) into the atmosphere, which are major contributors to air pollution. The most benign chemical cleaning substances are aqueous-based solutions, and the trend in industry is to move away from solvent-based cleaning to aqueous cleaning that, if free of hazardous substances, does not pose an environmental problem for (re)manufacturers. Typical aqueous cleaning consists of a washing, rinsing, and drying step.[26] The cleaning capability generally comes from the temperature and/or chemistry of the

Figure 3 Part cleaning using ovens at Chinese remanufacturers.

water, whether it is acidic, alkaline, neutral, and/or emulsion,[27] and the mechanical form of application (e.g., immersion, spray, or both mechanisms). Additional differences include the following:[21,28,29]

- Emulsions can be chemically or mechanically induced.
- Immersion can be accompanied by part, liquid, and/or ultrasonic wave agitation.
- Spraying can vary by water pressure, spray configuration, and/or pattern.
- Drying can be carried out with steam, air, centrifuges, and/or vacuum.

Aqueous cleaning arose as a "greener" alternative to solvent cleaning because it does not use flammable, VOC producing, ozone-depleting, and/or hazardous substances.[30] However, it can consume more water and energy, and produce more wastewater and energy-related air emissions. The most efficient way to use water-based solutions is to have a closed-loop system that recycles the cleaning solution and reduces the need for make-up water and detergent. If the components to be cleaned have hazardous substances that will end up in the solution, such as heavy metals, a closed-loop system is the only option.

Biological cleaning combines aqueous emulsion cleaning with bioremediation.[27] Basically, oils and greases are removed by the emulsions, which are then consumed by bacteria in the bath. Since the cleaned oils and greases are consumed and oxidized into CO_2 and H_2O by the bacteria, the only waste is a sludge consisting of dead organisms.[31] Additional benefits include the ability to recycle surfactants (leading to a longer life for the cleaning solution), minimal downtime, and relatively small water, energy and cleaning agent consumption.[27] A study conducted by the EPA on a biological cleaning system implemented by a manufacturing company found approximately 50 percent savings in water consumption, 80 percent savings in wastewater generation, and annual cost savings of $86,000 by the company due to the system.[32] Chemical CO_2 cleaning of metal parts is also possible and can be carried out in two ways: liquid or supercritical. In both cases, once the CO_2 dissolves the contaminants from the parts, the CO_2 can be easily separated and reused. The only waste produced with CO_2 cleaning is the contaminant itself, which can be easily treated, recycled, or disposed. However, it has high capital costs, and raises safety concerns regarding the high-pressure components and potential increase in CO_2 levels in the worker area.[27]

Whether thermal cleaning or chemical cleaning is used, components also need to be abrasively cleaned to remove rust and scale, as well as to improve surface finish and appearance. Abrasive cleaning is used by most remanufacturers to obtain like-new appearances, which is very important in this business. Most shops that remanufacture mechanical automotive parts (such as clutches, drive shafts, and engines) use airless centrifugal steel-shot abrasion technologies, whereas remanufacturers of electrical parts, such as starters and alternators, use air-blasting units with glass beads, aluminum oxide, and zinc oxide.[33] Airless centrifugal steel-shot abrasion technologies are self-destructive, which create high maintenance and repair costs during the life of the equipment and induce higher work-in-process inventories due to down time. In addition, they create noise and dust pollution. An alternative to using separate chemical and dry abrasive cleaning processes is to use wet-blasting, which is a surface treatment process that performs simultaneous degreasing, deburring, surface cleaning, and descaling in one operation, without the use of chemicals. Machines from, for example, Vapormatt, use a recirculated water/abrasive media slurry with a high-volume vortex pump that feeds a nozzle or gun. Compressed air can be added at the nozzle to increase the effectiveness or aggressiveness of treatment, to create a specific surface condition. A problem with abrasive cleaning technologies (e.g., using airless centrifugal steel shot) is that the technology can be too aggressive, and extra work is created when the core components are damaged.

One of the biggest problems with abrasive (and other) cleaning processes is that spent media will many times contain concentrations of heavy metals that exceed local and/or federal toxicity limits. Depending on the components being blasted, threshold limit may be exceeded and spent media will need to be disposed of as hazardous waste. The administrative costs make hazardous waste disposal very costly.

Finally, abrasive cleaning shot retention plays a big role in the life of remanufactured products. For example, in the case of remanufactured engines,[22] 67 percent of engine wear occurs when it is first started after being remanufactured. The media left inside the engine can generate considerable wear. Shot-media retention is one of the biggest concerns for engine remanufacturers. Abraded parts that are not completely dry from a previous chemical cleaning process are especially prone to retaining shot media. Even compressed air blasting is susceptible to having wet media and subsequent shot retention, because of the moisture in the air lines. In high-humidity areas, moisture will cause a constant problem. For electrical components, when air blast media are propelled through the air, static electricity is generated, which can ruin electrical components. To prevent this problem, remanufacturers make sure their systems are completely grounded. One way remanufacturers deal with shot retention is by using tumblers and vibratory units that blow, shake, or wash out shot after the abrasion operation. One alternative to avoiding shot media retention is dry ice (CO_2) blasting. The dry ice dissipates into carbon dioxide gas after impacting the surface being abraded, thus only the heavy metals or contaminants are left over. Due to its expense, it is not used widely. Xerox has used it for cleaning copier parts.

3.4 Refurbishing Processes

In addition to simply replacing a damaged part (which is quite effective for cheap and commonly available parts), two basic strategies exist for refurbishing damaged or worn high-value and less-available parts: (1) cut out the damaged or worn area and/or (2) add (new) material to the damaged, worn, or cutout area. Often a combination is used in conjunction with smart selection of cheap replacement parts.

Cutting out a damaged or worn area is one of the most frequently used approaches, especially in refurbishing metal parts that have surface damage only. For example, damage to cylinder head surfaces (e.g., caused by cracked cylinder head gaskets) can often be removed by milling down the surface. The increase in compression ratio is often insignificant and/or can be offset by a slightly thicker head gasket. Damage to races in bearing housings can be removed by milling, turning, or other metal-appropriate cutting processes. The increased diameter can be offset by installing a new (replacement) bearing with a larger outer diameter. Here, the cost of the new bearing is much lower than the housing. Similarly, damage to crankshaft journals is often removed by turning processes and diameter differences are (again) offset through changes in bearing sizes. In this case, however, surface treatments typically also need to be performed because the original (hardened) surface has been removed. In some cases, surfaces may need to be refurbished using electroplating. This may be done on site or outsourced, due to environmental permitting issues.

Parts that have internal material failures, such as fatigue cracks, require much more extensive refurbishing processes. If the part is valuable enough, remanufacturers go through great lengths to refurbish it. For example, Caterpillar can repair cylinder head cracks by cutting out the material containing a crack completely and filling the cavity using welding processes, almost analogous to dental cavity repair. Field testing and in-house experiments have verified this to be a viable refurbishment process.[10]

Adding weld material is one process option for *adding* material to worn or damaged surfaces. Another process option is to use *thermal spraying* technology where molten material is sprayed onto a surface. Once the desired thickness has been achieved, the surface is machined to correct surface tolerances. Thermal spraying is a process of particulate deposition in which molten or semi-molten particles are deposited onto substrates, and the micro structure of the coating results from the solidification and sintering of the particles.[34] Various coatings can be achieved by using different combinations of equipment and consumables. Basic thermal spray systems typically consist of a spray gun, a power supply or gas controller, and a wire or powder feeder. There are several thermal spraying methods: flame spraying or combustion flame spraying; atmospheric plasma spraying; arc spraying; detonation-gun spraying; high velocity oxy-fuel spraying; vacuum plasma spraying; and controlled atmosphere plasma spraying. Arc spraying is one of the fastest, simplest, and most energy efficient and inexpensive thermal spraying method.[35] Arc spraying machines consist of a wire feeder that pushes two wires through the arc spray gun. The heat zone created by the electric arc melts the wires, which are consumable electrodes. Compressed air blows the molten particles onto the substrate. Thermal spray equipment is used to apply coatings for building up worn areas; salvage improperly machined parts; improve characteristics of finished parts; and apply wear-resistant coatings. Depending on the material selected for resurfacing applications, the coating can result in parts that will outwear the originals by factors of two or more.[35]

3.5 Reliance on Employee Skills versus Automation

Remanufacturing relies heavily on human labor and skill due to the variation in product cores, condition, and subsequent uncertainties in processing. Employee skill can be a predominant issue in inspection, refurbishing, and reassembly.[20] Much of this is due to the diversity of unique products that the employee must be familiar with, and the different assembly and disassembly techniques required for each. The process of inspection can be significantly affected by the availability of the operator to identify which quality standards the specific part must measure up to. This skill is extremely important when specifications are not available. Many aftermarket remanufacturers must often define their own part specifications, as these specifications are mostly not available from the manufacturer.

Consequently, the use of factory automation beyond basic material handling is typically low for third-party remanufacturers. Although automated (e.g., robotic) disassembly has been tried and experimented with, a skilled employee equipped with air tools and expertise to simultaneously inspect and sort parts while disassembling is hard to beat. OEMs can have higher levels of factory automation, especially when remanufactured parts are mixed with new parts in the (new) product assembly process. Automation, however, is pursued for processes where danger to human health may exist (e.g., in metal spraying and cleaning operations). Fuji-Xerox, for example, uses a small robotic cleaning cell for washing its copier chassis. The robot's chassis-specific cleaning program is initiated through reading a specific barcode that is embodied in a specific location on each chassis.[10] Caterpillar has used robotic metal spraying booths that shield workers from metal dust.

4 OVERARCHING DESIGN PRINCIPLES AND STRATEGIES ENHANCING REUSE

Products can be designed to facilitate remanufacturing. Prior to worrying about designing for facility-level remanufacturing processes, however, manufacturers should ensure that the actual product is even a candidate for reuse. Hence, the *first* step in effectively designing a remanufacturing process is to enhance the overall reusability of the original product (or specific components). In this context, market requirements are just as important as technical requirements. This section highlights a number of overarching issues that may enhance or hinder reuse and remanufacture.

4.1 Product or Component Remanufacture?

Remanufacturing should be part of a larger business strategy. As such, products should not be designed "just" for remanufacturing, but also for functionality, initial manufacturability, and so on. Conflicts with other design guidelines can occur, and detailed design analyses may need to be performed.

When designing a product, keep in mind that remanufacturing the entire product may not be the best strategy and is more often an exception than the rule. Rather, remanufacture of certain product *subassemblies* is often more appropriate. A rather trivial example of this is an automobile. Powertrain components are commonly remanufactured, but interiors and bodies are not.

Similarly, remanufacturing entire products can be bad for the environment. Consider the fact if appliances and automobiles from the 1950s were kept in service as-is through remanufacturing. We would have much higher energy consumption due to their older and less efficient technology. Clearly, remanufacturing has its limitations. Leading OEMs who have internalized remanufacturing as part of their business, therefore, will spend significant time designing a product architecture that allows for technology upgrades. Fuji-Xerox, for example, looks five years ahead to see what technology may need to be incorporated in its copier systems and identifies which components should be designed as replaceable by upgrades versus which components should be designed for reuse. Also in manufacturing equipment, we see that such "upward" remanufacturing is done by adding new control systems. Hence, 100 percent reuse of all components is typically not feasible, or even desirable. Finding what to reuse and what to replace by upgrades and how to design the architecture around that is the first major challenge for OEM designers.

4.2 Product Architecture Design Guidelines

Products become obsolete and are replaced because of five primary factors:

1. Degraded performance, including structural fatigue, caused by normal wear over repeated uses
2. Environmental or chemical degradation of (internal) components
3. Damage caused by accident or inappropriate use
4. Newer technology, prompting product replacement
5. Fashion changes

In general, the first three categories tend to be driving product returns and remanufacturing of mechanical engineering products. Replacement of information technology products (e.g., computers) is mostly caused by rapid technology changes. Consumer electronic products (e.g., cell phones) are examples where products are simply being replaced due to newer technology and/or changes in fashion. The replaced products are often fully functional and well within their operating specifications. In such cases, the remanufacturing process may collapse to a simple "collect, test, and resell or discard" operation.

To achieve a high degree of product and/or component reuse, designers and manufacturers must find a way to counter these factors. Components and subassemblies that are good candidates for reuse, therefore, have the following characteristics:

- Stable technology—not much change is expected in the product's lifetime.
- The product is resistant to damage.
- Aesthetics and fashion are (largely) irrelevant.

Given that we often do not know future technology or fashion demands, a critical issue is therefore the "openness" of the product design to future modifications and upgrades. Upgradeable products allow for a larger percentage to be salvaged. In creating product designs that address these characteristics, designers should set the following goals:

- *Strive for open systems and platform designs that have modular product structures to avoid technical obsolescence.* Platform design attempts to reduce component count by standardizing components and subassemblies while at the same time maximizing product diversity. Designing the product in *modules* allows for upgrading of function and performance (e.g., computers) and replacement of technically or aesthetically outdated modules (e.g., furniture covers). As mentioned before, Fuji-Xerox develops multi-year upgrade plans and associated product modules for its copier design. More information on modular design can be found[36] where a method is described to design products with consistent modularity with respect to life-cycle viewpoints such as servicing and recycling. The authors define *modularity* with respect to life-cycle concerns, beyond just a correspondence between form and function.
- *Strive for a "classic" design to avoid fashion obsolescence.* Aesthetically appealing and timeless designs are usually more desirable (higher priced), better maintained, and have greater potential for long life spans and multiple reuse cycles. This is more in the realm of industrial design than mechanical design, but designing a product that does not become uninteresting or unpleasing quicker than its technical life will reduce the product's obsolescence and increase its desirability and potential for reuse.
- *Strive for damage-resistant designs.* Although this sounds like basic good engineering, lighter-duty materials and smaller, more optimized, part sizes and geometries are engineering design aspects that potentially reduce the number of service cycles, and can become problems in various facets of remanufacturing. Both are directly related to design, as current designs are being optimized primarily to reduce weight, space, and cost. A good example is the reduction in wall thicknesses between cylinders in engine

Figure 4 Damage to clutch pressure plate ear.

blocks. This reduces mass, but it also affects remanufacturability because damage due to, for example, scoring in the cylinder walls cannot be removed using machining. Instead a sleeve may have to be inserted, but this may not be possible due to the thin walls. Clearly, this practice benefits the manufacturer, but can cause difficulty for remanufacturers. Figure 4 shows a clutch pressure plate that has a broken ear. Rough handling (e.g., by dropping it) in shipping or removal may have cause this failure. The plate (cast iron with machined surfaces) can only be salvaged using (expensive) welding and testing processes. This type of accidental failure will only increase if parts are designed closer to strength and endurance limits.

4.3 Product Maintenance and Repair Guidelines

The service life of products can be extended in two basic ways: (1) make the product stronger and more durable, and (2) allow for good maintenance. Overall reliability and durability is enhanced by following solid engineering principles in developing a sound design and avoiding weak links. Methods such as failure mode and effect analysis are effective approaches to check the design.

Although maintenance is needed for many products, *incorrect* maintenance can have disastrous results. For example, car owners may add the wrong oil type to their automotive engines and transmissions. The design team can choose whether to allow for (user) maintenance and run risk of unintended failures due to poor maintenance, or to design the product so that it is either maintenance free or can *only* be maintained through specialized (OEM) personnel. In general, the latter is preferable when it is known that nonqualified personnel (such as users) will attempt prescribed maintenance operations. Maintenance by OEM personnel also adds a new business dimension for the OEM's overall business strategy.

Given that qualified personnel are available for maintenance, *designs should allow for easy maintenance and repair where needed.* Product design should follow available design for serviceability guidelines. Again, the best strategy is typically to design the product such that it needs little or no maintenance, or only maintenance by expert personnel. If maintenance has to be done by users, it should be designed absolutely fool-proof. Some strategies for achieving easy maintenance follow:

- Indicate on the product how it should be opened for cleaning or repair.
- Indicate on the product itself which parts must be cleaned or maintained (e.g., by color-coding lubricating points).
- Indicate on the product which parts or subassemblies are to be inspected often due to rapid wear.
- Make the location of wear detectable so that repair or replacement can take place on time.
- Locate the parts that wear relatively quickly close to one another and within easy reach.
- Make the most vulnerable components easy to dismantle.

Plus, provide clear maintenance and repair manuals and communication. Consider including vital information on the actual product itself, too. Good examples are stickers or labels with tire pressure ratings and oil type requirements placed in cars, which also aids service personnel.

4.4 Design for Reverse Logistics

If the remanufacturing process is part of an OEM's integrated strategy, core collection and reverse logistics also become crucial processes that can be aided by design. Core collection can be done by independent core managers or core brokers, through third-party subsidiaries, or suppliers/customers (e.g., single-use cameras through photofinishers and automotive parts through parts stores), or through direct channels (e.g., direct mail-in of toner cartridges to OEMs). Although often overlooked, the design of easy-to-use and protective single or bulk packaging can greatly increase core returns. Good examples are toner cartridges that come in returnable boxes with prearranged return addresses and shipping labels.

4.5 Parts Proliferation versus Standardization

Product diversity (or part proliferation) is a significant problem, especially in automotive parts remanufacturing. In automotive remanufacturing, the term *part proliferation* refers to the practice of making many variations of the same product, differing in only one or two minor areas. However, these differences (such as electrical connectors) are distinct enough to prevent interchanging these similar products. For example, for a given model year, a car line may have one or more different alternators for each variation of the vehicle—the alternator for the two-door model would not be able to be used to replace the alternator for the four-door model. Not only can they not be used within the car line, but no other car line made by the manufacturer can use the part, either.

Problems arising from this practice range from having to keep a large inventory of replacement parts to having to keep track of several, nonstandardized assembly and disassembly processes. An increase in the variety of assembly and disassembly processes also results in an increase of the number of process set-ups that have to be made, causing a reduction in throughput. Employee training also becomes a significant issue as a result, as they must be familiarized with all of the various, unique parts and the processes for each new product.

It is interesting to note that the trend of parts proliferation in the automotive sector started in the early 1980s. Consider the following numbers from an Atlanta-based large automotive remanufacturer. In 1983, there were approximately 3,400 different part numbers for brake products. By 1995, there were approximately 16,500 different part numbers! This coincides with the move of major U.S. automakers to a platform organization and a move toward lean production. Between 1982 and 1990, Japanese automakers nearly doubled the number of models on the road, from 47 to 84 models. Reacting to this condition, U.S. automakers also increased their models on the road from 36 to 53 in the same period of time.[37] Furthermore, the independence of individual platforms within an automaker's organization seems to have led to a reduction of shared components among automotive models, resulting in decreased standardization and increased parts proliferation.

A good design practice to counter part proliferation is to use standard parts. Standardization always supports remanufacture, and also manufacturing operations, and should be pursued wherever possible. Among other things, standardization reduces the number of different tools needed to assemble and disassemble, increases economies of scale in replacement part purchasing, and eases warehousing. Different product aspects can be standardized:

- *Components:*. Use as much as possible standard, commonly and easily available components. Use of specialty components may render remanufacture of assemblies impossible if these specialty components cannot be obtained anymore.
- *Fasteners:*. By standardizing the fasteners to be used in parts, the number of different fasteners can be reduced, thus reducing the complexity of assembly and disassembly, as well as the material handling processes.

- *Interfaces:*. By standardizing the interfaces of components, fewer parts are needed to produce a large variety of similar products. This helps to build economies of scale, which also improves remanufacturability. The PCI interface standard in computers is a good example of a standard interface.
- *Tools:*. Ensure that the part can be remanufactured using commonly available tools. The use of specialty tools can also degrade serviceability.

4.6 Hazardous Materials and Substances of Concern

A critical issue is to avoid hazardous substances and materials of concern. Products that contain hazardous materials require specialized processing equipment (higher capital costs) and will be in lower demand, resulting in low(er) profit margins. Plastics that contain halogenated flame retardants are a good example of this in the material recycling domain. Although a large volume of these exist suitable for recycling, recyclers cannot find markets for these plastics. Sometimes, hazardous materials can be removed and retrofitted using nonhazardous materials during remanufacturing. Air-conditioning and refrigeration systems that used freon are examples where a new refrigerant can be substituted. Performance, however, may degrade slightly because the product design was not necessarily optimized for the new refrigerant. Regardless of the ability to retrofit, one should always strive to reduce the number of parts that contain environmentally hazardous materials. Also, machining or otherwise processing of parts with (heavy) metals like chromium, zinc, or lead may trigger EPA Toxic Release Inventory (TRI) reporting and require special air-handling equipment, as per federal and local regulations, adding to remanufacturing costs.

4.7 Intentional Use of Proprietary Technology

Use of technology that is proprietary or difficult to reverse engineer will block/limit the number of independent entrepreneurs remanufacturing OEM parts and products. This practice has started to emerge as certain OEMs have realize the value of remanufactured products and how third-party remanufacturers can take away market share of OEM product and component sales. In the inkjet printing industry, some OEMs include chips that can only be reset by an authorized remanufacturer. Similarly, Kodak's single-use cameras have become more difficult to disassemble with common available tools in order to counter third-party film reloading and reuse. This strategy is counter to what many academics say should be done regarding product design for remanufacture, but this practice clearly makes sense from a higher-level business strategy where an OEM wants to retain market share and sales.

4.8 Inherent Uncertainties

Last, but not least, in remanufacture, the number and range of uncertainties are higher than for original manufacture and logistics because many of the concerns are out of the control of the OEM and the designers. Some sample product uncertainties encountered follow:

- How long is a typical use and/or life span?
- What is its state after its each use?
- What changes have been made during use and throughout its life?

This affects organizational uncertainties:

- How many will be available for take-back, and when?
- How long will it take to reprocess the product?
- What is the demand?

Some remanufacturing operations have throughput yields as low as 40 to 60 percent (unheard of in manufacturing), due to a combination of poor quality cores and poor processing. Designers and product-realization teams should be aware of these uncertainties, and should ideally try to manage or even eliminate the uncertainties by smart product and process design. For example, changes can be avoided if the product design eliminates the possibility of user tampering.

5 HARDWARE DESIGN GUIDELINES

In the preceding, some specific design guidelines were given that enhance the overall suitability of remanufacturing a given product. In this section, some specific component and machine-design type guidelines will be given that primarily facilitate the facility level remanufacturing processes (as discussed in Section 3). Clearly, this discussion is not exhaustive; you are encouraged to use your own engineering insights as well to identify design guidelines for remanufacturing operations and product designs.

5.1 Basic Sources and Overviews

There are relatively few publications and sources with general design for remanufacturing guidelines in existence. The emergence of WEEE and ELV take-back directives from the European Union,[3,38] however, has resulted in a number of design for recycling guidelines—some of which are applicable to remanufacturing. General design-for-recycling guidelines were formalized in the German engineering standard.[2] This guideline also contains directional criteria for the design of remanufacturable products. According to VDI 2243[2] and other sources,[4,5,7,14,39] remanufacturable assemblies should be designed with special emphasis on the following:

- *Ease of disassembly:* Where disassembly cannot be bypassed, by making it easier, less time can be spent during this non-value-added phase. Permanent fastening such as welding or crimping should not be used if the product is intended for remanufacture. Also, it is important that no part be damaged by the removal of another.
- *Ease of cleaning:* Parts that have seen use inevitably need to be cleaned. In order to design parts such that they may easily be cleaned, the designer must know what cleaning methods may be used, and design the parts such that the surfaces to be cleaned are accessible, and will not collect residue from cleaning (detergents, abrasives, ash, etc.).
- *Ease of inspection:* As with disassembly, inspection is an important, yet non-value-added phase. The time that must be spent on this phase should be minimized.
- *Ease of part replacement:* It is important that parts that wear are capable of being replaced easily, not just to minimize the time required to reassemble the product, but to prevent damage during part insertion.
- *Ease of reassembly:* As with the previous criteria, time spent on reassembly should be minimized using Design For Assembly guidelines.[40] Where remanufactured product is assembled more than once, this is very important. Tolerances also relate to reassembly issues.
- *Reusable components:* As more parts in a product can be reused, it becomes more cost effective to remanufacture the product (especially if these parts are costly to replace).

In the following section, we will focus on a number of guidelines in more detail. Clearly, the inherent and underlying assumption is that the products are being designed for remanufacture by an OEM or friendly third party. Otherwise, there is no incentive to follow any of these design guidelines.

5.2 Sorting Guidelines

Sorting is the first step in any remanufacturing process. Mostly it is coupled with an initial inspection as well. Figure 5 is illustrative of how cores are received by many third-party remanufacturers. The container in Figure 5 contains boxed and unboxed starters, alternators and brake shoes of varying types, shapes, sizes, and conditions. In such cases, worker knowledge and expertise are key in the sorting process. Product and part design can facilitate the sorting process by following some guidelines:

- *Reduce product and part variety.* The less different parts need to be sorted, the less time it costs. This can also be achieved by remanufacturers themselves through specialization on specific products and cores. This also

Figure 5 Cores arrive at automotive remanufacturer.

implies for internal components. Standardization of fasteners, bearings, pulleys, and so on will greatly speed up the initial core as well as subsequent part sorting.

- *Provide clear distinctive features that allow for easy recognition.* If different parts have to be used, make sure they are easily recognizable. For example, having two housings being exactly the same except for one different-sized hole may not be the best strategy because the sorter/inspector has to distinguish based on small size differences. A binary yes/no type distinction is much easier to do and can be achieved by, for example, changing the hole pattern.
- *Provide (machine) readable labels, text, barcodes that do not wear off during the product's service life.* Most products and parts have labels. Those that are exposed to the environment, however, tend to wear off during life unless they have been stamped, casted, or molded in. Even riveted serial plates and numbers can shear and wear off. Internal parts fare better, provided they have part numbers. Some companies are experimenting with radio-frequency identification (RFID) tags to facilitate sorting, but that is rather the exception than the rule.

5.3 Disassembly Guidelines

A phrase often heard is, "If a remanufacturer can take a product apart, it can be remanufactured." At first, this statement would seem to indicate that the design should focus on disassembly to ensure that the product can be remanufactured. However, there is a hidden assumption in this statement. A more correct statement is, "If a remanufacturer can take a product apart *without damaging important parts*, it can be remanufactured." The two key ideas that designers should extract from this statement are nondestructive disassembly and preventing key parts from being damaged.

In remanufacture, the objective is to reuse cores and components. That means that (in contrast to material recycling), destructive disassembly techniques like shredding are not an option. Manual disassembly, supported by pneumatic or other handheld mechanized means, is the general norm of the industry—for better or for worse (see, for example, Figure 6). Proper design can make disassembly easier so that less time can be spent during this non–value-added phase, but the goal of remanufacture is to salvage cores and components of value, and any damage must be repaired. Speedy disassembly is desired, but not at the expense of damaging cores. Avoiding and preventing damage, therefore, is often the more important objective than increasing speed. Given this, we can define number of simple overarching *guidelines for fasteners*:

Avoid and prevent damage:

- Avoid permanent fasteners that require destructive removal (such as rivets, welds, crimp joints, etc.).
- If fasteners require destructive removal, ensure that their removal will not result in damage to core and other reusable parts by incorporating breakpoints or appropriate strong lever points.
- Reduce number of fasteners prone to damage and breakage during removal (e.g., snap fits). For example, Phillips/Blade/Torx fasteners are more easily

Figure 6 Rivets being drilled out a clutch pressure plate—a cause for damage and high reject rates.

prone to head damage and removal difficulties than hex and allen bolts. Molded-plastic snap fits often break due to the aging of the plastic, either causing a need for repair or resulting in the whole part being scrapped.

- Increase corrosion resistance of fasteners, where appropriate. This reduces damage and facilitates removal.

Increase speed:

- Reduce total number of fasteners in unit.
- Reduce the number of press-fits, which do not have "push-out" capability.
- Reduce number of fasteners without direct line of sight.
- Standardize fasteners by reduce the number of different types of fastener (hex/phillips/allen/torx, metric/SAE, etc.). Reducing the number of different *size* fasteners (i.e., length, diameter) will speed up reassembly and allow for larger economies of scale in purchasing fasteners.

5.4 Design for Reassembly

Reassembly, the last process in a typical remanufacturing process, is basically identical to assembly in manufacturing. To design for reassembly, follow common design for assembly (DfA) guidelines. Table 2 contains common DfA guidelines that can be found in the general literature.

Manufacturers tend to use design for assembly and manufacturing processes that make it difficult for parts to be reused or remanufactured. For example, solenoids for starter motors are crimped into their housings. Not only is it difficult to remove the crimps in order to remanufacture the solenoid, but crimped fasteners cannot be re-crimped without degrading the strength of the crimp.

5.5 Cleaning

Parts that have seen use inevitably need to be cleaned. In order to design parts such that they may easily be cleaned, the designer must know what cleaning methods may be used, and must design the parts such that the surfaces to be cleaned are accessible and will not collect residue from cleaning (detergents, abrasives, ash, etc.). The following guidelines capture the basic aspects:

- *Protect parts and surfaces against corrosion and dirt.* The best strategy is to minimize cleaning wherever and whenever. Proper corrosion coating and dirt protection will support this. However, also consider that any coating (e.g., paint) may have to be removed if damaged. Hence, a balance may have to be found between protection and ease of removal.
- *Avoid product and/or part features that can be damaged during cleaning processes, or make them removable.* For example, when thermal cleaning is used, make sure all materials can withstand the heat without adverse effects. Abrasive cleaning methods can gouge surfaces.

Table 2 Common Design for Assembly Guidelines

1. Overall component count should be minimized.
2. Minimize use of fasteners.
3. Design the product with a base for locating other components.
4. Do not require the base to be repositioned during assembly.
5. Design components to mate through straight-line assembly, all from the same direction.
6. Maximize component accessibility.
7. Make the assembly sequence efficient.
   - Assemble with the fewest steps.
   - Avoid risks of damaging components.
   - Avoid awkward and unstable component, equipment, and personnel positions.
   - Avoid creating many disconnected subassemblies to be joined later.
8. Avoid component characteristics that complicate retrieval (tangling, nesting, and flexibility).
9. Design components for a specific type of retrieval, handling, and insertion.
10. Design components for end-to-end symmetry when possible.
11. Design components for symmetry about their axes of insertion.
12. Design components that are not symmetric about their axes of insertion to be clearly asymmetric.
13. Make use of chamfers, leads, and compliance to facilitate insertion.

- *Minimize geometric features that trap contaminants over the service life.* A sharp concave corner is an example of a geometric feature that traps contaminants. If a rib or plate is expected to trap dirt or grease, consider making it removable.
- *Reduce number of cavities/orifices that are capable of collecting residue (abrasives, chemicals, etc.) during cleaning operations.* Any orifice that can collect dirt or cleaning debry will have to be plugged or cleaned afterward.
- *Avoid contamination caused by wear.* Internal components can become dirty due to wear of other components. For example, oil seals may wear and the resulting leakage will cause contamination of other parts. Proper shielding or design out such sources of wear can reduce the cleaning effort required.

5.6 Replacement, Reconditioning, Repair

In general, remanufacturing tends to avoid the replacement of parts, but there are trade-offs as to whether to spend money to buy a new part or spend money to repair the part. For commonly available parts like bearings and fasteners, the choice is easy, but the higher the part price, the more incentive for refurbishment instead of replacement. The cost of replacement can be reduced by the following guidelines:

- Reduce number of parts subject to wear.
- Avoid materials that degrade through corrosion.
- Reduce numbers of parts to be removed to gain access to damaged parts to be replaced (or refurbished).
- Reduce number of independently functioning parts that are inseparably coupled.
- Reduce number of special parts (including aesthetic features).

As discussed in section 3, there are a number of basic strategies for repairing damage and refurbishing surfaces. Proper material selection can aid remanufacturing, as can surface protection. An interesting problem with surface protection such as heavy-duty paint or powder-coating is that it protects a part, but can cause significant cleaning problems in remanufacturing when the coating needs to be removed for renewal. For such surface reconditioning, consider the following guidelines:

- Reduce number of parts whose surface finish cannot be refinished through commonly available and conventional means.
- Minimize number of orifices that must be masked prior to painting.
- Reduce the number of (exterior) parts that must be removed prior to painting.
- Minimize the number of parts that can retain dents/deformations.

5.7 Inspection and Testing

Inspection and testing can be facilitated by reducing the number of different testing and inspection equipment pieces needed, as well as reducing the level of sophistication required. Although not in the realm of product design per se, good testing documentation and specifications should be provided to ensure that the correct specifications are achieved and tested for. This assumes (again) OEM involvement in the remanufacturing process.

6 DESIGN FOR REMANUFACTURING CONFLICTS

It should be noted that in some cases, design for remanufacturing can conflict design for manufacturing, and even be not in the best interest for the environment. For example, increasing longevity by adding material can increase part weight, causing more upfront material expenditures (and cost) and potentially more fuel consumption and emissions in transportation systems. Some differences also exist between design for disassembly (DfD) versus DfA. For example,

Table 3 DfD Guidelines and Effects on Assembly

Positive Effect on Assembly

1. Reduce the number of components.
2. Reduce the number of separate fasteners.
3. Provide open access and visibility for separation points.
4. Avoid orientation changes during disassembly.
5. Avoid nonrigid parts.
6. Ensure that disassembly can be done with common tools and equipment.
7. Design for ease of handling and cleaning of all components.

Negative Effect on Assembly

1. Design two-way snap fits or break points on snap fits.
2. Use joining elements that are detachable or easy to destroy.
3. Design for ease of separation of components.
4. Use water-soluble adhesives.

DfD Guidelines Having Relatively Little Effect on Assembly

1. Design products for reuse.
2. Eliminate need to separate parts.
3. Reduce number of different materials (for recycling).
4. Enable simultaneous separation and disassembly.
5. Place components in logical groups according to recycling group and disassembly sequence for modular design.
6. Identify separation points and materials.
7. Facilitate the sorting of noncompatible materials (for recycling).
8. Use molded-in material name in multiple locations to provide cut points (for recycling).
9. Provide techniques to safely dispose of hazardous waste.
10. Select an efficient disassembly sequence.

Note: From Refs. 43 and 44.

complete nesting can slow disassembly by not providing a location for the disassembler to reach, grasp, or otherwise handle.[41] Table 3 broadly presents the influence of DfD on assembly, compiled from literature.[7,39,41,42] The guidelines are divided into three categories according to their hypothesized influence on assembly: (1) positive effect on assembly; (2) negative effect on assembly; and (3) relatively little effect an assembly. Most DfD guidelines affecting the product structure are placed in the "positive effect" category concerning assembly. The four negative effects on assembly for the most part deal with making easily separable joints. This would negatively affect assembly in the sense that the purpose of the assembly step could be easily negated during product use. A compromise solution would be to design joints that are very hard to disassemble during product use but are easy to dismantle after the customer use or for the purpose of servicing a product. The DfD guidelines with relatively little effect on assembly for the most part deal with material selection and identification.

Different DfD strategies will have different effects on the overall processing time, especially if coupled with reassembly processes. A study on single-use cameras suggested that a modular design was slightly more effective in improving disassembly efficiency than parts consolidation, and much more effective than reducing orientation changes during disassembly.[44,45] *Clearly, as indicated in section 4, good design for remanufacturing processes should take a life-cycle perspective—both from economical as well as environmental points of view.*

7 DESIGN DECISION SUPPORT TOOLS

Although some OEMs have gained significant experience in remanufacturing and have developed associated specialized computer tools, relatively few design for remanufacture decision support tools exist in the general domain. See Ref. 46 for a simple worksheet that can be implemented in a spreadsheet. A more sophisticated spreadsheet based tool is described in Hammond and Bras (1996),[47] in which a nondimensional metric is used to quantify and ultimately compare product designs for remanufacturability. A sample of the results worksheet is given in Figure 7.

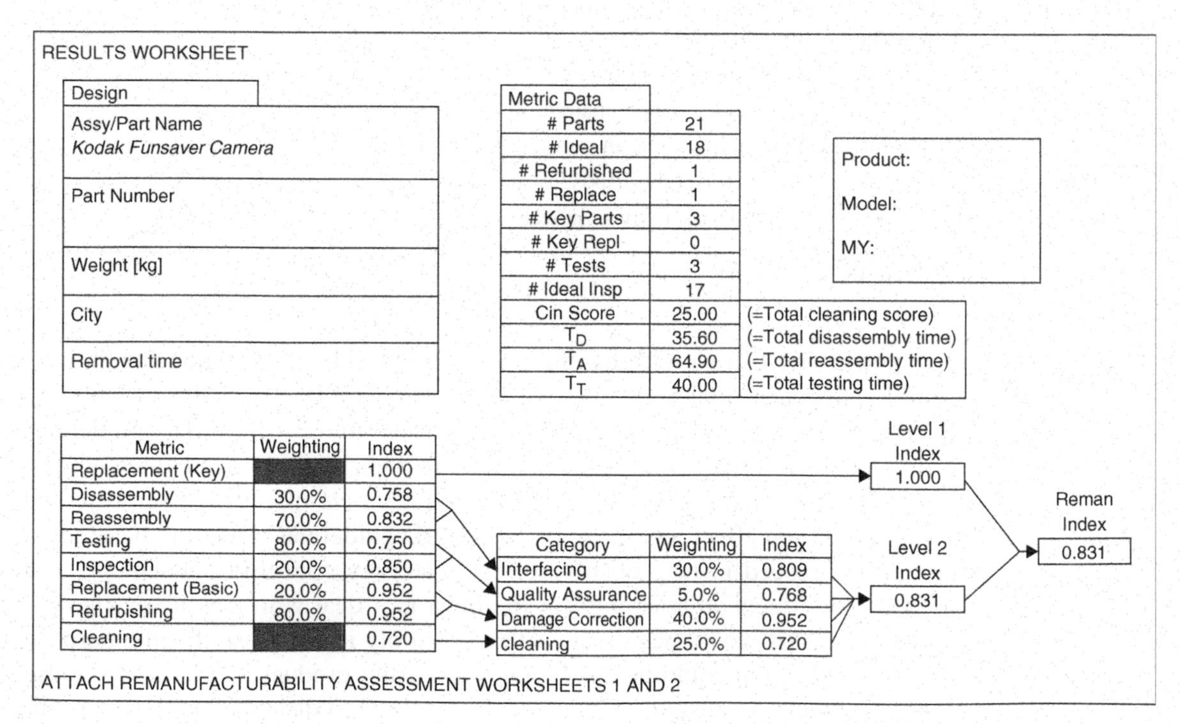

RESULTS WORKSHEET

Design	
Assy/Part Name	*Kodak Funsaver Camera*
Part Number	
Weight [kg]	
City	
Removal time	

Metric Data		
# Parts	21	
# Ideal	18	
# Refurbished	1	
# Replace	1	
# Key Parts	3	
# Key Repl	0	
# Tests	3	
# Ideal Insp	17	
Cin Score	25.00	(=Total cleaning score)
T_D	35.60	(=Total disassembly time)
T_A	64.90	(=Total reassembly time)
T_T	40.00	(=Total testing time)

Product:

Model:

MY:

Metric	Weighting	Index
Replacement (Key)		1.000
Disassembly	30.0%	0.758
Reassembly	70.0%	0.832
Testing	80.0%	0.750
Inspection	20.0%	0.850
Replacement (Basic)	20.0%	0.952
Refurbishing	80.0%	0.952
Cleaning		0.720

Category	Weighting	Index
Interfacing	30.0%	0.809
Quality Assurance	5.0%	0.768
Damage Correction	40.0%	0.952
cleaning	25.0%	0.720

Level 1 Index	Level 2 Index	Reman Index
1.000	0.831	0.831

ATTACH REMANUFACTURABILITY ASSESSMENT WORKSHEETS 1 AND 2

Figure 7 Spreadsheet-based remanufacturability assessment (From Ref. 47).

A model for assessing how product design characteristics, product development strategies, and different business conditions impact remanufacturing viability in terms of Net Present Value for an OEM interested in integrated manufacture-remanufacture is discussed in McIntosh and Bras, 1998.[17] The model is not focused on the analysis of a single-period and/or single product, but focuses on the interplay between multiple products over multiple time periods. It illustrates the benefits of designing and remanufacturing a *family* of products with shared components.

It is encouraging to see that more and more research, including work in business schools, is focusing on providing OEMs the ability to assess (1) whether their business conditions are capable of implementing remanufacture and (2) how their decisions impact the success of remanufacture.

8 SUMMARY

In this chapter, design for remanufacturing guidelines were presented and discussed. An overview of the industry, including typical facility-level processes was given in order to better understand the rationale for the given design guidelines. It is encouraging to see that more and more research, including work in business schools, is focusing on providing OEMs the ability to assess whether their business conditions are capable of implementing remanufacture and how their decisions impact the success of remanufacture.

REFERENCES

1. N. Nasr, "*Remanufacturing—From Technology to Application*," Global Conference on Sustainable Product Development and Life Cycle Engineering, Berlin, Germany, Uni-Edition, September 29–October 1, 2004.
2. VDI, Konstruieren Recyclinggerechter Technischer Produkte (Designing Technical Products for ease of Recycling), VDI-Gesellschaft Entwicklung Konstruktion Vertrieb, Germany, 1993.
3. European Union, "Directive 2000/53/EC of the European Parliament and of the Council of September 18 2000 on End-Of Life Vehicles," *Official Journal of the European Communities*, L 269/34–42 (2000).
4. R. T. Lund, "Remanufacturing," *Technology Review*, **87**, 18–23.
5. H. C. Haynsworth and R. T. Lyons, "Remanufacturing by Design, The Missing Link," *Production and Inventory Management* (Second Quarter), 25–28 (1987).
6. U.S. EPA, "Remanufactured Products: Good as New," *WasteWi$e Update*, U.S. Environmental Protection Agency, Solid Waste and Emergency Response (5306 W), 1997.
7. U.S. Congress, *Green Products by Design: Choices for a Cleaner Environment,* Office of Technology Assessment, 1992.
8. B. K. Fishbein, L. S. McGarry, and P. S. Dillon, *Leasing: A Step toward Producer Responsibility*, New York, NY, Inform, Inc., 2000.

9. ELA, *Industry Research: Overview of the Equipment Leasing & Finance Industry*, Equipment Leasing Association, http://www.elaonline.com/industryData/overview.cfm, 2004.
10. T. G. Gutowski, C. F. Murphy, D. T. Allen, D. J. Bauer, B. Bras, T. S. Piwonka, P. S. Sheng, J. W. Sutherland, D. L. Thurston, and E. E. Wolff, "Environmentally Benign Manufacturing," International Technology Research Institute, World Technology (WTEC) Division, Baltimore, MD, 2001.
11. D. T. Allen, D. J. Bauer, B. Bras, T. G. Gutowski, C. F. Murphy, T. S. Piwonka, P. S. Sheng, J. W. Sutherland, D. L. Thurston, and E. E. Wolff, "Environmentally Benign Manufacturing: Trends in Europe, Japan and the USA," *ASME Journal of Manufacturing Science*, **124**(4), 908–920 (2002).
12. R. T. Lund and F. D. Skeels, "Start-up Guidelines for the Independent Remanufacturer," Center for Policy Alternatives, Massachusetts Institute of Technology, 1983b.
13. R. T. Lund and F. D. Skeels, "Guidelines for an Original Equipment Manufacturer Starting a Remanufacturing Operation," Center for Policy Alternatives, Massachusetts Institute of Technology, 1983a.
14. V. J. Berko-Boateng, J. Azar, E. DeJong, and G. A. Yander, "Asset Recycle Management—A Total Approach to Product Design for the Environment," in *International Symposium on Electronics and the Environment*, 1993. Arlington, VA: IEEE.
15. B. Paton, "Market Considerations in the Reuse of Electronic Products," IEEE International Symposium on Electronics & the Environment, San Francisco, CA, IEEE, 1994.
16. M. W. McIntosh and B. A. Bras, *Determining the Value of Remanufacture in an Integrated Manufacturing-Remanufacturing Organization*, 1998 ASME Design for Manufacture Conference, ASME Design Technical Conferences and Computers in Engineering Conference, Atlanta, Georgia, September 14–16, 1998.
17. M. McIntosh and B. A. Bras, "Product, Process, and Organizational Design for Remanufacture—An Overview of Research," *Robotics and Computer Integrated Manufacturing—Special Issue on Remanufacturing*, **15**, 167–178 (1999).
18. W. R. Stahel, "The Utilization-Focused Service Economy: Resource Efficiency and Product-Life Extension, in B. R. Allenby and D. J. Richards (ed.), *The Greening of Industrial Ecosystems*, Washington, D.C., National Academy of Engineering, 1994, 178–190.
19. D. S. Rogers and R. S. Tibben-Lembke, *Going Backwards: Reverse Logistics Trends and Practices*, Pittsburgh, PA, Reverse Logistics Executive Council, 1999.
20. R. Hammond, T. Amezquita, and B. Bras, "Issues in Automotive Parts Remanufacturing Industry: Discussion of Results from Surveys Performed Among Remanufacturers," *Journal of Engineering Design and Automation, Special Issue on Environmentally Conscious Design and Manufacturing* **4**(1), 27–46 (1998).
21. A. J. Niksa, "*Cleaning Equipment and Methods—Overview Presentation*," 16th ASM Heat Treating Society Conference & Exposition, 1996.
22. D. Brooks, "Ovens," *Automotive Rebuilder*, **31**, 38–43 (1994).
23. A. Belkind, S. Zarrabian, and F. Engle, "Plasma Cleaning of Metals: Lubricant Oil Removal," *Metal Finishing*, **94**(7), 19–22 (1996).
24. W. Petasch, B. Kegel, H. Schmid, K. Lendenmann, and H. U. Keller, "Low-pressure Plasma Cleaning: A Process for Precision Cleaning Applications," *Surface and Coatings Technology*, **97**, 176–181 (1997).

25. P. P. Ward, "Plasma Cleaning Techniques and Future Applications in Environmentally Conscious Manufacturing," *SAMPE Journal*, **32**(1), 51–54 (1996).
26. J. Brown, *Advanced Machining Technology Handbook*. New York, NY, McGraw-Hill, 1998.
27. C. Heaton, C. Northeim, and A. Helminger, "Pollution Prevention: Opting for Solvent-Free Cleaning Processes," *Chemical Engineering*, **111**(1), 13 (2004).
28. J. B. Durkee II, "Environmentally Conscious Cleaning for the Millennium," *SAE International*, 1–8 (2000).
29. C. Nelson, *Parts Cleaning Systems: Using Water and Steam as Cleaning Agents*. FabTech International, 1993.
30. D. B. LeBart and R. Sivakumar, "Aqueous Cleaning Technologies: Present and Future," 16th ASM Heat Treating Society Conference & Exposition, 1996.
31. J. B. Durkee II, "What's in Your Cleaning Tank?" *Metal Finishing*, **103**(2), 46–48 (2005).
32. G. Eskamani, D. Brown, and A. Daniels, "Evaluation of BioClean USA, LLC Biological Degreasing System for the Recycling of Alkaline Cleaners," *Environmental Technology Verification*, Environmental Protection Agency, 2000.
33. B. Bissler, "Blast & Tumble," *Automotive Rebuilder*, **31**, 44–47 (1994).
34. L. Pawlowski, *The Science and Engineering of Thermal Spray Coatings*. West Sussex, John Wiley & Sons.
35. C. Howes, "Thermal Spraying: Processes, Preparation, Coatings and Applications," *Welding Journal* (April), 47–51 (1994).
36. P. J. Newcomb, B. A. Bras, and D. W. Rosen, "Implications of Modularity on Product Design for the Life Cycle," *Journal of Mechanical Design*, **120**(3), 483–490 (1998).
37. J. Womack, D. Jones, and D. Roos, *The Machine hat Changed the World: The Story of Lean Production*, 1991, New York: Harper Perennial.
38. European Union, "Directive 2002/96/EC of the European Parliament and of the Council of 27 January 2003 on Waste Electrical and Electronic Equipment," *Official Journal of the European Communities*, L 37/24–38 (2003).
39. W. Beitz, "Designing for Ease of Recycling—General Approach and Industrial Applications," in *9th International Conference on Engineering Design*, 1991, The Hague, Zurich, Switzerland: Heurista.
40. G. Boothroyd, and P. Dewhurst, *Product Design for Assembly*, Wakefield, MA: Boothroyd and Dewhurst, Inc., 1991.
41. R.M. Noller, *Design for Disassembly Tactics*, Assembly (January), 24–26, 1992.
42. G. Boothroyd, and L. Alting, *Design for Assembly and Disassembly*, Annals of CIRP, **41**(2), 625–636 (1992).
43. J. F. Scheuring, B. A. Bras, and K.-M. Lee, "Effects of Design for Disassembly on Integrated Disassembly and Assembly Processes," Fourth International Conference on Computer Integrated Manufacturing and Automation Technology, Rensselaer Polytechnic Institute, Troy, New York, October 10–12, IEEE, 1994a.
44. J. F. Scheuring, B. A. Bras, and K.-M. Lee, "Significance of Design for Disassembly in Integrated Disassembly and Assembly Processes," *International Journal of Environmentally Conscious Design and Manufacturing*, **3**(2), 21–33 (1994b).

45. J. F. Scheuring, *Product Design for Disassembly,* M.S. Thesis, George W. Woodruff School of Mechanical Engineering, Georgia Institute of Technology, Atlanta, Georgia, 1994.
46. T. Amezquita, R. Hammond, and B. Bras, "Design for Remanufacturing," 10th International Conference on Engineering Design (ICED 95), Praha, Czech Republic, Heurista, Zurich, Switzerland, August 22–24, 1995.
47. R. C. Hammond and B. A. Bras, "Design for Remanufacturing Metrics," First International Workshop on Reuse, Eindhoven, The Netherlands, November 10–13, 1996.

CHAPTER 9

MATERIALS SELECTION FOR GREEN DESIGN

I. Sridhar
School of Mechanical and Aerospace Engineering
Nanyang Technological University, Singapore

1 INTRODUCTION

Failure of an engineering component is related to one or more of the key factors: design analysis (e.g., not considering all possible failure mechanisms during the analysis stage), material and process selection, and service environmental conditions. Different periods of human civilization, called Stone Age, Bronze Age, and Iron Age, clearly reflect the role played by materials in improving the quality of life. Materials are the key ingredients for design and, hence, the performance of a product depends on the type of materials used to fabricate it. There are more than 100,000 materials available for the designer to select in realizing a product. Every material cannot be a right choice for a given application; therefore, there is a need for suitable material selection. Depending on the selection of a

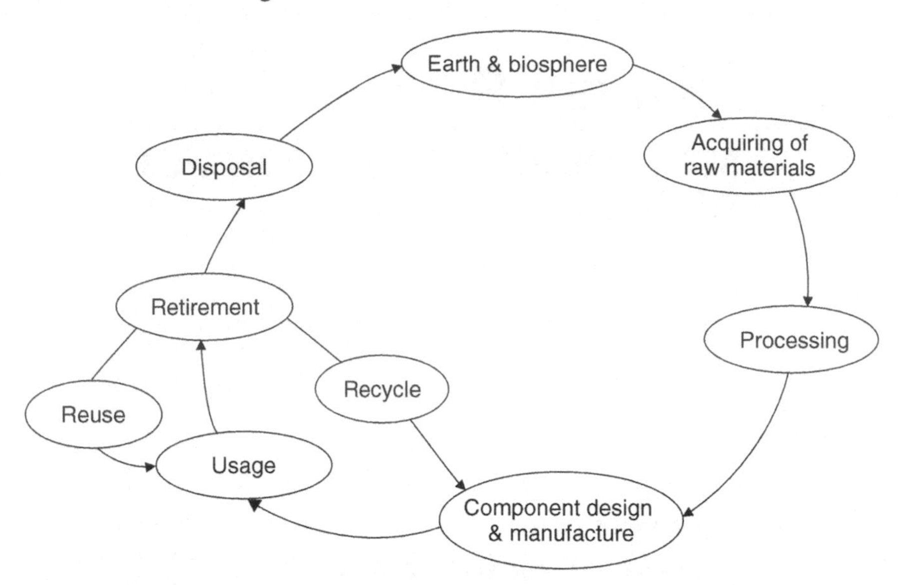

Figure 1 The life cycle or circuit of a material. (From Ref. 1.)

material, the design, processing route, cost, and performance of the product all can change.

Usage of materials has certain degree of environmental effect or impact. Figure 1 illustrates material life cycle.[1] As shown by the arrow marks, materials are predominantly extracted from earth and after their useful lifetime are ultimately returned to it. When a product's useful life is over, its components are disposed of, reused, or recycled. For many products, no thought is given beyond disposal. For example, even though single-use camera is disposable after use, Kodak has recycled 40 millions of its cameras. Similarly, Helwett-Packard and Xerox reuse or recycle 90 percent of parts and assemblies from the toner cartridges they manufacture. Each of the material cycle activity is associated with energy inputs or costs. Needless to say that these costs need to be reduced and nature or environment has to be preserved by minimum use of materials.

In material-intensive designs such as civil constructions, mechanical components, and consumer goods, the material cost is almost 50 percent of the product cost. For material-insensitive designs such as microelectronic components or biomedical prosthesis the material cost is within the range of 5 percent to 20 percent of total product cost. But they require high quality materials. Generally, material-intensive products are driven by cost, whereas material-insensitive designs are determined by their performance. Because the volume of material used in material-intensive designs is very high, there are several opportunities for the designer to explore material selection to minimize its environmental impacts. The designer is responsible for the resources used in the product. This chapter addresses basic fundamental procedures that can be adopted to reduce the environmental impact with judicious selection of materials in engineering design.

2 THE NEED FOR MATERIAL SELECTION

Engineering materials are broadly classified into five categories: natural materials, engineering alloys (metallic alloys), polymers, ceramics, and composites. In the beginning of the Industrial Revolution, ferrous alloys (and, to some extent, aluminum) were used extensively for various engineering designs. With the technological advancements, new materials and processing routes are being developed and are growing faster than ever before. Currently, a variety of polymers, ceramics, and fiber-reinforced plastics (composite materials) are replacing metals in commercial and industrial designs to obtain lightweight products. With the increase in customer demand for higher performance, quality, and reliability, competition among companies has increased.

Governments all over the world are passing new laws for a better environment and are encouraging sustainable product development concepts. For example, automakers are encouraged to develop vehicles with fuel efficiency of 14 km per liter of gasoline (about 33 miles per gallon). In light of global needs and environmental awareness, lightweight materials are gaining importance in various industry sectors. It is estimated that for every 10 percent reduction in vehicle weight, there is a 6 to 8 percent improvement in fuel economy. This is equivalent to a reduction of about 17 to 20 kg of carbon dioxide (CO_2), the main greenhouse gas, per kg weight reduction over the lifetime of the vehicle.

In general, material selection for consumer products is driven by cost and ease of handling: the automotive market is driven by energy or fuel efficiency; the aerospace and sport industry requires high performance and lightweight materials; the marine industry needs lightweight and corrosion-resistant materials; and the electronics industry requires Si with highest purity.

This interest in lightweight and high-performance materials has led to the development of composites technology. Composite materials can provide solutions to fulfill the needs of various industrial sectors such as aerospace, marine and sports goods. Current research in materials technology includes development of metallic foams, lattice structures, functionally graded materials, nanostructured materials, and components made by tough metallic glasses.

Need for material selection arises out of one or more of the following reasons:

1. As in adaptive design, the designer works on an existing product to improve its performance.
2. Material is needed for a new device or product or application.
3. Sometimes a new application demands a new material.

4. A changed market requirement, environmental concerns, government legislations, or global trade issues necessitates new material.

Material substitution leads to altogether change in the product process planning. The components design will change to obtain maximum performance (real or perceived).

3 THE GREEN DESIGN PROCESS

Design problems are open ended and can have multiple solutions. Design procedure can't be defined very easily. It can be thought of a combination of technical design and industrial or artistic design. Technical or scientific design takes care of the primary and secondary functions that need to be carried out by the product. The industrial aspect of the design improves the appearance, feel, and style or contours of the product.

Any design process in general consists of the following steps, as illustrated in the flow chart in Figure 2:[2,3]

1. Identify the need and problem specifications or requirements.
2. Classify the requirements into product functions, objectives, constraints to be met, and variables to be identified in the problem to obtain an optimum solution.
3. Develop conceptual solutions to achieve the required functions and select the optimum concepts.
4. In the embodiment design stage, refine the selected preliminary layout of the conceptual design to obtain a more robust, reliable, better design by reducing component mass and power consumption. The designer employs the principles of functional clarity, simplicity, and safety to improve the design performance.

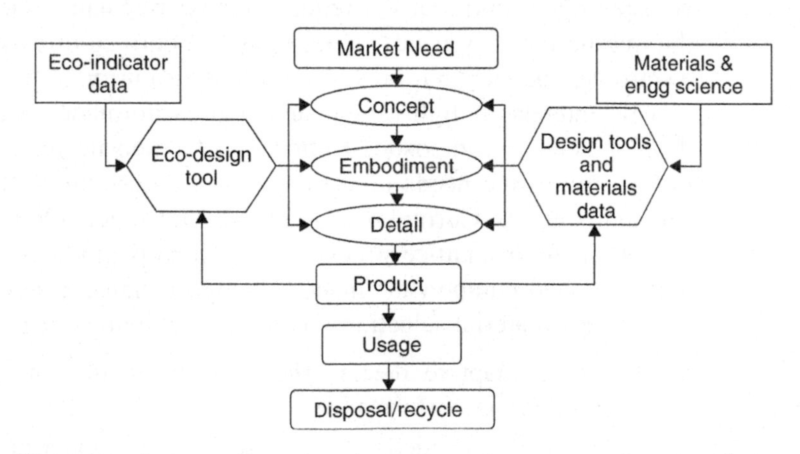

Figure 2 A design flow chart depicting various stages of product design.

5. In the detailed design stage, exact dimensions of various components are obtained by optimizing the design with the help of high-end computer aided design (CAD) tools.

In the green design,[4] in addition to the traditional aspects of function, material selection, process identification, reliable life, and cost the designer should take into account the long term environmental impact of the designed product. In many industries, a group of experts are employed instead of a single person to design a product. The group may include mechanical engineers, material specialists, production engineers, industrial engineers, customer representative, financial analysts, and a company management or legal expert. This type of teamwork allows concurrent engineering to occur and imposes upon designers the need for simultaneous consideration of product design, function, manufacturing, and cost while also taking into account later-stage considerations such as reliability, quality, and, more recently, environmental priorities.

Kinematics, stress analysis, and mechanical control system design will be handled by the mechanical engineer. The production engineer produces plans for manufacturing process flow in a product as it is designed. The material scientist helps in selecting appropriate materials for various components for improved performance. The financial specialist delves into cost and availability issues, and the systems engineer prepare logistics plan. Design tools often involve functional analyzers, high-end computer-aided design (CAD) packages for product drawings, finite element analysis software packages for thermal or mechanical stress analysis, and cost modeling softwares.

Design for the environment (DfE) includes carrying out a life-cycle assessment (LCA), a "cradle-to-grave" or "earth-to-earth" evaluation of environmental and energy impacts of a given product design (see Chapter 1). In principle, LCA covers *all* stages in the life cycle of a product system, from earth to earth. This includes extracting resources, processing them into materials and fuels, producing usable components, manufacturing a product, using and maintaining the product, and disposing of it. Figure 3 shows a standard LCA framework as developed by the Society of Environmental Toxicology and Chemistry (SETAC). It essentially breaks down a product life-cycle inventory into inputs and outputs for material and energy, as well as environmental releases. The core idea is to analyze the estimated life-cycle cost to the environment due to the product design. Environmental impact assessment (EIA) is related to LCA and is described through the flowchart in Figure 4. EIA is intended to structure the study of impact of emissions so that potential major problems are identified. Coalition for Environmentally Responsible Economics (CERES) Principles (formerly known as Valdez Principles, www.ceres.org) are designed to aid companies in monitoring their environmental performances. They include 10 specific guidelines a design company should adopt.

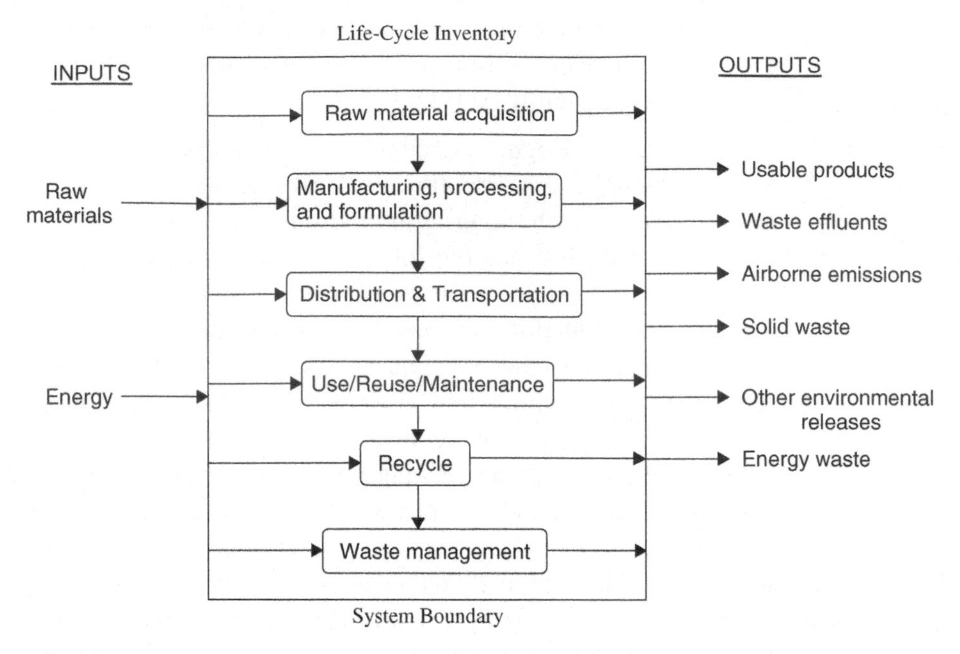

Figure 3 The life-cycle assessment (LCA) framework suggested by SETAC.

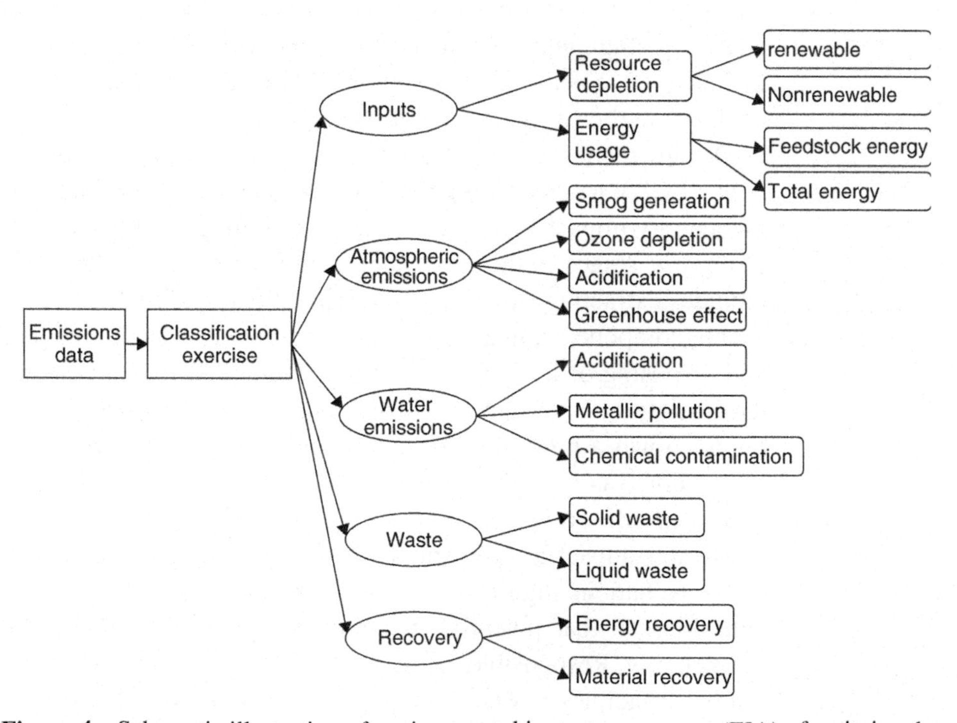

Figure 4 Schematic illustration of environmental impact assessment (EIA) of emission data. (From Ref. 4.)

The CERES Principles (Formerly known as Valdez Principles)

1. *Protection of the biosphere.* Companies will minimize the release of any pollutant that may endanger air, water, or earth.
2. *Sustainable use of natural resources.* Companies will make sustainable use of renewable natural resources, including the protection of wildlife habitats, open spaces, and wilderness.
3. *Reduction and disposal of waste.* Companies will minimize waste and recycle wherever possible.
4. *Wise use of energy.* Companies will use environmentally safe energy sources and invest in energy conservation.
5. *Risk reduction.* Companies will minimize environmental health risk to employees and local communities.
6. *Marketing of safe products and services.* Companies will sell products or services that minimize adverse environmental impacts and are safe for the consumers' use.
7. *Damage compensation.* Companies will take responsibility through clean-up and compensation for environmental harm.
8. *Disclosure.* Companies will disclose to employee and community incidents that cause environmental harm or pose health or safety hazards.
9. *Environmental directors.* At least one member of the board will be qualified to represent environmental interests and a senior executive for environmental affairs will be appointed.
10. *Annual audit.* Companies will conduct an annual self-evaluation of progress in implementing these principles and make results of an independent environmental audit available to the public.

Materials data enter at various stages of the design process, but the level of accuracy required on material property information differs at each stage. In the initial stage of the design process (conceptual stage), approximate data for wide range of materials are gathered and design options are kept open. Essentially all classes of materials are considered here. In the embodiment stage of design, an approximate analysis of the design is carried out and a specific class of material is selected. For example, if the design has to work at 300°C, polymers and fiber-reinforced polymer composites should be rejected and only metals or ceramics should be considered. In the final stage of design—detailed design—we consider further only a subset of materials from the choice of material in the embodiment design. The exact composition, required post treatment to the material is identified. Finally, a sample specimen is requested from the material supplier and tested for its properties. These properties' values are used in the analysis and performance evaluation calculations.

In traditional design function, the selection of material and process interact with the external shape of the component. In green design, the design procedure

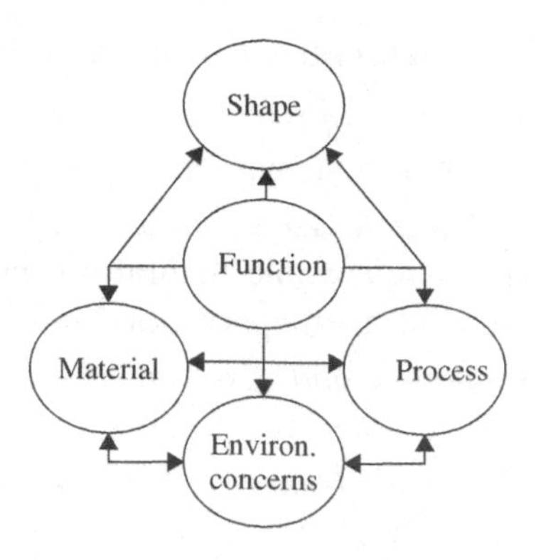

Figure 5 A diagram depicting the interaction between product function, material, synthesis, component geometry, and environmental concerns. (From Ref. 4.)

further includes the integrated concerns of environmental issues, as shown in Figure 5. Some natural materials, such as wood, synthetic composites, and foams, do contain microscopic shape (cellularic) associated with them. If these materials are employed in design, they enhance the design performance when compared with a solid shaped structure.[5,6] Function specifies the choice of both material and the component macroscopic shape. Then, material and component geometry identify the appropriate manufacturing process based on materials formability, weldability, machinability, and so on. When compared with ferrous alloys, the manufacturability of wood is very much limited. The autoclave technology is also limited to fabricate fiber-reinforced polymer composites. It is also to be noted that the specification of shape restricts the choice of material and process, but equally, the specification of manufacturing process limits the materials and shape. Both the material and process interact with the environmental consequences.

4 FACTORS IN MATERIAL SELECTION FOR GREEN DESIGN

Broadly, the factors for material selection can be divided into three categories: material properties profile, manufacturing profile, and environmental issues, including after-use recycling aspects. In the following sections, these three factors are discussed briefly.

4.1 Material Properties

The performance of engineering design depends on a combination of material properties based on the competing failure modes under the applied loading conditions. These properties can be classified into intrinsic (namely, physical properties, chemical properties, mechanical properties, and dimensional properties) and extrinsic (eco-impact, energy content, etc.). These properties should be considered with reference to the operative environment. Density is the most important physical property in the design of lightweight structures. When two or more materials are joined in a design, then their coefficient of thermal expansion is an important property that needs to be considered if the design is subjected to a variation in temperature. The casings of hand phones have to be nonmagnetic. The checklist for physical properties is quite long.

Chemical properties strongly depend up on the atomic bonding and structure. One of the chemical properties of concern is the environmental resistance of various materials. Metals in their native state are prone to dry oxidation and wet corrosion. Polymers can be damaged due to UV rays and other types of radiation. Generally, ceramics are chemically inert.

Mechanical properties influence the designs to a large extent. Appendix A provides definitions of commonly used material properties. Components subjected to fatigue load demand high endurance limit. High-fracture toughness is an important consideration for structures such as bullet-proof vests or armor vehicles subjected to impact loads. Compressive strength should be the design criterion for short struts, and high bucking strength is needed for slender columns. Materials for energy-absorbing devices demand constant plastic stress over a long range of plastic strain. For high-strength applications such as turbine blades, excellent creep strength is desirable.

Finally, the designer has to specify the exact dimensions of the components and allowable tolerances. Specimens subjected to fatigue loading demand mirror like surface finish. Also, in the green design, we should include extrinsic properties related to the environmental impact of the chosen material.

Energy Content

The energy content of a material is an approximate estimate of the energy used to make it from its naturally occurring ores, feed stocks, plus the energy content of the source material itself. The energy content of the source material is small, except, for example, when the source is oil. The energy content of aluminum is determined by the electric power absorbed in its extraction from bauxite ore. For polymers, for which the feed stock is crude oil, the energy content is the energy contained in the oil itself plus that of the subsequent processing. For wood, it is the energy content of wood plus the energy required to harvest it. The energy content values for some engineering materials are listed in Table 1.

Eco-Indicators

The *eco-indicator* of a material or process is a number that indicates the environmental impact of a material or process. The environmental effects are those that damage ecosystem or life. The general environmental effects included in the eco-indicator are listed in Table 2. The eco-indicator is expressed in millipoints (mPts) per functional unit and is calculated according to the Eco-indicator 99 method (as developed by Pre Consultants bv, Netherlands, www.pre.nl). The

Table 1 Energy Content and Eco-indicator Values for Various Engineering Materials. (From Ref. 5.)

Material	Energy (MJ/kg)	Eco-indicator (millipoints/kg)
Metals		
Aluminum and alloys	230 to 300	10 to 18
Copper and alloys	50 to 120	60 to 85
Magnesium and alloys	360 to 420	20 to 30
Iron	20	
Carbon steels	50 to 60	4 to 4.3
Stainless steels	100 to 120	16 to 18
Cast irons	50 to 250	3 to 10
Titanium and alloys	500 to 550	80 to 100
Polymers		
Nylons	170 to 180	12 to 14
Polyethylene		
Low density	80 to 100	3.7 to 3.9
High density	100 to 120	2.8 to 3.0
Polypropylene	110 to 115	3.2 to 3.4
Polystyrene	100 to 140	8.0 to 8.5
PVC	70 to 90	4.2 to 4.3
Synthetic elastomers	120 to 140	13 to 16
Ceramics		
Glasses	10 to 25	2.0 to 2.2
Cement	4 to 8	1.0 to 2.0
Concrete	3 to 6	0.6 to 1.0
Composites		
Wood	2 to 4	0.6 to 0.8
Reinforced concrete	8 to 20	1.5 to 2.5
Glass-fiber reinforced polymers	90 to 120	12

higher the indicator, the greater the environmental impact; a low value indicates lower environmental effect. For materials, this functional unit is 1 kg, so the indicator is valid for the production of 1 kg of a material.

Some effects that are not included in the estimation of eco-indicator are:

- Toxic substances that are only a problem in the workplace but scarcely occur in the outside environment because they decompose rapidly
- Exhaustion (depletion) of raw materials
- Quantity of waste (The effects of waste processing are included)

Calculation of the Eco-indicator

To calculate an eco-indicator, a life-cycle inventory (LCI) is completed, to collect the environmental impact data. The data are used to calculate the eco-indicators according to the Eco-indicator 99 method as detailed below.[5]

Table 2 Environmental Effects Included in the eco-indicator of Engineering Materials

Greenhouse effect	The increasing concentration of gasses such as CO_2, CH_4, and CO that increase the atmospheric temperature by restricting heat radiation by the Earth
Ozone layer depletion	The increase in ultraviolet radiation on Earth caused by high-altitude decomposition of the ozone layer
Acidification	Degradation of forests by, for example, acid rain due to the presence of gases such as SO_x and NO_x in the air
Eutrophication	The disappearance of rare plants that especially grow in poor soils, as a result of the emission of substances, that has the effect of a fertilizer and the changes in aquatic ecosystems
Heavy metals	Health damage caused by heavy metals (Cd, Hg, Pb) in the soil, water, and air
Carcinogenity	Carcinogenic substances such as polyaromatic hydrocarbons cause cancer in people
Winter smog	Emission of dust and SO_2 create smog in the winter mist
Summer smog	Smog forming, with peaks in summers, due to increases in the ozone concentration
Pesticides	Groundwater quality deteriorates considerably due to the leaching of pesticides

First, for every environmental effect a score is calculated, and these scores are normalized by comparing them with the current level of the environmental effect per person per year. Normalization reveals which effects are large and which are small in relative terms. Second, the effects are multiplied with a weighting factor to reveal the relative importance of the effect. These subjective weighting factors are determined by considering the seriousness of an effect and the distance between the current level and the target level. Finally, the weighted normalized contributes are summed up to provide the eco-indicator value. Table 1 lists eco-indicator values approximated to a European nation; these are only an approximate measure with low resolution of the eco-burden. Eco-indicator values are useful in the initial screening of materials (see Section 6.1). It is to be noted that the eco-indicator is not meant for use as a marketing tool or an environmental feature. The eco-indicator can be used in two ways: (1) to analyze products or ideas, with the aim of finding the most important causes of environmental pollution and finding opportunities for improvement, and (2) to compare products, semi-finished products, or design concepts, after which the least environmentally polluting components can then be chosen.

4.2 Manufacturing Issues

It is to be understood that materials have different levels of manufacturability. Commonly employed manufacturing processes are described in Appendix B. From an environmental perspective it is desirable to produce components to near net shape so that the scrap can be reduced. The manufacturing process selection depends on the material, component geometry (including shape and size), production batch size, and required dimensional tolerances. Powder consolidation techniques may be more appropriate than sand casting to produce components with good surface finish. Under green manufacturing, the designer should seek manufacturing processes that consume minimum energy and produce minimum atmospheric emissions, liquid, and solid waste.

4.3 Environmental Issues

Environmental considerations were once seen as personal or moral obligations on the part of the designer. Now in the green design process, they are integral part of the product design. Several countries are considering legislation on the usage of materials in various designs. It may not be legal to use a particular material unless it is on the approved list of some regulatory body. In the United States, the Food and Drug Administration (FDA) regulates the materials that are to be used in the construction of biomedical prosthesis such as hip joints, knee joints, and heart pacemakers. Many material-processing technologies liberate various byproducts. Under the U.S. Clean Air Act, various product manufacturers have to follow the National Emission Standard for Hazardous Air Pollutants (NESHAPS). The release of chemicals such as asbestos, benzene, mercury, beryllium, lead, inorganic arsenic, copper, nickel, and radio nuclides should be minimized. Also, airborne particles known as aerosols are an increasing focus of environmental regulators. It is noteworthy that in 1952, thousands of children and elderly people died in London due the inhalation of soot and sulfur dioxide. The environmental protection agencies of several countries restrict the release of particles less than 10 mm in diameter (referred to as PM10) because they are not likely to be expelled by the human respiratory system, if inhaled by chance.

Recycling

The design team should carefully consider the disposal of the product after its reliable life. Recycling is becoming a vital issue in some countries due to the widespread use of consumer goods. Some localities are running out of places for landfills, and hence recycling is an attempt to reduce the used materials. Currently, about 15 percent of the solid waste in landfills is metals and 20 percent is plastics. Most of this material is reusable or recyclable. If a product cannot be designed to be recycled or reused, at least it should be biodegradable. The designer should have the knowledge of percentage of biodegradable material in a product and the time it takes to degrade. Recycling can lead to substantial energy savings and can help in the reduction of pollution emissions. Materials that are highly inert in their environmental surroundings are most appropriate for recycling. In the following, we will discuss the recycling issues of various engineering materials.

Among metals, aluminum alloys are good examples of recycles, as a low ratio of energy is required to refine recycled aluminum relative to its virgin material (see Table 3). Similarly, several ferrous and nonferrous alloys are recyclable. The quality of alloys that are recycled tends to diminish with each cycle. Also components having more than one engineering alloy are difficult to recycle. Examples include bimetallic strips used in sensors and galvanized steel cans.

Thermoplastic materials such as polyethylene (PE) and polypropylene (PP) are easier to recycle. The extensive cross-linking thermosetting polymers such as epoxies make them unsuitable for recycling. Whereas, elastomers such as rubber, once vulcanized, act as a thermosetting polymer and is therefore generally not recyclable. Therefore, mixing plastics in an assembly poses a recycling problem. Similarly, decorating plastic components with metallic coatings such as magnesium or gold can create a recycling problem. Currently, an elaborate recycling code is incorporated on a variety of polymeric products, especially food and beverage cans. Table 4 presents the recycling code numbers adopted in the United States of America and their associated materials.

In general, crystalline ceramics are not recycled, whereas amorphous glass containers are well known for recycling. Glass-container manufacturers routinely

Table 3 Approximate Embodied and Recycling Energies for Metals

Material	Energy Content, Virgin Material (MJ/kg)	Energy to Recycle Material (MJ/kg)
Aluminum	190	35
Magnesium	356	38
Carbon steel	24	14

Table 4 Generally Adopted Identification Numbers for Engineering Polymers for Recycling Purposes

Recycle Code	Polymer Name
1	Polyethylene terephthalate (PET or PETE)
2	High-density polyethylene (HDPE)
3	Polyvinyl chloride (PVC)
4	Low-density polyethylene (LDPE)
5	Polypropylene (PP)
6	Polystyrene (PS)

incorporate a combination of recycled glass (known as *cullet*) and raw materials of sand, sodium carbonate, and calcium carbonate. This use of recycled glass provides increased production rates and reduced pollution emissions. Also, glass containers with polymeric coatings are difficult to recycle due to dissimilar material combinations.

Currently several composites materials are in the market. They include metal matrix composites (aluminum matrix with carbon fiber reinforcement), polymer matrix composites (carbon or glass-fiber reinforced epoxy composite), and ceramic matrix composites (carbon and carbon composites). These composites, in general, are a fine-scale combination of different material components and are difficult to recycle.

The microelectronic circuit boards have a combination of materials and are therefore difficult to recycle. Regulatory agencies scrutinize semiconductor fabrication operations to maximize both solid and chemical waste recycling.

5 SOURCES FOR MATERIAL DATA

Materials properties data could be in unstructured format or structured format. Materials property information can be obtained from handbooks in unstructured format, wherein the information about various classes of materials is described in narration. Whereas in structured format the material properties information is listed in tabular form which can be accessed easily. Sources of material information include the following:

- Published literature (referred journals and conference proceedings)
- Material suppliers' brochures
- Handbooks of societies such as American Society of Metals (ASM) and American Society of Testing of Materials (ASTM)
- Compact discs from commercial companies
- Internet-based World Wide Web pages (for example www.matweb.com)

In the published literature, usually a range is indicated for most of the material properties. The precision with which the property range (uncertainty) is available to the engineer helps in predicting the reliability of design. More detailed information on material property with high precision is required during the final design process. (In-house testing of selected materials should be preferred here.) Materials properties should be listed after testing them according to the well-established procedures of American Standards of Testing of Materials (ASTM). Cambridge Engineering Selector (CES), developed by Granta Design, Cambridge, United Kingdom, provides databases of both intrinsic and extrinsic properties of materials, including eco-properties. Several data sources for extracting material properties are listed in ASM handbook of materials selection[7] and by Ashby.[5]

6 STEPS IN THE MATERIAL SELECTION PROCESS

There are basically three major steps—namely, screening, ranking, and final short listing,[5] involved in narrowing the list of materials to a suitable material of choice.

6.1 Screening

Functional requirements dictate the relevant material properties desirable for the designed component. For example, the functional requirement of a spring is to store elastic energy and that of a building column is to resist elastic buckling. In most of the applications the designers consider all possible mechanisms of material or component failure. There may be several benefits a material, can offer, but some requirements are critical to the application. For example, weight may not be critical for a consumer product, but it is critical for an aerospace part. Material density is of importance on weight-based design. High specific heat is a requirement for storage heating elements, and low coefficient of thermal expansion is needed in the design of precision measurement devices. Properties requirements also dictate the screening procedure: Only the materials that could do the function well will remain in the selection. For example, if a component has to withstand temperatures higher than 300°C, we eliminate polymers and composite materials based on polymer matrices. Similarly, electric power transmission lines must be good conductors. So only engineering alloys are considered for power transmission cables. Here in the screening stage, materials that cannot meet the primary property profile are eliminated.

6.2 Ranking of Materials

In general, the performance of a redesign depends on functional requirements (e.g., load-carrying capacity of a tie rod, pressure contained in a pressure vessel, etc.), geometry, and material properties. Material performance index is a function of material properties, which can be used to maximize the design performance meeting the imposed constraints. Keeping the functional requirements and geometrical details constant, the screened materials can be ranked using material performance index. Section 7.5 elaborates on this procedure. The materials with high ranks based on the material performance index are short listed as the most appropriate materials for the design. It is desirable to set minimum performance required in ranking the materials.

6.3 Determination of Candidate Materials

The short-listed materials are further evaluated based on their past usage, failure behavior and analysis, availability and supply, environmental concerns, cost, and economic issues. After selecting the candidate materials for the designed component, prototype parts are made using the appropriate processing route and then it is tested to validate the design performance against benchmark calculations. Depending on the seriousness of an application, the number of parts to be tested should be decided. Aerospace, nuclear power plant, automobile, and military design parts generally require more tests to ensure that the part functions safely and reliably under various service conditions. Minimum number of tests are sufficient for consumer products, as sudden breaking of the part may not result in severe damage.

For critical applications, a large property database is created to explain the behavior of the part under various service conditions. The purpose of all these tests is to generate reliable performance data under the conditions expected during the service life of the product.

7 MATERIAL SELECTION METHODS

The materials used in a component determine its quality. However, different materials have different environmental implications. Material selection guidelines attempt to guide designers toward the environmentally preferred material. *Use appropriate design to minimize materials in products, in processes, and in service.* Here are some guidelines:

- Component design mass can be minimized by choosing appropriate cross-sectional geometry (shape) for bending, twisting, and buckling loading conditions.
- Where thin sheets of metal or plastic are used, increase stiffness and strength by providing supporting bosses (protruding studs included for reinforcing holes or mounting subassemblies) and ribs.
- Where needed, employ more smaller supporting ribs rather than a few large ribs.
- Gussets (supporting members that provide added strength to the edge of a part) can aid in designing thin-walled housing.
- Design bosses so that they play useful roles in supporting the main structural elements of the product.
- In joints design, make fasteners accessible and easy to release; use adhesives sparingly and make them water soluble if possible.
- *Choose abundant, nontoxic, nonregulated materials where possible.* If toxic materials are required for the manufacturing process, try to generate them on site rather than have them made elsewhere and shipped.
- *Choose natural materials (e.g., cellulose) rather than synthetic materials (e.g., chlorinated aromatics).* Note that this is always better. For example, the amount of atmospheric emissions in the production of 1 kg of synthetic

plastic is more than the harvesting of 1 kg of wood. But, eco-designers should also consider the amount of energy needed in drying wood, the paint required to preserve wood, and sawing losses. As noted previously in Section 4.3, thermoplastics can be recycled whereas wood cannot.

- Minimize the number of material systems or combinations used in a product.
- Try to use materials that have an existing recycling infrastructure.
- Use recycled materials where possible, rather than the materials obtained through raw materials extraction.

In addition to these generic guidelines, companies such as IBM and DaimlerChrysler are developing and using specific materials-selection guidelines for environmentally sound product development, which describe in detail the materials that should be used for specific applications.

Label advisors are generally marks on materials or products that reveal information about the material content relevant to materials handling and waste management. For example, the plastic bottles used in many consumer products usually have a plastic identification symbol that can be used in plastics resorting and recycling efforts (see Table 4). Even labels are subject to green design. Embedding labels into a material is preferable to a separate label attached with an adhesive.

There is no standard technique used by the designer to select a right material for an application.[5,8–11] Sometimes, a material is selected based on what worked before in similar conditions or what a competitor is using in his or her product. In the current competitive market, this *short-cut* method may cause engineers to overlook new emerging technologies and may put the product in a less-competitive position. The job of materials and design engineers is to consider all possible opportunities to utilize new material systems and technologies for the reduction of manufacturing cost and weight for the same or increased performance. This section presents some material selection methods for the evaluation of a suitable material.

7.1 Cost versus Property Analysis

Cost is a crucial factor in a material selection process, as low cost is often a requirement. The current method compares the material cost for equivalent material property requirements. The property could be tensile strength, compressive strength, flexural strength, fatigue, creep, impact, or any other property that is critical to the application. The method determines the weight required by different materials to meet the desired property. Note that mass is proportional to cost and environmental impact. For structural applications, the volume of the material required to carry the load is determined, and then the volume is multiplied by the density to determine the weight. For equal volumes, steel is three to four

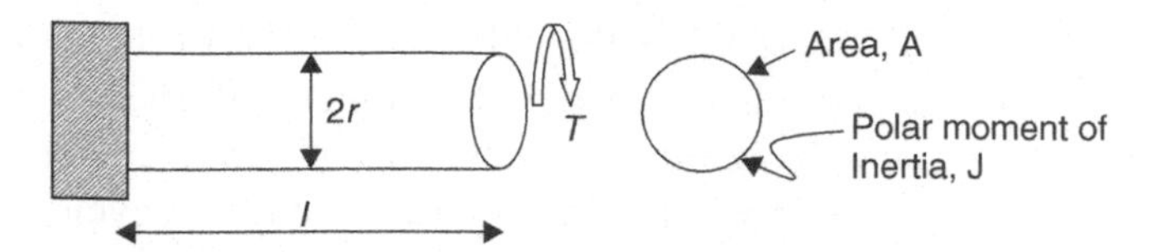

Figure 6 A shaft of radius r, length L subjected to an external torque T.

times heavier than carbon fiber reinforced epoxy composites. Thus, composites can offer significant weight savings as compared to steel if properly designed and manufactured.

Because cost is very important in any green design, it is most appropriate to consider total life-cycle cost as the basis for material selection. Life-cycle cost includes not only raw material and manufacturing costs, but also the cost for serviceability, maintainability, and recycling.

In this section, only material cost is compared for equal performance requirement. For an application, the most critical performance requirement might be tensile strength, compressive strength, fatigue strength, impact, or creep, or any combination of these. If tensile strength is the determining factor for the selection of a material, then the weights required by different materials of choice are determined to have the same tensile strength values.

Suppose there are two solid circular shafts of length L made of materials a and b and are subjected to a constant torque T, as shown in Figure 6. The shear strengths are τ_a and τ_b; densities are ρ_a and ρ_b and their radii are r_a and r_b, respectively. If the shafts have to carry the toque without plastic failure, then the minimum cross-section radii required is calculated using the following formula:

$$r_a = \sqrt[3]{\frac{2}{\pi}\frac{T}{\tau_a}} \tag{1}$$

and that of shaft b will be:

$$r_b = \sqrt[3]{\frac{2}{\pi}\frac{T}{\tau_b}} \tag{2}$$

For equal lengths L, the shafts' mass ratio in terms of their cross-sectional areas A_a and A_b will be

$$\frac{m_a}{m_b} = \frac{\rho_a L A_a}{\rho_b L A_b} = \frac{\rho_a A_a}{\rho_b A_b} \tag{3}$$

If the unit cost (cost per kilogram material) of materials a and b are C_a and C_b, respectively, then the ratio of the total material costs $T\$_a$ and $T\$_b$ will be:

$$\frac{T\$_a}{T\$_b} = \frac{C_a m_a}{C_b m_b} = \frac{C_a \rho_a A_a}{C_b \rho_b A_b} = \frac{C_a \rho_a}{C_b \rho_b}\sqrt[3]{\frac{\tau_b^2}{\tau_a^2}} \tag{4}$$

If the ratio of $T\$_a$ to $T\$_b$ is less than unity, it indicates that material a is preferable to material b.

Table 5 Pugh Decision Matrix Method

Material Property	Material A	Material B	Material C
Density	+	+	−
Tensile Strength	+	−	+
Modulus	+	S	−
Energy content	S	−	−
Eco-impact	+	+	+
Σ+	4	2	2
Σ−	0	2	3
ΣS	1	1	0

For various other design constraints such as stiffness, similar relationships can be determined. In addition, several other properties (e.g., chemical resistance, corrosion resistance, wear resistance, durability, and energy content) could be important to an application. These properties need to be given sufficient consideration in material selection. Providing a coating to the structure can incorporate other secondary features such as corrosion resistance or wear resistance, with further increased cost of the product.

7.2 Pugh's Decision Matrix Method

The Pugh method[12] is one of the simplest decision-making technique used, especially to replace or substitute current material in the design. Material selection is based on the qualitative or quantitative comparison of the alternative materials with the datum or current material choice. In this method, a decision matrix is constructed as shown in Table 5. The first column lists the material property that need to be considered and the remaining columns contain alternative materials listed in the decision matrix. Each of the properties of a possible alternative new material is compared with the corresponding property of the currently used material, and the result is recorded in the decision matrix as + if more favorable, − if less favorable, and *S* if it is the same.

Then, at the end of the decision matrix, create three more rows with the row headers being Σ+, Σ−, and ΣS. The decision on whether a new material is better than the currently used material is based on the analysis of the result of comparison (i.e., the total number of +, −, and S). New materials with more favorable properties than unfavorable ones are selected as serious candidates for replacing the existing material to redesign the component.

7.3 Weight Factors

Weight factors is one of the traditional methods used in selection problems, whether it is conceptual design layout selection or material selection. Several problems can arise while trying to determine the optimum material when a number of required properties and a number of materials meet these different properties in different ways. The degree of importance of each property could be different in a given application. The issue is then to determine the material that achieves the best balance of properties. The weight factor method, or *merit index method*, involves assigning a relative merit value for each of the relevant properties based on its importance in the design. Because properties are measured in different units, each property is normalized to obtain the same numerical range. This is done by a scaling method, based on either property maximization or minimization requirement.

Material properties such as strength, stiffness, and percentage elongation are desired in a structure to be a maximum value. These properties are scaled to a value of less than or equal to 1 in the following way:

$$\alpha = \text{scaled property} = \frac{\text{Numerical value of a property}}{\text{Highest value in the same category}} \tag{5}$$

Similarly, there are properties, such as cost, density, and wear rate, that are required as a low value in a design. Such properties are scaled as follows:

$$\alpha = \text{Scaled property} = \frac{\text{Lowest value in the same category}}{\text{Numerical value of a property}} \tag{6}$$

But, there are properties, such as wear resistance, corrosion resistance, repairability, machinability, weldability, and recyclability, that cannot be quantified as a numerical value. Such properties are given subjective ratings, A = 1 (outstanding) to E = 0.1 (poor), as shown in Table 6.

Once material properties are scaled, the cumulative weight index of a material is determined by $\gamma = \sum_i w_i \alpha_i$, where w_i is weighting factor, α is scaled property, and i is the summation of all the properties under consideration.

The weight factor w_i for various properties should lie between 0 and 1, chosen such that they add up to 1 (i.e., $\Sigma w_i = 1$). Engineers should use good judgment or past experience in choosing appropriate values of w_i for various properties. Usually, the most important property is given the largest w, the second most important, the second largest, and so on.

For example, in the design of an aircraft wing, material density, yield strength, Young's modulus, fracture toughness, and eco-indicators are considered. If all these properties are equally important, then equal weights (0.20) should be assigned to each property. By contrast, in the design of an electric bulb filament, material properties such as melting point, electrical conductivity, strength,

Table 6 Scaling of Nonquantitative Property

Property\Material	A	B	C	D	E
Dry oxidation resistance (Subjective rating)	Poor	Good	Outstanding	Satisfactory	Very good
Scaled property (≤ 1)	0.1	0.6	1.0	0.4	0.8

Table 7 Pairwise Comparison Method (Digital Logic Method) to Assign Weighting Factors

Material Property	Number of Positive Decisions $N = n(n-1)/2$						Positive Decisions	Weighting Factor
	1	2	3	4	5	6		
Density	0	1	0				1	0.167
Strength	1			1	1		3	0.500
Energy content		0		0		1	1	0.167
Eco-impact factor			1		0	0	1	0.167
						Total	6	$\Sigma w_i \approx 1$

and ductility are important. The design analysis also reveals that melting point is the most important property, followed by electrical conductivity, and then comes tensile strength and ductility. As a result, we might assign weighting factors to melting point as 0.4, conductivity as 0.3, strength as 0.2, and ductility as 0.1.

One of the methods used to avoid bias in assigning weighting factors to the considered materials properties is digital logic method, or pairwise comparison method. In this procedure, evaluations are arranged such that only two properties are considered at a time. In comparing two properties or performance indices, the more important goal is given numerical one (1) and the less important one is given zero (0). When all the comparisons are made, add the ones for each property, and total the results. The weighting property for each property is the sum of the ones it received, divided by the total number of decisions made. The total number of possible decisions is $n(n-1)/2$, where n is the number of properties under consideration. Table 7 shows the pairwise comparison made for an automobile spring. Here, the properties are listed in the left-hand column, and comparisons are made in the columns to the right. The obtained weighting factor is given in the last column.

Example 1 Evaluate alternative materials given in Table 8 for their suitability as a leaf spring in an automobile for maximum elastic energy storage capacity per unit weight. Assume that the energy content and eco-impact of the material should be minimized.

Solution: Springs are commonly used to store elastic energy in variety of devices. Table 8 lists 304 stainless steel (SS), 6061 aluminum, titanium alloys (Ti-6Al-4V) and unidirectional carbon-fiber reinforced epoxy-based composite as possible materials for the design of automobile leaf spring. Herein, we wish to maximize the elastic energy storage capacity of the spring per unit weight and at the same time minimize the energy content and eco-impact of the spring material. Based on the experience, allocate 0.45, 0.30, and 0.25 as the weight factors for energy storage, energy content, and eco-impact variables. The last column in Table 8 gives the cumulative weight index for each of these materials. For the

Table 8 Properties of Candidate Materials for an Automobile Leaf Spring

Material	Density (Mg/m^3)	Young's Modulus, E (GPa)	Yield Strength, σ_y (MPa)	$\sigma_y^2/(E\rho)$	Energy Content, MJ/kg	Eco-indicator, mp/kg	Performance Index
304 steel	7.85	200	320	65.22	110	17	0.5098
6061 Al	2.70	70	120	76.19	265	14	0.3791
Ti-6Al-4V	4.40	120	950	1709.28	525	90	0.2055
CFRP	1.35	100	1000	7407.41	215	22	0.7626

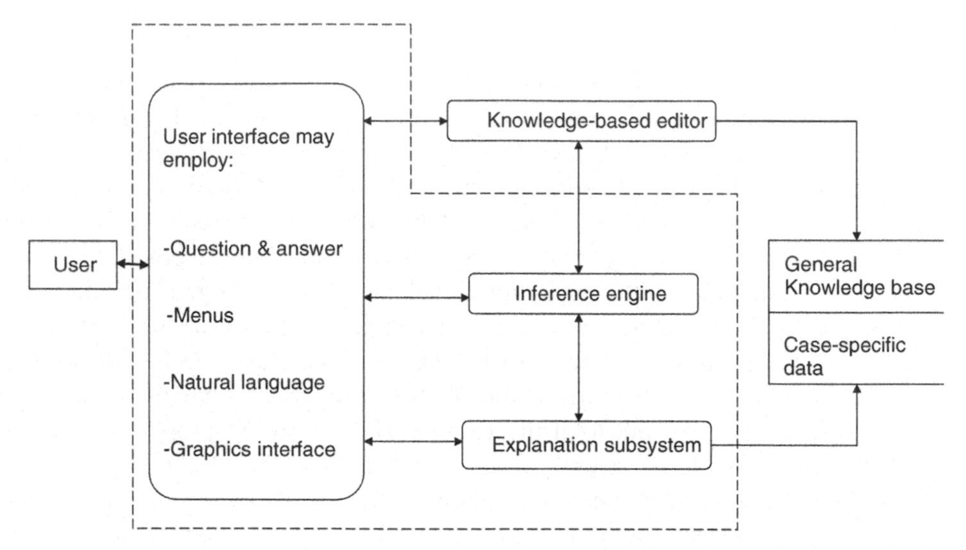

Figure 7 Architecture or block diagram of a typical expert system. (From Ref. 13.)

current design example, CFRP has the highest performance index and is thus the most suitable material for the leaf spring, followed by 304 stainless steel.

7.4 Expert System for Material Selection

Thousands of material choices are available to an engineer to assist in appropriate material selection. This reveals the need for an *expert system* for the selection of alternative materials for a given application. Expert systems are finding many applications in industry, including the areas of design, trouble-shooting, failure analysis, manufacturing, and materials selection. Expert systems, also called *knowledge-based systems*, contain a domain specific knowledge base combined with an inference engine that processes knowledge encoded in the knowledge base to respond to a user's request for advice. The classical structure of a typical expert system is shown in Figure 7.[13]

Expert systems rely on heuristics, or rules of thumb, to extract information from a large knowledge base. Expert systems typically consist of three main components:[10]

1. The knowledge base contains facts and expert-level heuristic rules for solving problems in a given domain. The rules are normally introduced to the system by domain experts through a knowledge engineer.
2. The inference engine provides an organized procedure for sifting through the knowledge base and choosing applicable rules in order to reach the recommended solutions. The inference engine also provides a link between the knowledge base and the user interface.
3. The user interface allows the user to input the main parameters of the problem under consideration. It also provides recommendations and explanations of how such recommendations were reached.

A commonly used format for the rules in the knowledge base is of the following "if–then rules" form:

IF (condition 1) and/or (condition 2)

THEN (conclusion 1)

In an expert system, the user feeds in the service property requirements, usually a range with minimum and maximum values (e.g., operating temperature range, eco-indicator, energy content, chemical resistance, fluid exposure, percent elongation, fracture toughness, strength, etc.). Then, subject to the available material database, the expert system output provides a list of materials satisfying the design requirements.

In the case of material selection problems, expert systems based on the user's requirements will search large in-built databases and provide impartial recommendations meeting the requirements. Here are a few examples of currently available experts systems:

- *The Chemical Corrosion Expert System produced by the National Association of Corrosion Engineers (NACE) in the United States.* The system prompts the user for information on the environmental conditions and configuration of the component and then recommends candidate materials.
- *Materials and Process Design Advisor developed by Rockwell International.* It consists of expert systems for corrosion protection, adhesive selection, heat treatment of metallic materials, conformal coating selection, and selecting soldering processes.
- *The Cambridge Material Selector (CMS) developed by Granta Design, Cambridge, United Kingdom.* It has a large database of almost all classes of engineering materials.

7.5 Performance Index Method

Ashby and co-workers from University of Cambridge have pioneered this systematic approach of material selection.[5,6,8] Material-selection methods such as weight fraction method are, to some extent, based on human judgment, and not much depends on engineering analysis of the problem at hand. The databases are the hard core of the selection procedure: Up to a certain point, they can be cast in a standard format. For specialized design problems, a more rigorous approach based on function, objective, constraints, and free variables is needed. But a database as used in Expert Systems does not compare the output, hence evaluation tools to compare and rank different materials are needed. Simple mathematical modeling allows such a tool to be built, but the price to be paid for using this method is that dimensioning of the structure is very crude.

One has to keep in mind that the aim is to identify the materials for which accurate structural mechanics calculations will have to be performed later on. Each set of requirements has to be structured in a systematic manner:

1. What is the objective of the design or the function to be performed by the component?
2. What are the free and the imposed (fixed) variables?
3. What are the design constraints?

Example 2 For instance, one might look for a tie rod (the function of a tie is usually to carry uniaxial tensile forces effectively) for which the length L is prescribed and the section area A is free, *which shouldn't yield under a prescribed load P* (constraints), and which should be of minimum environmental impact (objective). The stress should not exceed the material tensile yield stress σ_y.

Solution: Note that the environmental impact of tie rod material is measured in terms of the energy content of the material, e, which is measured kJ per unit weight of the material.

We wish to minimize the total energy content of the designed tie rod, given in terms of the mass (m) of the tie rod by

$$E = em = e(\text{volume})(\text{density}) = (LA)e\rho \tag{7}$$

The objective equation (7) is to be minimized subject to the design constraint of strength. The constraint that the tie rod material should not yield under load P imposes a minimum value for the cross-sectional area, A, of the tie-rod.

Using the definition of normal stress, we have

$$\frac{P}{A} \leq \sigma_y \rightarrow A = \frac{P}{\sigma_y} \tag{8}$$

Combining equations (7) and (8), we obtain the total energy content of the tie-rod:

$$E = (LP)\frac{e\rho}{\sigma_y} \tag{9}$$

Therefore, the material that will minimize the energy content of the tie rod would be the one that *maximizes* the *performance index* M_1, given by

$$M_1 = \frac{\sigma_y}{e\rho} \tag{10}$$

The performance index is essentially the specific strength divided by the energy content of the material. M_1 can be maximized either by maximizing the tensile strength or by minimizing the density and content of the material. Materials with the same M_1 should have the same energy content and hence same environmental effect.

In this example, if the design constraint was changed to stiffness such that the tie-rod material shouldn't elongate more than a prescribed limit δ under an applied load P, the material selection analysis for minimum environmental impact is as follows:

If the deflection δ is specified, then for an applied load P, the stiffness $S = P/\delta$.

From Hooke's law we have,

$$\text{Load, } P = \frac{EA}{L}\delta \tag{11}$$

Now the free-variable area A can be expressed as

$$\text{A} = \frac{P}{\delta}L\frac{1}{E} = SL\frac{1}{E} \tag{12}$$

Substituting the expression for the sectional area A in equation (7) we obtain:

$$E = SL^2\frac{e\rho}{E} \tag{13}$$

The environmental impact of tie-rod material can be minimized by maximizing the performance index M_2:

$$M_2 = \frac{E}{e\rho} \tag{14}$$

This is essentially the specific modulus divided by the energy content of the tie-rod material. Index M_2 can be maximized either by maximizing the Young's modulus or by minimizing the density and energy content of the material. Materials with the same M_2 should have the environmental impact.

The above two very simple derivations illustrate the prescription one should follow for obtaining performance indices: Write the constraint and the objectives, eliminate the free variable, and identify the combination of materials properties that measures the efficiency of materials for a couple (constraints/objectives). These performance indices have now been derived for many situations corresponding to simple geometry (bars, plates, shells, beams) loading in simple modes (tension, torsion, bending), for simple constraints (do not yield, prescribed stiffness, do not buckle), and for various objectives (minimum weight, minimum volume, minimum cost). They have been extended to thermal applications. The way to derive a performance index for a real situation is to do the following:

- Simplify the geometry and the loading
- Identify the free and fixed (design) specified variables
- Develop a mathematical expression for objective including free and fixed variables
- Express design constraint mathematically again having free, fixed, and functional variables
- Eliminate the free variables between the constraint and the objectives

Table 9 provides some standard performance indices currently used in mechanical design. Ashby and Cebon have derived such performance indices for a wide

Table 9 Examples of Performance Indices to Minimize Environmental Impact

Objective	Component	Loading	Constraint	Performance Index
Minimum energy content	Tie	Tension	Stiffness and length prescribed, section free	$\frac{E}{e\rho}$
Minimum CO_2 emissions	Beam	Bending	Stiffness and length fixed and side length of the square cross-section free	$\frac{E}{[CO_2]\,\rho}$
Minimize eco-indicator points	Plate	Compression	Buckling load, length and width fixed. Thickness is free	$\frac{\sqrt[3]{E}}{I_e\rho}$
Minimum energy content	Cylinder	Internal pressure	Imposed maximum elastic strain, thickness of the cylinder free	$\frac{E}{e\rho}$
Minimum energy content	Tie	Tension	Strength and length prescribed, section free	$\frac{\sigma_e}{e\rho}$
Minimum CO_2 emissions	Beam	Bending	Strength, length fixed and side length of the square cross-section free	$\frac{\sigma_e^{2/3}}{[CO_2]\rho}$
Minimize eco-indicator points	Plate	bending	Strength, length, width fixed and thickness free	$\frac{\sqrt{\sigma_e}}{I_e\rho}$
Minimum energy content	Cylinder	Internal pressure	Under applied pressure the shell material doesn't yield, thickness of the cylinder free	$\frac{\sigma_e}{e\rho}$

range of applications.[8] A simple way to use the performance index is with the *selection maps* shown in Figures 8 and 9. On a logarithmic scale, the lines corresponding to equal performances are straight lines whose slopes depend on the exponents entering the performance index. Intuition says that material lying in the left upper-hand portion of the chart is the best material for the energy-conscious design. Figure 9 shows one of these maps used for stiff components at minimum energy content. Materials for stiff ties should maximize $E/(q\rho)$, materials for stiff beams should maximize $\sqrt{E}/(q\rho)$, and materials for stiff plates should maximize $\sqrt[3]{E}/(q\rho)$.

One of the drawback with material selection charts is that the property values are assumed to be constant throughout the life cycle of the component. But, in reality, the designs are usually analyzed for a finite lifetime. In fatigue design, the quasistatic strength is replaced with materials endurance limits, a fatigue property. But in designing for creep resistance or corrosion resistance, a new set of

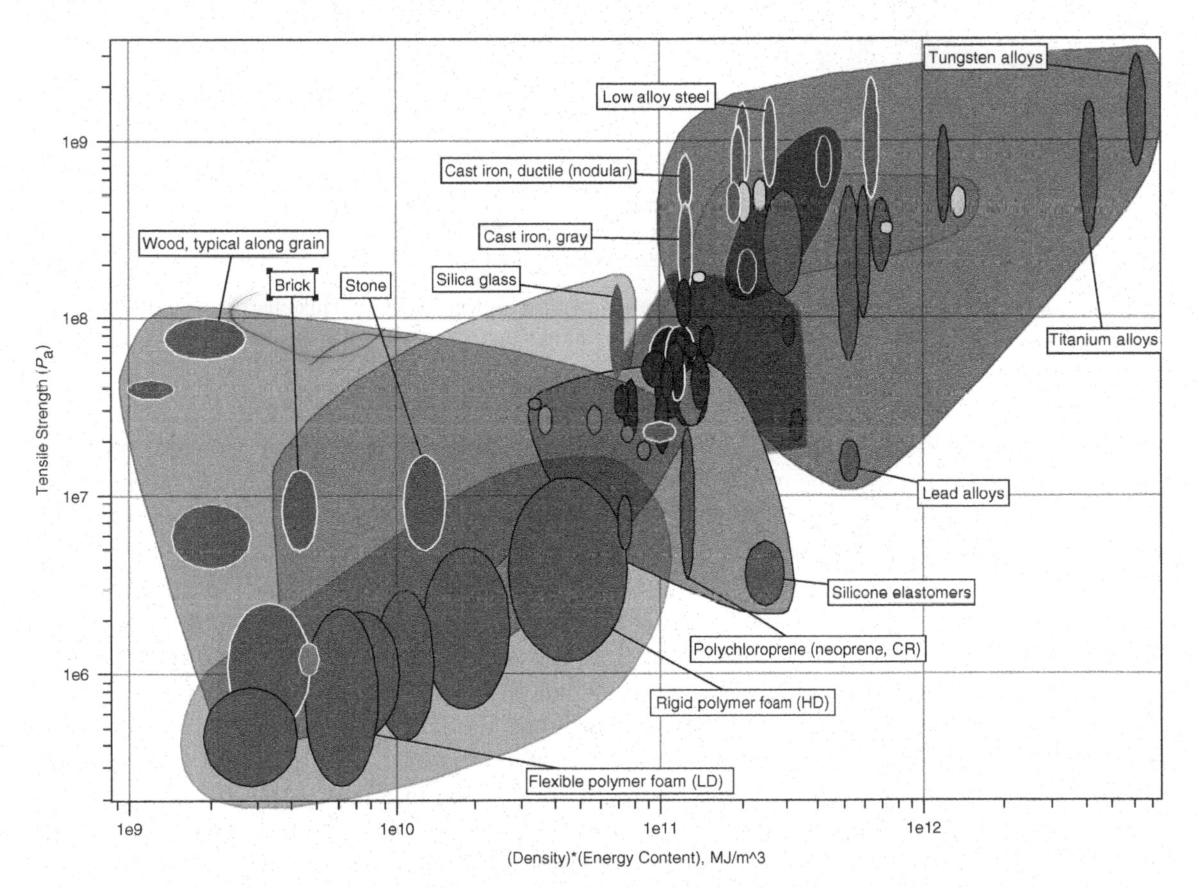

Figure 8 The tensile strength versus energy content material selection chart. This chart is useful in evaluating materials for strong structures at minimum energy content (generated using Cambridge Engineering Selector V6 Software).

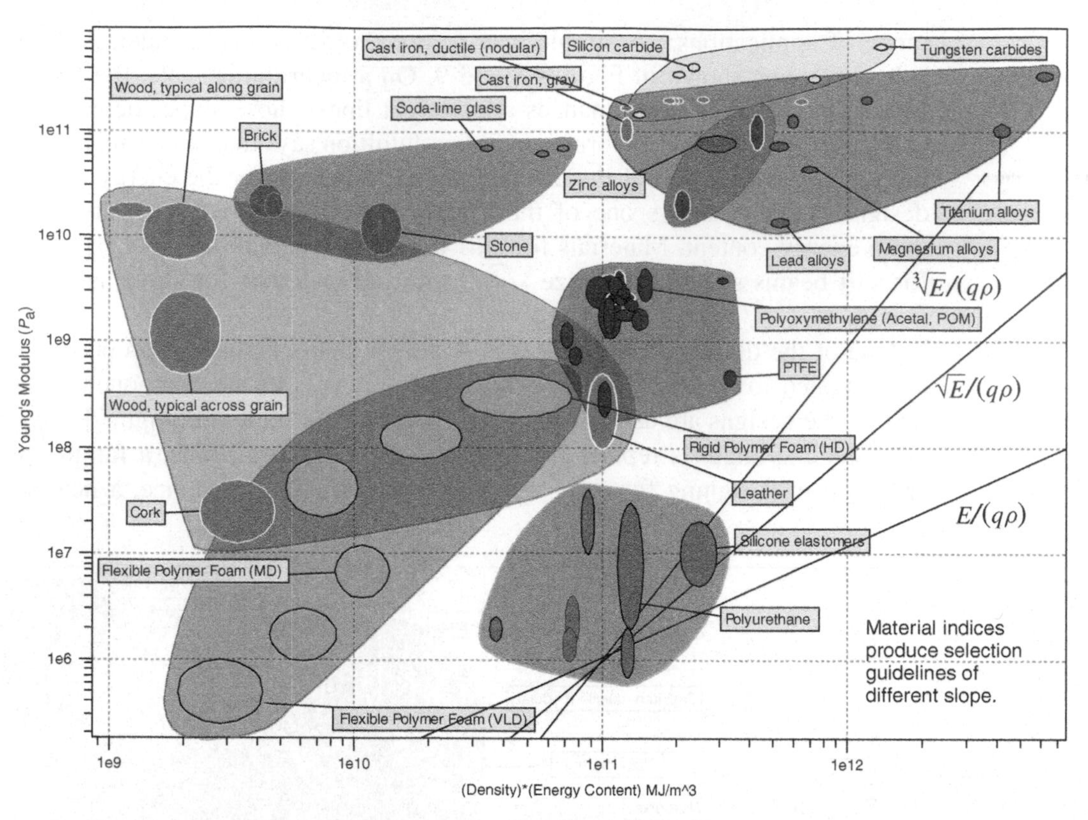

Figure 9 The Young's modulus versus energy content material selection chart. This chart is useful in selecting materials for stiff structures at minimum energy content (chart is generated using Cambridge Engineering Selector V6 software).

performance indices involving rate equations (for creep or corrosion) has to be developed. The performance indices then depend not only on the materials properties, but also on operating conditions such as applied load, specimen geometry, and life span of the component.

Design problems have multiple objectives and/or multiple constraints. For a single objective and a multiconstraint problem (such as designing a component that should be strong and at the same time stiff), the problem is to identify the limiting constraint. Detailed information of the component geometry is required to identify the limiting constraint. Otherwise, using a systematic method called *coupling equations* (see Chapter 7 of reference 5), one can solve the problem by simultaneously satisfying both constraints. In a multiobjective problem (e.g., designing a component for minimum weight and minimum cost), designers need to identify an *exchange coefficient* between the two objectives—for instance, how much the user is ready to pay for saving 1 kilogram of weight.[6] These exchange coefficients can be obtained either from a value analysis of the product or from the analysis of existing solutions. They allow engineers to compute a value function, which is the tool needed to rank the possible solutions. Both the value analysis and the coupled equation method provide an objective treatment of the multiple criteria optimization. However, they require extra information compared to the simple performance index method. Kampe has used this procedure to determine energy efficient beam material over its usable lifetime.[14]

8 CONCLUSIONS

In the engineering community there is an increasing awareness of environmentally friendly designs. Along with the traditional tools of engineering practice, design for environment (DfE) has become an integral part of overall professional engineering. The extrinsic properties of the materials should be considered in the selection of materials. This chapter elaborates on several methodologies for material selection in order to design components with minimum environmental impact.

REFERENCES

1. M. Cohen, "Societal Issues in Materials Science and Technology," *Materials Research Society Bulletin*, **19**, 3–8 (1994).
2. G. E. Dieter, *Engineering Design: A Materials and Processing Approach*, McGraw-Hill, New York, 1983.
3. G. Pahl and W. Beitz, *Engineering Design*, 2nd ed., trans. K. Wallace and L. Blessing, Springer-Verlag, Berlin, 1997.
4. L. Holloway, D. Clegg, I. Tranter, and G. Cockerham, "Incorporating Environmental Principles into the Design Process," *Materials & Design*, **15**(4), 259–266 (1994).
5. M. F. Ashby, *Material Selection in Mechanical Design*, 3rd ed. Elsevier Butterworth Heinemann, Oxford, United Kingdom, 2005.
6. M. F. Ashby, "Multi-objective Optimization in Material Design and Selection," *Acta Materialia*, **48**, 259–369 (2000).
7. J. H. Westbrook, "Sources for Materials Property Data and Information," *ASM Handbook*, vol. 20: *Materials Selection and Design*, Volume Chair George Dieter, ASM International, Materials Park, OH, 1997, pp. 491–506.
8. M. F. Ashby and D. Cebon, *Case Studies in Material Selection*, Granta Design Limited, Cambridge, United Kingdom, 1996.
9. F. A. A. Crane, J. A. Charles, and J. Furness, *Selection and Use of Engineering Materials*, 3rd ed., Butterworths, London, 1997.
10. M. F. Farag, *Selection of Materials and Manufacturing Processes for Engineering Design*, Prentice Hall International, Englewood Cliffs, NJ, 1989.
11. M. Kutz (ed.), *Handbook of Materials Selection*, John Wiley & Sons, Inc., New York, 2000.
12. S. Pugh, *Total Design: Integrated Methods for Successful Product Development*, Addison-Wesley, Reading, MA, 1991.
13. V. Weiss, Computer-Aided Materials Selection," *ASM Metals Handbook*, vol. 20, Volume Chair George Dieter, ASM International, Materials Park, OH, 1997, pp. 309–314.

14. S. L. Kampe, "Incorporating Green Engineering in Materials Selection and Design," Proceedings of the 2001 Green Engineering Conference: Sustainable Environmentally-Conscious Engineering, Virginia Tech's College of Engineering and the U.S. Environmental Protection Agency, Roanoke, Virginia, July 29–31, 2001.

APPENDIX A MATERIAL PROPERTIES

Some of the commonly used intrinsic material properties are defined below:

1. Strength is a microstructural sensitive property. Its value depends up on temperature and loading rate.
   a. *Yield strength.* The stress a material can withstand without permanent deformation.
   b. *Rupture strength.* The stress on the material at the time of breakage is known as rupture strength.
   c. *Ultimate tensile strength.* The maximum stress a material can withstand without tearing or fracturing in a tension test.
   d. *Endurance limit.* This is the maximum safe stress that can be applied to a material for infinite number (usually 10^7) of fatigue cycles.
   e. *Creep strength.* Defined as the ability of a metal to resist slow deformation due to stress, but at a stress level less than that needed to reach the yield point. Creep strength is usually stated in terms of time, temperature, and load.
2. *Ductility.* The ability of a metal to deform plastically without breaking, but with sustainable reduction in cross-sectional area.
3. *Hardness.* It is defined as resistance to permanent deformation and it influences ease of manufacture.
4. *Fracture toughness.* It is defined as the resistance to crack propagation and measured in terms of energy needed to create a unit surface area.
5. *Coefficient of Thermal Expansion (CTE).* It is ratio of the change in length of a specimen per unit length when its temperature is raised by 1°C.
6. *Thermal conductivity.* It is a measure of the ability of a metal to transmit heat under steady-state heat transfer conditions.
7. *Critical temperature.* As a metal is cooled, it passes through distinct temperature points where its internal structure and physical properties are altered. The rate of cooling will greatly influence the ultimate properties of the metal.
8. *Corrosion and oxidation resistance.* It indicates how well a metal can resist the corrosive effects of the operative enviroment.
9. *Heat treatability.* A measure of how the metal's basic structure will vary under an operation, or series of operations, involving heating and cooling

of the metal while it is in a solid state. Ferritic, austenitic, and martensitic steels all vary as to their heat treatability.

10. *Thermal shock resistance.* The maximum sudden temperature change a material can withstand without fracture.

APPENDIX B MANUFACTURING PROCESSES

For completeness some of the most commonly used metalworking procedures are listed below:

1. *Casting.* An age old process in which a molten state is allowed to solidify in an open or closed mold to produce a component to near net shape.
2. *Forging.* A process of plastically deforming metal to a finished shape by hammering or pressing in a closed die. It can be achieved under cold or hot conditions based on the operating temperature relative to the metals melting point.
3. *Extrusion.* Metal is pushed through a die to form various cross-sectional shapes.
4. *Machining.* A process whereby metal is formed to a desired shape by removing material in the form of chips. Various procedures include drilling, thread cutting, milling, grinding, and other unconventional cutting procedures involving laser beams and electron beams.
5. *Welding.* A process of joining or fusing two pieces of metal together by locally melting part of the material through the use of arc welders, plasmas, lasers, or electron beams.
6. *Electrochemical machining (ECM).* ECM is accomplished by controlled high-speed deplating using a shaped tool (cathode), an electricity-conducting solution, and the workpiece (anode).
7. *Powder consolidation.* Metal powders are blended, pressed, sintered (a process of fusing powder particles together through heat in a reducing atmosphere), and then coined out of the pre-alloyed powders.
8. *Protective surface coatings.* These include plating by means of electrical and chemical processes, or by painting. Surface treatments for increased wear may include processes such as nitriding, cyaniding, carburizing, diffusion coating, and flame plating.
9. *Shot peening.* A plastic flow or stretching of a metal's surface by impregnating round metallic shot thrown at high velocity. This is done to increase fatigue life.
10. *Heat treatment.* A process to impart specific physical properties to a metal alloy. It includes normalizing, annealing, stress relieving, tempering, and hardening.

11. *Inspection.* Part inspection is critical for quality assurance. Non-destructure inspection methods include visual inspection, magnetic particle and dye penetrant inspection, X-ray inspection, and inspection by devices using ultrasound, light, and air.

CHAPTER 10

EMPLOYING TOTAL QUALITY MANAGEMENT/SIX SIGMA PROCESSES IN ENVIRONMENTALLY CONSCIOUS DESIGN

Robert Alan Kemerling
Ethicon Endo-Surgery, Inc., Cincinnati, Ohio

1 WHAT IS TOTAL QUALITY MANAGEMENT?

1.1 Traditional Quality

From World War II to recently, most companies have been partially organized around the model developed by the Department of Defense for the war effort. That model utilized a *quality department,* separate from production, to inspect and mark acceptable the output of the firm. This served well for the war effort and for many years after that, but business leaders began to question the model. Why should only the *quality department* be concerned with quality? Is it a good business strategy to hire a lot of people to inspect product? Why should a business tolerate scrap and rework from poor quality? Could another approach bring about a business advantage?

1.2 New Approach

The new approach, called, originally, *total quality management* (TQM), makes the case that quality is not just the responsibility of a certain department. Rather, it must be a responsibility of every part of the organization. This is necessary not only to avoid the large costs of scrap and rework, but also to focus on

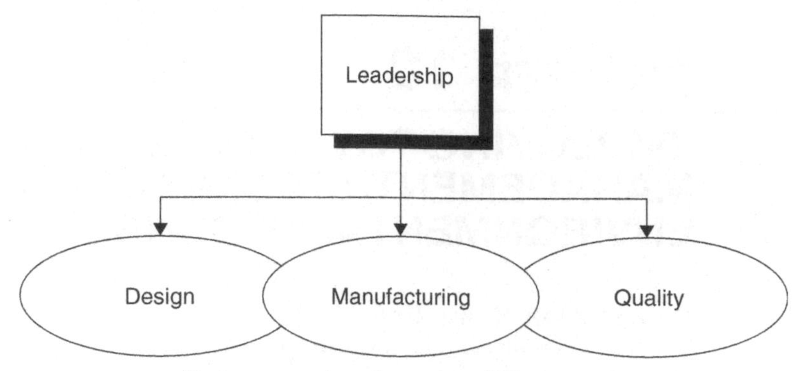

Figure 1 The interlocking responsibilities of total quality management.

satisfying customer needs. Both allow the firm to be competitive in the global marketplace. Figure 1 shows the interlocking responsibilities of functions within a TQM environment.

Obviously, mechanical engineers and managers will not be making or inspecting product. It will also give you, the mechanical engineer, exposure to the tools and knowledge to be successful with your processes and teams as you face the challenges to develop new products with a greater awareness of environmental responsibility.

Since the 1990s, a process publicly known as Six Sigma has effectively replaced TQM. Begun by Motorola and adopted by companies such as GE, Honeywell, Texas Instruments, and many others, Six Sigma strives to identify the key attributes of a product or process and employs a particular process order and tools to assure high performance with respect to the key attributes. The process has become known as the DMAIC process (pronounced duh-MAY-ick) after the following process steps:

1. *Define.* Define the key attributes of customer satisfaction or wants.
2. *Measure.* Determine how the key attributes of the product or process will be measured.
3. *Analyze.* Find the key functions in the design or process that control the important factors to the customer and that reduce variation.
4. *Improve.* Remove the sources of variation and/or set key process parameters for optimum output.
5. *Control.* Install controls to keep the process as designed and stable.

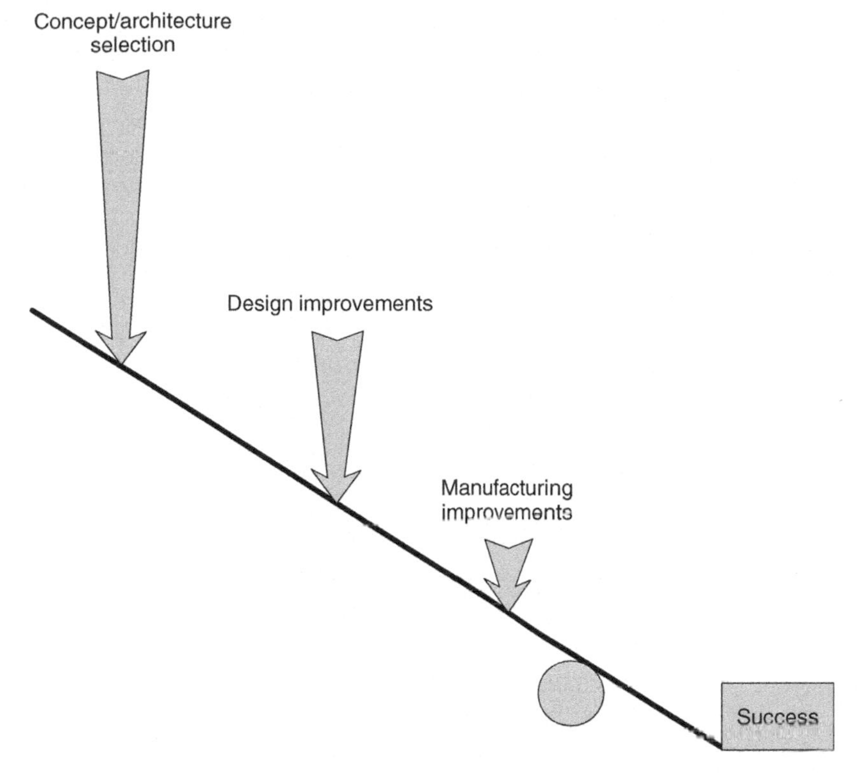

Figure 2 Relative impact on design success.

As the Six Sigma process took hold, practitioners looked for a way to extend it into the design and development processes. As shown in Figure 2, many thought that the leverage from design and development could potentially overwhelm the improvement opportunities available to manufacturing after the design is completed. What was born from this is called *Design for Six Sigma*. Since past literature uses both terms, this chapter will adopt the term *TQM/SS* for the remainder of the text.

1.3 Definitions of Quality

Expectations of quality have changed over time. The following are some definitions that have been used for quality:

- Freedom from defects
- Fitness for use
- The totality of features and characteristics of a product or service that bear on its ability to satisfy given needs
- The features and characteristics that delight the customer

As you can see, the progression of these definitions moves from a narrow assessment of defects, to the broader consideration of satisfying the customer. As a matter of fact, your customers will probably take into account their entire experience with your firm. Focusing on satisfying customer requirements is the *heart* of the success of TQM/SS. The TQM/SS process will cause your design team to *focus* on the important aspects of the customer and the environment.

2 THE IMPACT OF ENVIRONMENTALLY CONSCIOUSNESS ON DESIGN

It would appear that having to design in an environmentally conscious manner would complicate the design process. Actually, employing TQM/SS means this is less of a complicating factor than might be anticipated. For the designer, the concept of designing for the environment means that a new customer or type of customer will be providing some of the inputs of design requirements. Although a new customer might seem to be a complicating matter, design engineers are used to considering multiple customers for the final product. The TQM/SS process simply treats this as another input to the process.

2.1 A New Customer

There are many different reasons for considering the environment as a customer. These range from the logic that we must live in the world we create to the fact that environmental considerations are the law in many parts of the world.

The remainder of this chapter will *not* continue to discuss why the environment should be considered. It assumes that you are reading this because of personal commitment or requirement. The remainder of the chapter will focus on how the tools of TQM/SS will help you design with environmentally conscious considerations.

2.2 Advantages of Employing TQM/SS

As we consider the environmental impact of a new product, it is important that the new device and the processes that enable its manufacture leave the smallest footprint on the environment. One way that is accomplished is to ensure that the amount of scrap and loss created during manufacturing is the absolute minimum. This is also the general objective of TQM/SS. Without doing anything different, TQM/SS practices line up with environmental considerations in this aspect.

The other area where TQM/SS directly lines up with environmental consideration is in the area of optimizing the design for use. The aim of TQM/SS is to make the resulting device useful and efficient for the customer and the manufacturing processes. This often entails optimization work such as designs of experiments, statistical tests, process capability studies, and measurement systems analysis. All of these are key elements of TQM/SS. In the process of optimizing the design and supporting manufacturing processes, a significant part of the impact on the environment may be minimized.

3 THE TQM/SS APPROACH

Designing in the TQM/SS approach follows a logical flow of steps, often named after the acronym formed by the name of the steps:

1. Define
2. Measure
3. Analyze
4. Design
5. Verification (and Validation)

The name is *DMADV*, pronounced Duh-Mad-Vee. Some organizations, particularly in the medical device industry, refer to the process as DMADVV, making a distinction between engineering verification and final-user validation. In medical design, *verification* assures that the design was delivered correctly by the manufacturing steps. *Validation* is the assurance that the verified design can perform the intended function and is useful to the end user (clinician). For the remainder of this writing, the focus will be on the DMADVV process. For each step, some example tools will be discussed. In most cases, more information about the tools will be given. Due to space constraints, some alternative or adjunct tools will be mentioned only. Refer to material at the end of the chapter for more guidance. Commonly used tool names will be italicized when first used. Also examples for designing with environmental considerations will be discussed. Some practitioners may call the process steps by different terms, but at the heart of any Design for Six Sigma process, the steps will follow the same general flow.

3.1 Define

In the *define* phase of the design process, the design engineer must determine the expectations of the customers involved with the design. As already discussed, environmental concerns become part of the *customers' inputs* to the design of the finished product. You can anticipate that the requirements of environment as a customer will fall into one or more of four categories:

1. *Remove.* Find ways to remove harmful materials from the environment. This means both the device *and* the processes necessary to manufacture the device. In some cases, specific materials may be mandated for removal. At the time of this writing, MTBE has been targeted for removal as a gasoline additive due to the potential for ground pollution. In Europe, removal of lead from soldering processes is scheduled for phase-in.
2. *Reduce.* Reduce the impact on the environment from the device, its manufacture, packaging, and the final disposal of the device, packaging, and manufacturing materials. Use less material in the device, manufacturing, and packaging.
3. *Reuse.* Directly reuse the device, components, packaging, or manufacturing materials.

4. *Recycle.* Facilitate the ability to take the components out of the device that are not reusable and put their constituent materials into a form where they can be used in other devices as raw material.

Tools for the Define Phase

The key tools in the Define phase are not very technical tools, but their contribution to the success of the project cannot be overstated. These details must be defined in some form. The following are some recommended tools.

The most important tool in the Define phase is a *project charter*. This may take many forms and names in organizations. The different forms and names are not relevant to the tool. The *information and agreement* represented in the charter may be the key to the success of the project. The charter forms a contract between the organization's leadership, sometimes called *sponsors*, or simply *leadership*, and the design team. There are six elements of a charter:

1. A goal statement of the project
2. The process and project scope
3. The business case (preliminary return on investment estimate)
4. An opportunity statement
5. A preliminary project plan
6. The core team (some also include a *stakeholder analysis*)

Some may object to inclusion of the business case and project plan in the charter, saying correctly that it is very early to determine these. It is correct that it is early to have these elements in detail at this point, but they are still necessary to serve as initial screens. There is no reason that the charter remains a static document. These and other elements may be updated.

With regard to the *goal statement* portion of the charter, this must contain the environmental position of the new design. If the design aims to meet a new environmental regulation, this should be included here. If there is not a regulatory required value to achieve, the goal statement should define the environmental target value for the design. For example, "This design will reduce volatile hydrocarbon emissions during manufacturing by 20 percent."

Many experienced design engineers are familiar with *project creep*. Project creep comes about because new requirements are discovered, but it also comes about because of a type of "while you're at it" thinking. Two tools are useful to manage project creep. They are the *multi-generation plan* (MGP—also called the *multi-generation product plan*, or MGPP) and the scope tool.

A multi-generation plan is a simple table of four columns. The first column is a container for the elements of the plan. Columns two, three, and four are the next three generations. The table may look like Figure 3.

As you can see, the tool defines three increasing generations of a product. By defining the different generations, the design team provides a place to position

	Generation 1	Generation 2	Generation 3
Vision	General use	Those who expect more	Thought leaders
Key characteristics	Easy	More options	All automated
Technology platform	PC	PDA	Wireless phone
Market position	Basic	Upgrade	Premium
Environment	50% reduction in packaging	Lead-free solder	Easy breakdown for recyclers

Figure 3 Multi-generation product plan.

those "nice to have" elements into a next or third generation, avoiding arguments that the opportunity is lost forever if they're not included this time.

The *scope tool* is a simple exercise that the team can execute. On large paper or a white board, draw a large circle and label it "Design." Write the potential design elements, features, and options on sticky notes. Discuss each one as it comes up and decide as a team if it is in scope or out of scope. If it is in scope, it goes in the circle. Out of scope goes outside. As a further exercise, the out-of-scope elements may be assigned to one of the follow-on generations if desired. The reason these two tools are often joined is to make the initial decisions on the features and attributes of the final system. Having an MGP makes available a place holder for those features and attributes that will not be in the current design.

A final tool to employ in the Define phase is the *project plan*. This can take the form of a *Pert chart* or a *Gantt chart*. There are software products and extensive writings on project planning. The reader is encouraged to consult some referenced at the end of the chapter. Included with the project plan is a beginning *work breakdown structure* (WBS). A WBS is a logical decomposition of the entire project into logical task blocks that can be managed. These task blocks may be scheduled and resources assigned, breaking a large project into a set of smaller entities. This will aid in determining what resources are needed and when they will be needed. A work breakdown structure may be a separate document, but most organizations find it convenient to include as part of the Gantt or Pert charts. The project effort may also be displayed as part of a *network diagram*.

3.2 Measure

In the Measure phase, the team will put greater definition on the elements required for a successful project. A number of tools will be used to gather the *voice of the customer* (VOC). These will include requirements or targets for the environmental impact of the new design. In some cases, the requirements will be dictated by regulation. In other cases, the requirements may be softer, phrased as goals or *target values*. In either case, these must be captured, along with the VOC statements of the end users. All will become inputs of the design effort.

For the environmental elements of the design, it will be important to establish target or required values. For elements that are covered by regulation, values or thresholds may be set by law. For elements that are not regulated to specific values, perhaps, for example, target values to reduce wastewater discharge or packaging material, it will be important that the organization or the design team establish the target goal. End users may provide input due to their interest in environmental issues. For some design projects, addressing environmental issues may add value to users, increasing their support of the design.

Tools for the Measure Phase

Conjoint analysis is a technique not only to obtain the VOC on various development options for a new device, but also to help prioritize the options. Structured with estimated pricing, it can give an indication of the pricing that customers will tolerate for the different options. The different options available for a design are presented to customers in either pairwise format or a list of option sets. Based on the answers, customers indicate their preferences, including what they would consider paying for the new device. Customer selection will include indicating the more favorable option from the pairwise groupings or they will be asked to put the options in preferential order for lists. This specific response allows better quantification of the results.

With environmental considerations, some new designs are more costly until economies of scale can be achieved. Conjoint analysis can help the design team understand how much customers will accept additional cost (if any) in order to help with environmental issues. Of course, this assumes that the environmental development has some flexibility in response—that it is not mandated by law.

Rapid application development is a development approach used mostly in the software development practice, but the technique can be applied in some mechanical development. The development team will use rapid prototype techniques, usually some form of solid modeling by way of wax printing, stereolithography, machining, rapid tool development, metal sintering, or computer simulation to form prototypes for customer feedback. Depending on the quality and robustness of the prototypes, customers may even be able to try the device as a potential or actual solution in real-life usage. As feedback is obtained, the requirements are updated; new prototypes may be developed and used for further feedback.

Kano analysis is named after Dr. Noriaki Kano for his research and promotion of segregating customer features. Dr. Kano proposed that customers view features or characteristics as belonging to three broad categories. The lowest category is "Must-be," which denotes the features and attributes that must be part of the final design. The next category is "More-is-better" items. The highest category consists of "Delighters." The "More-is-better" features and attributes are often developed from good VOC research, where the customers describe what they would like to see as improvements and advancements to the current solution they are employing. To obtain understanding of Delighters, it may be necessary for experts to propose items beyond the current state of the art that could be added to the design as an attribute that is unexpected by the customer. When these are identified, the value of their inclusion in the design may be explored during customer research. A caution is warranted: Customers do not often describe Must-be elements because these are an unspoken given in the customers' minds. The design team must take care to assess the features of the current customer solution (if any). If features or attributes are to be removed, the wisdom of this decision should be explored with market research.

For many today, environmental elements in the design are Delighters. Soon, most will expect some form of environmental consideration in the design of the devices they use. Recent literature has begun to use the term GRINTECH for *green integrated technology* to describe the resulting expectation for design. In the report "Future R&D Environments: A Report for the National Institute of Standards and Technology," the environment is described as an increasingly strong pull factor, shaping research and development for the foreseeable future.[1]

As devices mature, elements that were once in the More-is-better and even the Delighter categories can move into the Must-be category. For example, car sound systems have generally moved from the CD player as a high-end option to a Must-be in most cars today. Currently, global positioning satellite (GPS) systems in autos are working their way down the Kano model. In premium vehicles, GPS is expected. In vehicles below premium they are becoming options or Delighters.

The best use of resources demands that a team focus on the important items in the process. Many design teams designate their top 5 to 10 must-hit requirements as *critical to quality CTQ*. These CTQs are held constantly before the team and are measured in the final verification and validation phase of the project. Later in the design phase, the use of CTQs will be expanded to a *CTQ Cascade*, where CTQs are related to subsystem and component features.

Most teams that are new to this process will want to discuss what it means to be *critical* to quality. There are many things that are critical if left out or damaged. The way to look at *CTQs* is to determine what parts of the process are difficult to do or difficult to control. For example, process parameters that have tighter tolerances than normal might be CTQs. Another candidate for designation of CTQ is something that is new to the process or is very important to the customer. CTQs may be developed from *quality function deployment* and *voice of the customer* research, discussed later in this section. Organizations might find it useful to put a display of the *design scorecard* in the development area, as shown in Figure 4. The design scorecard is usually a table of the CTQs and their target values, alongside the actual or predicted values for the CTQs. A design scorecard may also take the form of a graph, as in Figure 5.

Design Element	Target	Actual Mean	S.D.	Cpk
Weight	2800	2750	10	1.67
Mileage	40	43	1.2	0.83
Capacity	500	610	15	2.44

Figure 4 Design scorecard.

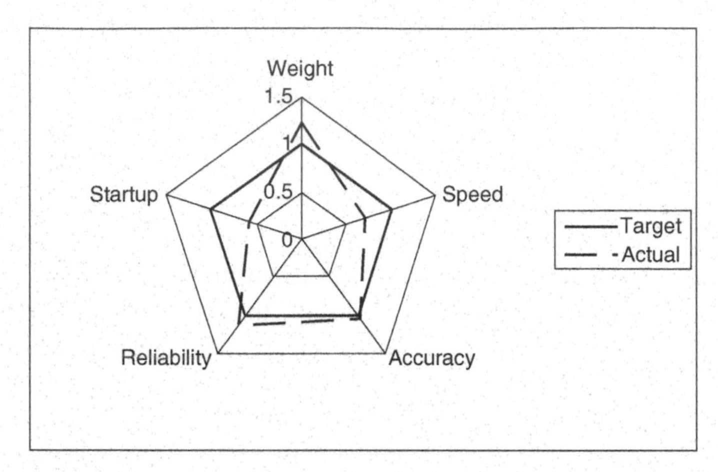

Figure 5 Design scorecard in radar chart form.

Measurement systems analysis (MSA) is a tool used in the measure phase and in the later phases of Verification and Validation. MSA defines the error coming from the measurement system and supports reduction in measurement error, so the resulting design and its manufacturing controls are not reacting to measurement error, but are rather reacting to actual variation. It is important to note that the final system variation is given by the following:

$$Variation_{\text{System}} = Variation_{\text{Device}} + Variation_{\text{MeasurementSystem}}$$

At this point, one might ask what measurements are available when the design has not been completed. This is a good question, with a good answer. At this point in the DMADVV process, the design team is establishing the metrics of *success*. It will be very important that this success be measured. If the issue is customer acceptance or parts per million of a particular substance, it would be most useful that these be measurable with some certainty prior to the company's commitment of resources to the effort.

There are a number of ways to obtain *Voice of the customer (VOC) research* data. Each has its own pros and cons. Although they may be called by different names, they fall into one or more of four different categories: interviews (phone or face-to-face); observations; focus groups; or surveys. Interviews, focus groups, and surveys are fairly well known so this chapter will not expand on them, but observations are of significant interest and are not as well known or used by design engineers.

In *observation research*, the design team has one or more *observers* watching different customers as they work in the proposed environment for the new device. This may be arranged in a *blind* format where the customers either don't know they are under observation or at least the observation is conducted in a way that is not very overt. The latter takes place when the observers are hidden, use cameras or one-way glass as examples. As the customers are watched, it may be easy for the observers to see difficulties with the current solution or even to observe the amount of stress, physical and mental, that the current solution requires. If the observations are recorded, it is also possible that very fine-grained timing of the steps may be obtained during the observations. These may be rich sources for input on the new design. For environmental considerations, observations may show opportunities for reduction of materials, including packaging. Also, better knowledge of the actual usage may show opportunities to reduce waste from initial overfilling or spillage during use or disposal. Obviously, observation requires that the user have some sort of solution currently in use, whether your company's or a competitor's.

After determining important features and needs in the product, the best tool to use for capturing these and relating them to the design elements is *quality function deployment (QFD)*. You will recognize the core form of QFD as a simple L-shaped matrix. QFD was initially applied in the 1960s in the Kobe shipyards of Mitsubishi Heavy Industries of Japan. It was refined through other Japanese industries in the 1970s. Dr. Donald Clausing first recognized QFD as an important tool. It was translated into English and introduced to the United States in the 1980s. Following publication of King's book, *Better Designs in Half the Time*, it has been applied in many diverse U.S. industries.[2]

At the heart of applying QFD are one or more matrices. These matrices are the key to QFD's ability to link customer requirements (referred to as VOC or customer *WHATs* in QFD literature) with the organization's plans, product or service features, options, and analysis (referred to as *HOWs*). The first matrix used in a major application of QFD will usually be a form of the A-l matrix.[2] This matrix often includes features not always applied in the other matrices. As a result, it often takes a characteristic box-shape with a roof and is called the *house of quality (HOQ)* in QFD literature. Figure 6 presents the basic form of the HOQ.

The A-l matrix starts with either raw (verbatim) or restated customer WHATs, along with corresponding priorities for the WHATs. Restated customer WHATs are generally still qualitative statements, but with more specificity. For example, if the customer requires high mileage, the design team might translate this (with some input from customers), into a highway mileage target of 40 miles per gallon. The priorities of these WHATs are usually coded from 10 to 1, with 10 representing the most important item(s) and 1 representing the least important. The

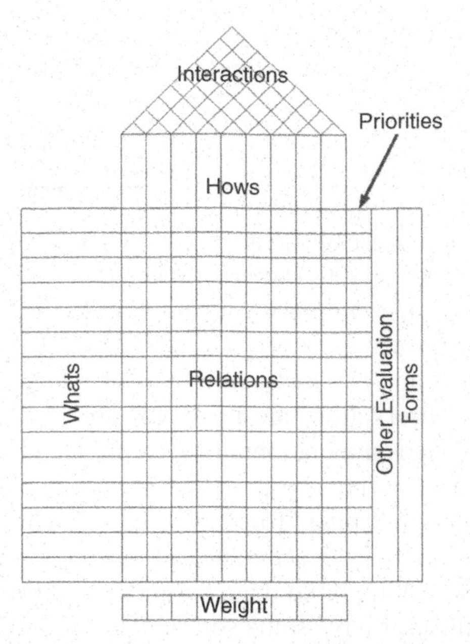

Figure 6 Traditional QFD house of quality (HOQ).

WHATs and their priorities are listed as row headings down the left side of the matrix. Frequently, we find that customer WHATs are qualitative requirements that are difficult to directly relate to design requirements, so the project team will develop a list of substitute quality characteristics and place these as column headings on this matrix. If such substitute VOCs are used, these will still carry the priority indicated by the customer. The column headings in QFD matrices are referred to as HOWs in QFD literature. Some authors use the term *voice of the business* (VOB), as these represent the business strategy to meet the customers' WHATs. Continuing the car example, some options for the VOB for the high mileage car might be a gas-electric hybrid, a diesel-electric hybrid, or a pure electric vehicle. Substitute quality characteristics are usually quantifiable measures that function as high-Ievel product or process design targets and metrics.

The term *substitute quality characteristic* might appear ambiguous. The best way to think of this is to consider the fact that verbatim customer requirements may be stated in words that cannot be directly translated to equations, features, and tolerances. For example, when a user describes the need to make a kitchen appliance "easy to use," there is no way to put that as a specification and measure the output. However, it is still the *voice of the customer*. So, using QFD, it would be placed down the left column as a *WHAT*. The team would then place their strategies to make the device "easy to use" along the top as column headings. Examples of ways to achieve "easy to use" might be well-marked controls, no more than one knob, or automatic sensing of the appropriate setting. Each of these becomes a HOW and, if it relates to the WHAT, becomes a substitute quality characteristic for the "easy to use" WHAT.

The relationships between WHATs and HOWs are identified using symbols such as ● for high relation, ○ for medium relation, and △ for low relation. These are entered at the row/column intersection of the matrix. The convention is to assign 9 points for a high relationship between a WHAT and a HOW, with 3, 1, and 0 for medium, low, and no relationship, respectively. The assignment of points to the various relationship levels and the prioritization of customer WHATs are used to develop a weighted list of HOWs. The relationship values (9, 3, 1, and 0) are multiplied by the WHATs' priority values and summed over each HOW column. These column summations indicate the relative importance of the substitute quality characteristics and their strength of linkage to the customer requirements.

The other major element of the A-l matrix is the characteristic triangular top (an isosceles triangle) that contains the interrelationship assessments of the HOWs. This additional triangle looks like a roof and gives the QFD matrix the profile of a house—hence its nickname, the house of quality (look back again at Figure 6). The roof contains indicators that show the relationship between HOWs. The best way to think of this is to consider what would happen to the other design elements if each one is increased in turn. Consider, for example, a QFD for a car. In response to customer needs and wants, we intend high mileage and ease of operation. To achieve high mileage, we also intend to forgo power steering and automatic transmission. The latter decision would improve mileage, but it would have a detrimental effect on ease of operation for most drivers. The relationship between HOWs is noted in the roof by five symbols: ++ (strong positive relation); + (positive relation); – (negative relation); – – (strong negative relation); and blank (no relation). The positive relationships indicate that increasing the design attribute (HOW) will cause a corresponding increase in the connected HOW. There is no numeric analysis done with these relationships. These are informative for potential trade studies.

Other features that may be added to the A-l matrix include target values, competitive assessments, risk assessments, and others. These are typically entered as separate rows or columns on the bottom or right side of the A-l matrix.

The key output of the A-l matrix is a prioritized list of substitute quality characteristics. This list may be used as the inputs (WHATs) to other matrices. For example, in Figure 7, we show the HOWs of the program team-feeding requirements (WHATs) to the subsystem team and the subsystem team HOWs feeding requirements (WHATs) to the suppliers.

The *affinity diagram* is a widely used tool that is excellent for generating and grouping ideas and concepts. Teams often find the affinity diagram a great tool to explore the issues in a project or to consider the factors involved in implementation. Materials needed are simple. Most teams use several stacks of sticky-notes and pencils. Ideas are generated by each member of the team and

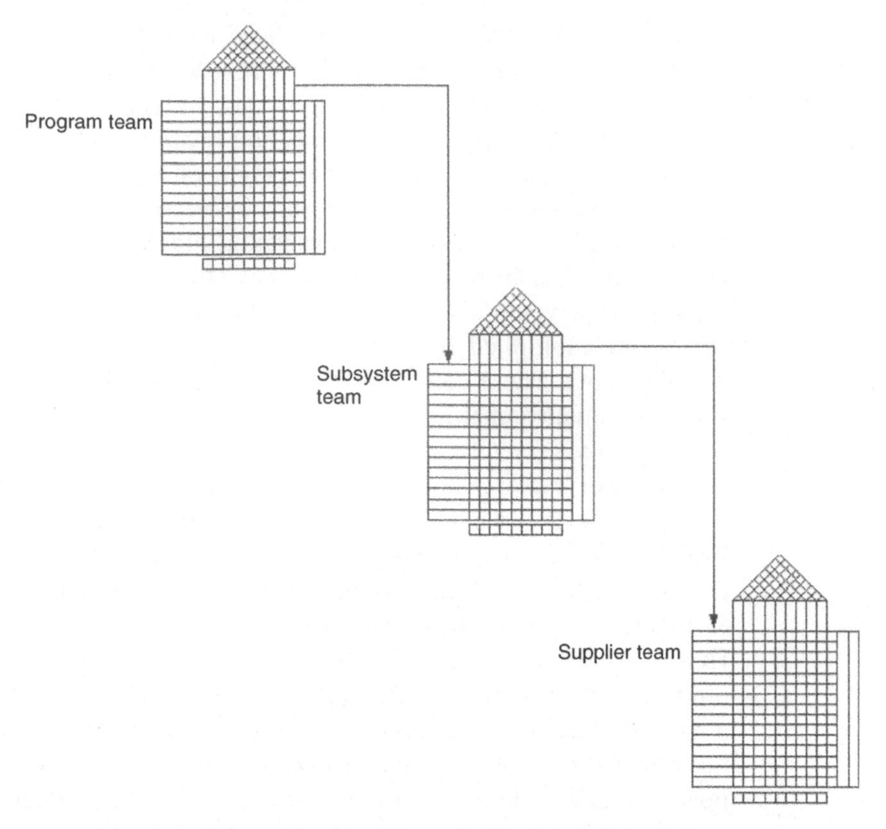

Figure 7 Flowdown of QFD requirements.

pasted on a white board or wall. These are then arranged into *affinity* groupings by the team and assigned a descriptive header. The affinity header identifies the key issue or consideration identified by the team. The number of cards under the header indicates the breadth of consensus by the team.

3.3 Analyze

Tools for the Analyze Phase

Concept reviews, sometimes called *concept selection*, are held to make decisions between two or more competing concepts for the design. Concept reviews can use the *Pugh Matrix* to help make the selection between competing concepts. A Pugh Matrix supports deciding between options with a complex mix of requirements. To utilize the Pugh Matrix, target values for each requirement must be established. Also, the design team must understand if *more is better* or *less is better* for each requirement. To evaluate the concepts, each concept is assessed against each requirement. If a concept exceeds a requirement, it receives a + symbol. If the concept's response to the requirement is the same as the requirement, it receives an S or 0. If it is less than the target, it receives a −. To analyze the results, each symbol is tallied. All other things being equal, the concept with the higher number of +s indicates that this concept will be the superior concept for trial.

Reliability testing may be performed in the analyze phase and then continued into the design and verification phases of the project. While employing environmentally conscious design, the design engineer may have to work with new materials or device configurations that were previously used. These must be tested to ensure that the required reliability is achieved in the hands of the customer. First, it is very important to define the required reliability as part of the analysis. The generally accepted definition of reliability is:

The probability that a device will perform its defined mission under defined conditions for a specified period of time.

The reason that reliability must be defined this way is to establish the requirements for design and testing. This definition will also help determine the appropriate demonstration for reliability testing.

In the analyze phase, reliability testing will often be a form of screening test. Here the testing will be used to make decisions on materials, design architecture, strength, or between concepts. The testing may not actually go as far as is necessary to demonstrate life performance. In some cases, deliberate overstresses may be employed to shorten the testing so as to come to a quicker decision between alternatives.

Reliability testing often employs Weibull analysis, which will be covered in the verification and validation section of this chapter. For the analyze phase, some other testing strategies will be discussed. One such strategy, especially useful for the mechanical environment, is to accelerate the usage of a device to simulate a lifetime of use and then test or measure the residual strength, wear surface, and so on that remain in the device. The distribution of residual strength as compared to the anticipated stress distribution is a measure of the probability that a device will perform to the required level for the mission. Figure 8 shows a graphic of the test approach. For example, environmental issues have required the removal of a protective coating from the device. You are evaluating a new type of coating. The expected life, in the customer's opinion, is five years. The coating could be subjected to accelerated aging such as UV, heat, and humidity. After a simulated time, the coating coverage is checked by cross-section in several places.

A very simple test can be constructed using nonparametric test methods. This test will easily detect if the new device is much better or much worse than the predecessor. If the two approaches are close to the same, the test will not be able to differentiate. The test consists of testing six old devices and six new devices to failure, counting the time, cycles, or actuations to failure for each device. If all six of the new devices survive testing longer than the old devices, there is only a 1.6 percent probability this could happen by chance. This would indicate statistically a low likelihood that the two designs are the same. The converse is that the new devices are more reliable than the older design.

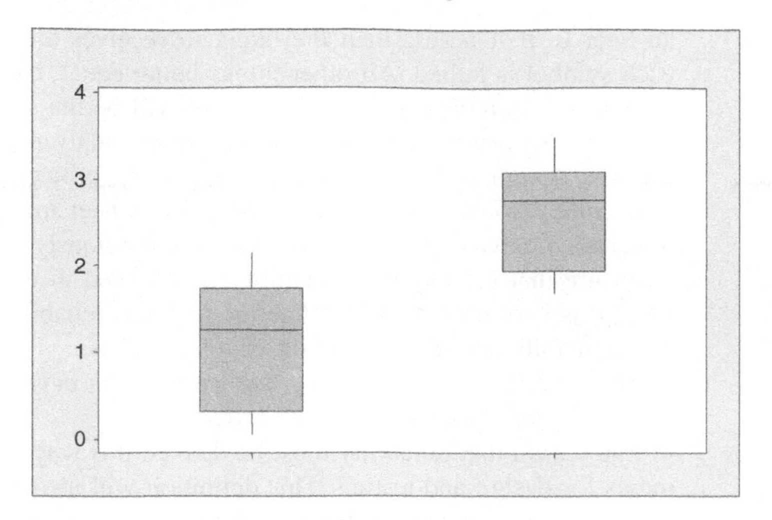

Figure 8 Accelerated aging results.

If testing will take a long time, deliberate overstressing can be used to shorten the test time. For example, all test devices will be run to an expected lifetime or a portion of the expected life. Then the stresses (load, temperature, vibration, etc.) may be raised by 10 percent. A specific method employed in some companies is to use *highly accelerated life testing*. This employs fast-cycling temperature, coupled with intense vibration on all six degrees of freedom, to expose potential weaknesses.

As with the previous testing, the time or cycles to failure will be noted for potential statistical comparison with other concepts or devices. A caution is warranted. It is a difficult and complicated problem to translate overstress testing to actual life or predictive testing. Such guidance is beyond the scope of this chapter, but overstress testing, properly managed, is useful to select between different design options or materials.

Brainstorming is fairly well known, so little of this chapter is spent on it. Primarily, rules must be enforced to allow everyone on the team to participate. They must have an equal chance to participate without their ideas being rejected. There are examples in business experience where the best ideas came from quiet process operators when they were finally encouraged to participate. Brainstorming may help with generating ideas for concepts. *TRIZ*, or *theory of inventive problem solving*, may be a rich source of ways that previous inventors have solved similar problems.[3]

3.4 Design

The design phase is where the details of the device and process design are worked out. Specifications, drawings, and process steps will be developed. The specific elements developed in the design phase should incorporate the key inputs and learnings from the previous phases in TQM/SS.

In the design phase, it is critical to develop the linkage from CTQs to the features and attributes of components and subsystems. These linkages, called a *cascade* or *flow down*, must be documented for the long-term maintenance of the design. The design tools help develop the relationship and document the linkage.

Design for manufacturability (DfM) is a discipline aimed at reducing the effort necessary to manufacture or assemble the device. Often DfM supports environmental concerns. This is achieved by making the device easier to manufacture and assemble, reducing labor, scrap, rework, special tooling, and special components. These will be seen in the following list of considerations for DfM:

- Reduce the number of parts required in the design.
- Foolproof the assembly design (*poka-yoke*) to prevent mis-assembly.
- Make the design so verification or testing is built in.
- Avoid tight tolerances beyond the natural capability of the manufacturing processes (see discussions on process capability and SPC).
- Design robustness in the product to tolerate component variability.
- Design for part orientation and handling to minimize non–value-added movement and difficulty in orienting parts.
- Use gravity to position and hold parts before fastening for ease of assembly and to avoid special fixtures. Use simple patterns of movement.
- Use common parts and materials to standardize handling, inventory, and tools.
- Use modular products to facilitate assembly and servicing.
- Use the minimum number of common fasteners supported by common tooling. Avoid using different types of fasteners where possible.

A key responsibility of a mechanical engineer is to obtain the required performance from a device, component, or process. This must also be done in the most efficient way possible for the company. This usually requires simulation, trade studies, or some form of experimentation with the possible input variables of one or more processes. Engineers are typically taught methods that include assumptions or approximations for the underlying equations. These may not be accurate enough to guide the engineer to the most efficient result.

Design of experiments (DoE) is the tool of choice for trade studies and design or process experimentation. A properly designed experiment will yield the most information possible from a given number of trials, fulfilling the engineer's fiduciary responsibility to the company. More importantly, properly designed experiments also avoid *misleading* results.

Instead of good, solid DoE work, some attempt to use a one-factor-at-a-time (OFAAT) approach where the engineer changes one factor, holding all others constant. This is repeated as the engineer works one at a time through all factors

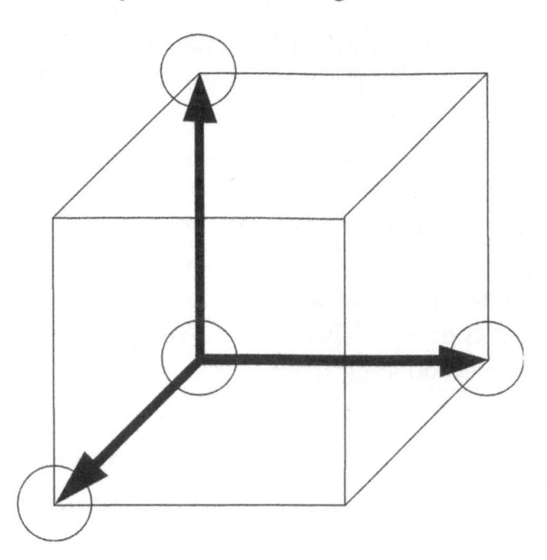

Figure 9 One factor at a time (OFAAT) experimentation.

of interest while monitoring the response(s). OFAAT has some appeal because of its simplicity. Unfortunately, OFAAT yields only linear, first-order responses. The engineer often knows there are interactions with the factors, or a factor's effect may be nonlinear (exponential or quadratic). OFAAT will not disclose this.

In Figure 9, a system space is shown consisting of three factors, each at two levels. Experimenting with OFAAT will only explore the circled corners, yielding no information about the remainder of the space. If there is any interaction between the factors, it will only show up at the four points that are *unexplored*. If there is any form of curvature to the response, we will need to experiment at some point within the interior space of the cube.

Another bad alternative to good DoE work is random experimentation. This takes place when the engineer changes more than one factor at a time, perhaps making multiple runs while trying different combinations. With random experimentation, desired results may be achieved, but the engineer will not know exactly why. The engineer may make a costly design or process change that is not necessary because the actual effect was created by one of the other factors that were randomly changed. Figure 10 shows a path of random experimentation. Like a random walk, this approach lacks an orderly approach to assessing the process environment.

As compared to OFAAT or random experimentation, well-planned DoEs systematically change factors according to a plan, measuring response(s) under known conditions. The experiment often starts with a multifunctional team agreeing on the likely important factors for the experiment. The team may use an affinity diagram or prioritization matrix to determine the priority of process factors. After determining what factors to use, the team must also decide how many

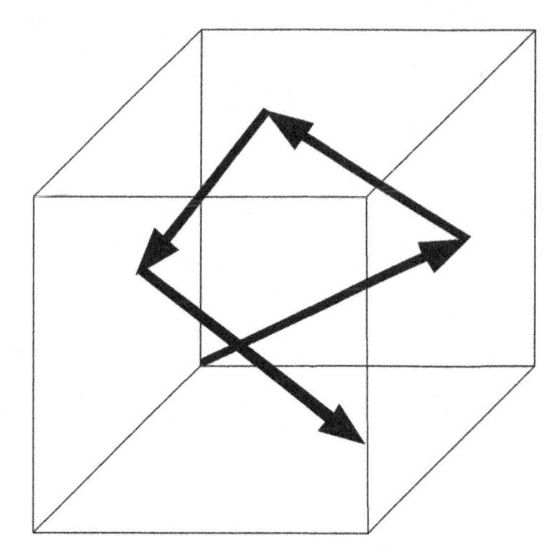

Figure 10 Random experimentation.

levels for each factor. More factors and levels drive more experimental runs, so the factors should be prioritized. Initial experiments often keep the factors at only two levels. This helps reduce the number of experimental runs and makes the analysis somewhat easier.

Types of Experimental Designs

There are many types of experimental designs, but they all fall into two major classifications:

1. *Full factorial.* An experiment where all possible combinations of factor levels are run, at least once. If there are n factors, each at two levels, this will require 2^n experiments for each replication. This type of experiment will yield all possible information, but may be more costly than the engineer or company can afford. Figure 11 shows the main effects output of a DOE of three factors, each at two levels. Figure 12 shows the interactions.
2. *Fractional factorial.* An experiment where a specific subset of the possible factor-level settings is run. A fractional factorial experiment provides only a subset of the information available from a fully factorial experiment. Even so, these designs are very useful if the subset available is planned carefully. Usually, a design is planned that does not identify higher-level interactions. These are *confounded*, or mixed in with other responses. If there are n factors, a half fractional factorial will require 2^{n-1} runs at a minimum. For example, considering an experiment with five factors, one run at each factor would require 32 runs. A half-fractional factorial would cut this to 16. Consult a DoE subject matter expert (SME) for help with fractional factorial experiments.

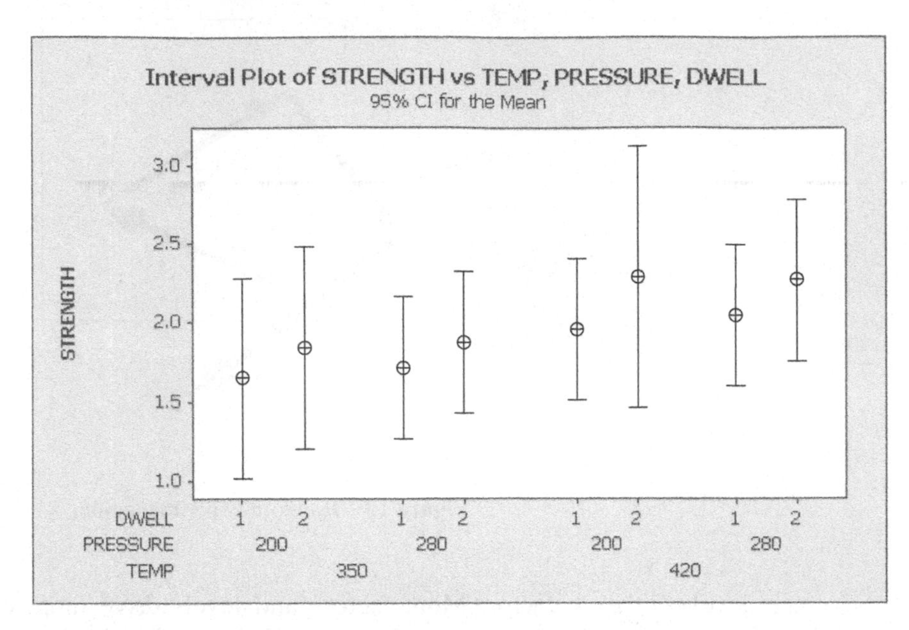

Figure 11 Output from a three-factor experiment (each at two levels).

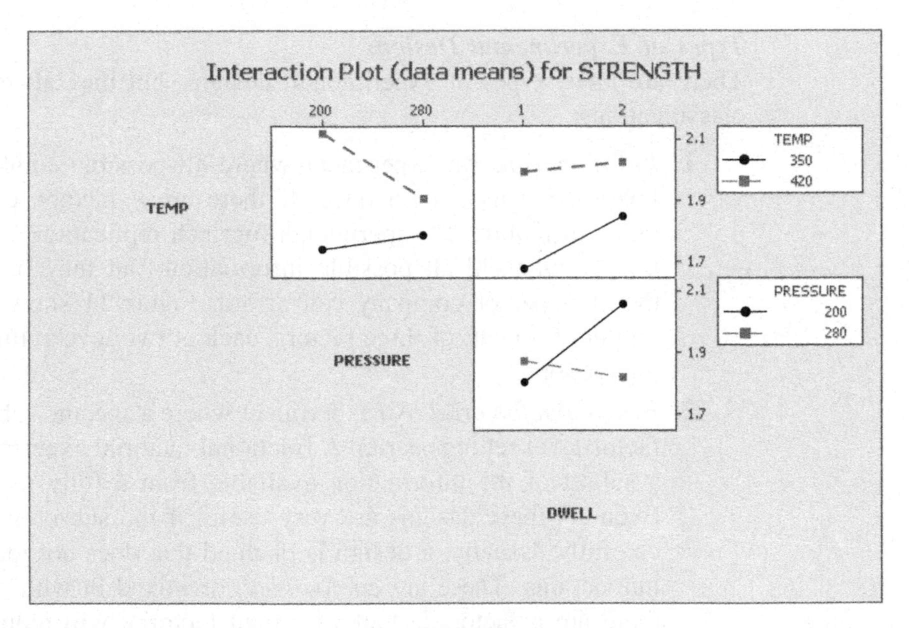

Figure 12 Interactions from a DoE—Different slopes or crossing lines indicate an interaction.

Several methodologies utilize these basic experimental design types. Classical DOE was developed by Sir Ronald Fisher in England and promoted by Box, Hunter, and Hunter in the United States starting in the 1970s. This type utilizes both full and fractional factorial designs. In the early 1960s, Dr. Genichi Taguchi began to promote in Japan, a form of experimental design that uses a special set of fractional factorial designs. Although the forms Dr. Taguchi used were not unique, his approach generated a dramatic increase in DoE usage, especially among engineers. Dr. Taguchi made three major contributions to the field of DoE. First, he developed a DoE methodology that offered clearer guidance to engineers than earlier approaches. Secondly, he promoted the concept of *robust design* and demonstrated how DoE could be used to obtain it. Finally, he promoted the application of something he called the *quality loss function*. This could express in dollars, how the enterprise and society in general are affected by variation from an optimal target.

Usually experiments are run at two levels. Occasionally, the engineer must experiment with factors at more than two levels. These may be attribute factors, such as different materials, or continuous variables, such as temperature, pressure, or time. DoE handles all these, but the planning and analysis get a bit more complicated.

No matter what experimental design is chosen, there are two key aspects of every experiment. The first is randomization. *Randomization* means to plan the experiment carefully, but run it in some type of random order. Using a random number generator, picking numbers out of a hat, or any other method, may accomplish this. The reason randomization is employed is to prevent some time-dependent factor from creeping into the experimental results. For example, a machine tool wears with use. If the experiment proceeds in a particular order with regard to the runs, the later runs will have the additional influence of tool wear. Randomization allows each factor level setting combination an equal chance to experience a time-related factor. The other part that must be considered is replication. It is rare that an experiment is only run once at each factor level setting combination. Even a full factorial experiment is usually run with at least two replications so sufficient information is obtained for good analysis.

While we've touched on the main types of experimental design, the author wishes to note that this has been a very rich field of research and innovation. As a result, there are several types of experimental designs we have not touched on that may be used for specific purposes (e.g., mixtures or the situations where output is nonlinear). These are discussed in the references provided for this chapter.

Statistical testing is often a tool employed in the design phase when materials or processes must be changed, as would be done for environmental reasons. The design engineer must assure that changing a material or process will still yield designed performance, or must adapt the design to the resulting performance. For example, a new material must be used in a particular structure. The specifications and even finite element analysis indicate that sufficient component strength may be delivered, but the engineer wants to assure that the new process will deliver the performance. The engineer may use a test of means to verify performance.

The t-test relies on a special statistical relationship formed if data are distributed normally or nearly normally. If two variables, distributed $N(\mu_1, \sigma_1^2)$ and $N(\mu_2, \sigma_2^2)$, are combined in a linear fashion (added), the result is distributed $N(\mu_1 + \mu_2, \sigma_1^2 + \sigma_2^2)$. If the two variables are subtracted, the result is $N(\mu_1 - \mu_2, \sigma_1^2 + \sigma_2^2)$. Note that the means are subtracted, but the variances are added. This is directly related to the root sum of squares (RSS) calculation familiar to most engineers. Translated from the statistical nomenclature, a linear combination result will be a combination (plus or minus) of the means, and the resulting variance will always be the sum of the variances. Variance is the standard deviation squared.

Applying this in testing, we can take two samples and form a Z-statistic from the formula $Z = (\mu_1 - \mu_2)/\sqrt{\sigma_1^2/n_1 + \sigma_2^2/n_2}$. The standard normal tables will give the probability that such a statistic would be observable from a normal distribution. Intuitively, if μ_1 and μ_2 are close together, the Z-statistic will be near zero and the probability will be higher that the two samples will perform about the same. If the means are far apart, it will signify a potentially significant difference. For example, if the design engineer needs to prove that the new material is at least as strong as previous material, the test hypothesis will be constructed with μ_1 representing the new material and μ_2 representing the older material. Two hypotheses will be formed:

$$\text{H}_\text{o}: \quad \mu_1 \geq \mu_2 \text{with } \alpha = 0.05$$

$$\text{H}_1: \quad \mu_1 < \mu_2 \text{with } \beta = 0.05$$

A critical Z value will be developed to guide the decision. If the $Z_{\text{observed}} \geq Z_{\text{critical}}$ the test is said to "fail to reject" the null hypothesis (H_o). If $Z_{\text{observed}} < Z_{\text{critical}}$, the test rejected the null hypothesis.

The Z-test may be performed when the sample sizes defining the two means and variations are large or the variation is well-known from previous testing. Often sample sizes of 25 to 30 are defined as the boundary of what statisticians call a large sample. If the variation is not well understood, or the two samples will be less than 25, the estimation of both the mean and standard deviation introduces a bias in the statistical test that may be overcome by using a slightly different and more complicated formula based on the Student t distribution.

$$t = \frac{\overline{x} - \overline{y}}{\sqrt{\frac{(n_x - 1)S_{\overline{x}}^2 + (n_y - 1)S_{\overline{y}}^2}{n_1 + n_2 - 2}\left(\frac{1}{n_x} + \frac{1}{n_y}\right)}}$$

where $S = \sqrt{\left(\frac{1}{n-1}\right)\sum(x - \overline{x})^2}$. The Student t distribution will have a critical decision value just as Z_{critical}, but will be a function of the degrees of freedom from the sizes of the two samples. The complication of employing the Student

t test might cause one to utilize the Z test under circumstances where it is not recommended. Besides the availability of statistical software, both the top commercially available and top open source spreadsheets offer an easy-to-use function called the "ttest," which performs the test and returns a probability rather than the t statistic.

A similar test is available to determine if the new material under consideration may have an effect on the variability of the design. This type of test, called the F-test, is a somewhat simpler test to construct. The observed variable is:

$$F = \frac{S_1}{S_2}$$

This is compared against an F_{critical}, which is a function of the size of the two samples in a variable called *degrees of freedom*. As before, F tests may be performed in statistical software or the two major spreadsheets available to users. The procedure name in the spreadsheets is "ftest."

As mentioned earlier in this chapter, CTQs are stated using the customer's language and do not appear to relate to process parameters. Using tools such as QFD or other analyses it is possible to relate customer CTQs to design features. This can be carried on down to components, component features, processes, and settings. This is called a *critical to quality cascade*. Development teams often find it useful to express these graphically in a tree diagram. A tree diagram graphic easily shows the people involved with a process how the parameters relate to customers. For example, the team might start with one CTQ, such as car mileage.

At the next level, aerodynamic design, efficient engine, and efficient transmission would be shown. Each of these subsystems could be further decomposed to their major elements (where this makes sense). This process can be continued to the logical place on each subsystem and component where key elements may be measured and controlled. At this point in the DMADVV process, it may not be possible to completely take a CTQ cascade to the lowest level of control, but the CTQ cascade can be updated as the team moves through the process. Figure 13 shows an example of a CTQ cascade tree with process control information.

A *process control plan* is another tool that can demonstrate this relationship. A process control plan is a process work instruction, generated in a word-processing document or spreadsheet. In this plan, it is convenient to show process settings in a tabular form. Linkages between a setting and a CTQ can also be shown here. Cascades can also be shown in a spreadsheet. Early proponents of QFD often proposed using two or more QFD matrices to do this linkage. This is an excellent analysis approach, but may be too difficult to maintain for a process work instruction. A process control plan fits that need very nicely.

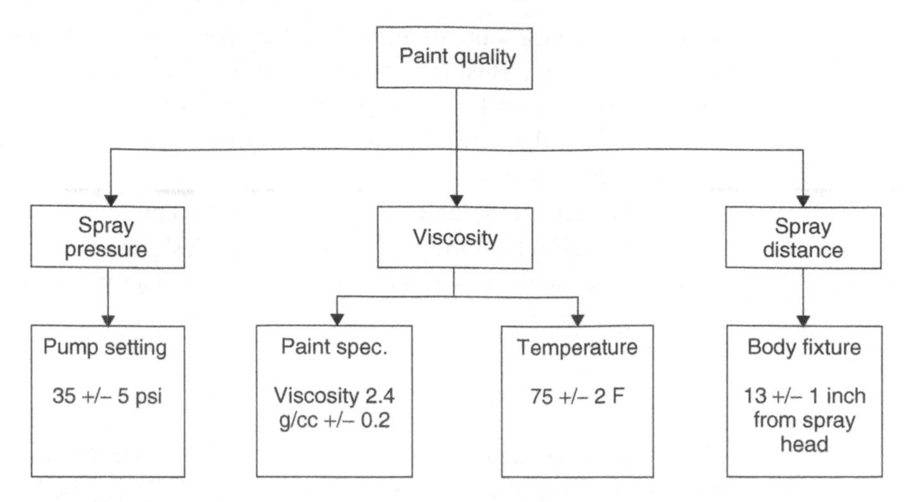

Figure 13 Cascade flowdown (tree) with process specifications for control.

3.5 Verification and Validation

A key function at the end of the design process is to verify that the design was properly translated by the manufacturing processes into components and devices that meet the design intent. This is done by a series of tests for both the components and the device to verify that the design requirements may be demonstrated in the device.

For medical devices and in some other industries, there may be a requirement to perform validation testing. Validation testing may be a good strategy in nearly all industries, even if it is not a requirement. In validation testing, product representing the design as translated by manufacturing is given to representative customers in order to determine if the design is an effective solution to the customer's problem. If the design effort is properly done, validation will be a confirmation that the device is useful. If there is a miss in customer requirements, the miss may be detected and perhaps addressed before the device is committed to the marketplace.

Tools for the Verification and Validation Phase

Process capability analysis or validation studies are properly descriptive from the name. These studies allow engineers and operators to assess the process to determine either the long-run or short-run performance of the process. Knowledge of process capability may aid in setting specifications or may support the prediction of scrap, rework, and throughput. If design engineers understand process capability and use that information to set specifications, there can be less wasteful conflict between design and manufacturing. It is critical that process capability analysis be applied to the design CTQs, as well as to the features and attributes linked to those CTQs on each component and subsystem of the design.

This will assure the long-term success of the design, as well as the reduction in scrap and rework so important to reducing the impact on the environment.

The best approach to process capability analysis is to take samples from some of the initial run of production or engineering pilot parts. Twenty-five subgroups of three to five parts taken from the run will be sufficient. Measure the critical features, as defined by CTQ flowdown or manufacturing need. Plot the average and range or average and standard deviation on an appropriate control chart. If the control chart shows the process is in control, the centerline of the averages is a good estimate for the process grand average. The process standard deviation may be estimated from the average range or average standard deviation of the in-control process. These values may be used to calculate C_{pk} or C_{p} using the following formulas:

$$C_{pk} = \frac{\min\{USL - \overline{\overline{X}}, \overline{\overline{X}} - LSL\}}{3^{*}\hat{\sigma}}$$

$$C_{p} = \frac{USL - LSL}{6^{*}\hat{\sigma}}$$

where USL = upper specification limit, LSL = lower specification limit, $\overline{\overline{X}}$ is the estimate of process mean, and $\hat{\sigma}$ is the estimate of process standard deviation.

What good is C_{p}? C_{p} demonstrates what the process can achieve if it is centered on the specification and it is controlled to stay there.

The capability index, C_{pk}, indicates how much room there is between the product specification (tolerance) limits and the expected output of the process. C_{pk} calculations and performance values are given in Figure 14. C_{pk} shows how many multiples of three standard deviations fit between the process output average and the closest specification limit. A C_{pk} of 1.0 indicates there are only three standard deviations between the process average and a specification limit. A C_{pk} of 1.33 indicates that there are four standard deviations at a minimum. Your company may have established a target value for C_{pk}. Some companies use 1.33 or 1.5 as the target value. Higher values of C_{pk} allow greater margin if the process slips out of statistical control.

What we call a Six Sigma process approach guides the team to set a process and product specification so that there are six standard deviations between the process average and the closest specification. A Six Sigma process would have a C_{pk} of 2.0 or greater (check the calculations to ensure this relationship is understood). Six Sigma strives for the extra margin to assure good output even if the process encounters a *shift*. Empirical observation shows that process may shift up to 1.5 sigma before the problem is detected and controlled. If the process is six standard deviations from the nearest limit and a process shift of 1.5 sigma occurs in that direction, the engineer will still have a process operating at 1.5 C_{pk}. Figure 14 also shows the effect of a 1.5 sigma shift under various initial C_{pk} values.

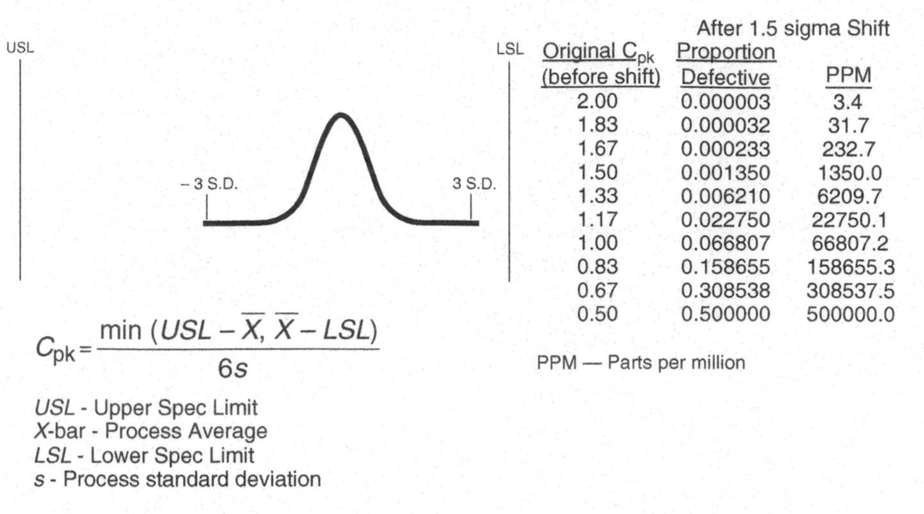

	After 1.5 sigma Shift	
Original C_{pk} (before shift)	Proportion Defective	PPM
2.00	0.000003	3.4
1.83	0.000032	31.7
1.67	0.000233	232.7
1.50	0.001350	1350.0
1.33	0.006210	6209.7
1.17	0.022750	22750.1
1.00	0.066807	66807.2
0.83	0.158655	158655.3
0.67	0.308538	308537.5
0.50	0.500000	500000.0

Figure 14 C_{pk} calculations and impact of process shifts.

What if a C_{pk} of 2.0 cannot be achieved? If this is not addressed, it will lead to long-term scrap and difficulty. Extra scrap especially will not help achieve environmental goals. The development team must work to either:

- Increase the design tolerance window
- Improve the process

This is not just a manufacturing problem. The development team must work with manufacturing to assure the process is robust. The general steps are:

1. If the process doesn't have SPC, apply it!
2. Get the process under statistical control, that is to say, make it predictable.
3. Assess the measurement system and improve it if necessary.
4. From the SPC chart, obtain estimates of the process average and standard deviation.
5. Assess the process C_{pk}.
6. Based on the resulting C_{pk}, determine whether to change the product specification or improve the process by applying the elements of the DMAIC process.

The last two steps are very important. If the process capability is not acceptable, the design or the process must be changed. If one or both of these are not done, your business must *live with the resulting low performance as long as the product is made*. The decision of which to address—product design, process, or both—is an economic one.

Another aspect of process assessment is the measurement system. If a significant portion of process variation comes from measurement, it may be easier to improve the measurement system than the process. Gage repeatability and reproducibility (GR&R) assessments are also a core part of the process.

Many companies utilize *design reviews* as a way to manage overall project risk. An independent team is assembled to review the development process and to assure that the design meets the original specified requirements. Generally, the team will assess the derivation of customer requirements, their translation into design intent, and the ability of the resulting manufacturing processes to deliver material meeting design intent.

Reliability testing in the verification and validation phases, takes on a different objective than it does in the analyze phase. Not all companies will use it, but in some industries it is mandated. In other industries, it is a good idea to assess the potential warranty costs, determine how many spare parts to stock for service, and understand if there are issues that must be addressed as the product design matures. Reliability here often employs more of demonstration tests, although it can still utilize structured tests using acceleration of environmental factors.

Much of reliability demonstration testing today utilizes some form of Weibull testing. The Weibull distribution is of the form

$$R(t) = e^{-\left(\frac{t}{\eta}\right)^{\beta}}$$

where t is time or cycles, η is called the location parameter, and β is called the shape parameter.

The Weibull distribution is often used in reliability work because of its ability to statistically model many different situations. For example, when $\beta < 1$, this situation is called an early failure region and will be marked by earlier failures, which diminish in rate quickly. When $\beta > 1$, the devices are said to be in *wearout*. The wearout region is characterized by increasing failure rate as the devices increasingly lose strength against the environmental stresses. When $\beta = 1$, the equation is the well-known exponential distribution. Using different parameters, the Weibull distribution may model what is known as the *reliability bathtub curve*, where the devices in the field go through the early failure, useful life, and wearout regions as they progress through use.

With regard to the different regions, reliability testing may be used to determine if the manufactured devices exhibit any early failures. If so, these may be addressed by discovering the root cause and improving the applicable manufacturing controls. For example, cycle testing of a mechanical device may show that some fasteners are backing off of the attachment bolts. Work on the root cause may find that the factory has a problem with torque variation of the nut drivers. Alternatively, the design may have to employ self-locking fasteners at key positions in the design.

Continuing reliability testing beyond early failure will allow the designers to determine the mean time between failures (MTBF) and to find out what components or subassemblies in the device will be failing first. If the MTBF is unacceptable and a number of common components fail early, these may be the prime candidates for improvement or targets for recommended preventive maintenance. A well-designed device that appropriately manages the usage stresses will show random failures of different components in the system. If the life of this device is not acceptable, the opportunities for improvement are more limited to a general redesign or reduction in usage stresses.

It is beyond the scope of this chapter to delve into reliability testing, but the Weibull distribution may be conveniently manipulated algebraically to a form where linear regression can be used to estimate η and β. With these estimates, the design engineer may predict the resulting device reliability. The linear regression also supports the application of confidence bounds such that the confidence of the estimate may be determined. The only statistical difficulty to Weibull testing is the assignment of probabilities in the form of order statistics to the failures as they occur. This may be done by Weibull software or tables available in many reliability handbooks.

Measurement systems analysis (MSA) is a process of assessing and, as necessary, improving the measurement process for a design or device. Since everything has variation at some measurable level, it is no surprise that measurement systems have variation also. As noted in the measure phase, system variation is the sum of the variation of the measurement system and the device or component. The tool most used for MSA is gage repeatability and reproducibility (GR&R), which isolates and quantifies variation from repeatability and reproducibility. These terms are defined as follows.

1. *Bias in the gage.* When a gage has a bias, it tends to indicate a reading that is above or below the true value. This is a function of the gage's calibration.
2. *Repeatability.* When a gage is not repeatable, it means that repeated measurements by the same operator, when the part is removed and replaced in the fixture or gage, show a large variation. Repeatability is influenced by gage design. Electrical noise, excess play in mechanical linkages, or a loose fit in fixture-retaining features can also influence repeatability.
3. *Reproducibility.* This pertains to the ability of a second operator to achieve the same result as a previous operator working with the same equipment and under the same conditions. Examples of factors that influence reproducibility include holding fixtures that are sensitive to operator technique and measurement instructions that give the operator significant discretion in how the part will be mounted and measured.

Some may be surprised at this level of detail for the process measurement systems. The reason is economic. If a process has lower than desirable yield, the issue may be with the process, measurement system, or both. The measurement system may be rejecting good parts and allowing nonconforming units to be sent to customers. First, it is often more economical to fix a measurement system than change a process. Second, if the measurements being taken have a large amount of uncertainty; it is likely that you are rejecting good parts, delivering parts that are not in conformance, or both.

Measurement systems analysis should be performed properly so the source of variation is identified. It is desirable that the parts exhibit variation covering the expected tolerance range, although this may be difficult for some processes. Most analysis involves approximately 10 parts and two to three operators. First, the gage is calibrated or the calibration record is checked. Second, each operator will measure and record the features of interest on each part two or three times. Parts will be run in random order to remove any time trending with the operator or measurement system. All measurements will be recorded with identification of the operator, part, and order of measurement. The statistical analysis then breaks down the sources of variation and to identify how much variation is coming from the parts, the repeatability of the gage, or the reproducibility of the gage.

Many companies place guidelines on the amount of measurement error they will tolerate in the system. Generally, less than 10 percent of the feature tolerance is an acceptable range. If the error is less than 30 percent, it may be tolerable, depending on the criticality of the feature. If the measurement error is greater than 30 percent of the tolerance range, the measurement technique should be improved.

Another aspect of measurement systems analysis is the comparison between gages. Often companies will rely on suppliers' measurements. If there is an

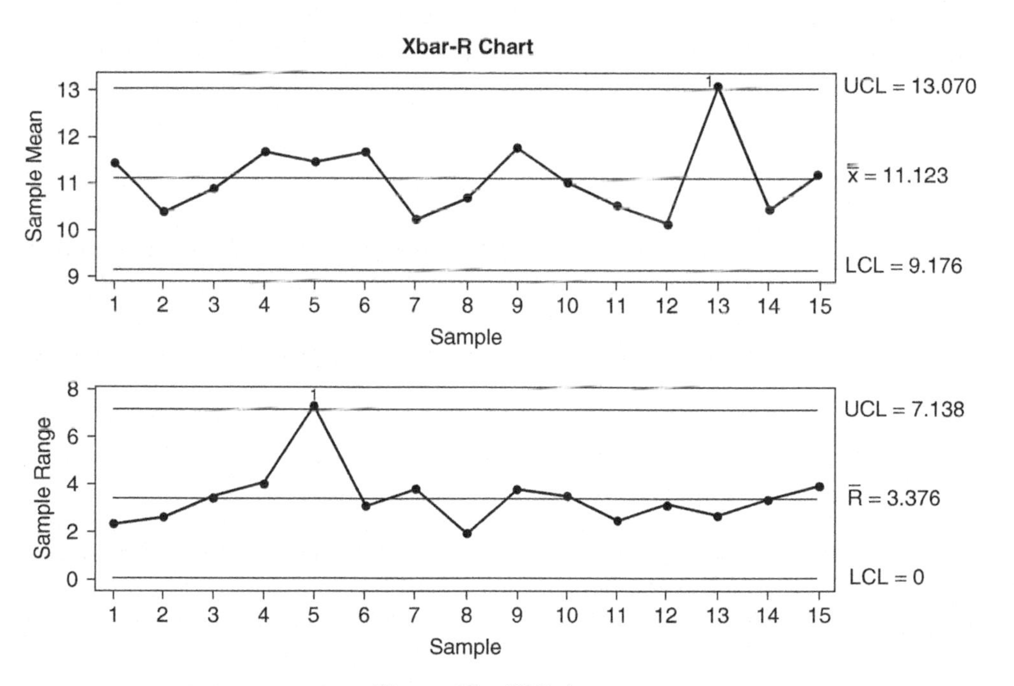

Figure 15 SPC charts.

issue, it is good to be able to assess parts at your facility and know your measurements will be similar to your suppliers' measurements. For more information on this extensive subject, the reader is referred to sources such as the Automotive Industry Action Group (AIAG).[4]

Statistical process control (SPC) is an important aspect of manufacturing. Design engineers have the responsibility to help manufacturing define the critical elements of manufacturing process outputs through CTQ cascades and to establish how processes may be controlled. This is done through designs of experiments and SPC. The focus in this chapter is on design; refer to the chapter references for further information on SPC. Figure 15 shows example SPC charts.

WEB RESOURCES

The following are very good online resources or represent organizations supporting TQM/SS concepts and tools. If the URLs change, a simple search will make the information available again.

American Society for Quality—numerous publications on quality, reliability, and applied statistics, available at www.asq.org.

APQC (formerly the American Productivity and Quality Center), available at www.apqc.org.

Automotive Industry Action Group—numerous publications covering SPC and process improvement, available at www.aiag.org.

National Institute for Standards and Technology (NIST) Engineering Statistics Handbook, available at http://www.itl.nist.gov/div898/handbook/

BIBLIOGRAPHY

M. L. Berryman, "DFSS and Big Payoffs," *ASQ Six Sigma Forum Magazine*, **2**(1), 23–28 (2002).

J. G. Bralla, *Design for Manufacturability Handbook*, 2nd ed., New York: McGraw-Hill, 1999.

M. Caroselli, *Quality-driven Designs: 36 Activities to Reinforce TQM Concepts*, Pfeiffer & Co, San Diego, 1992.

G. S. Day and D. J. Reibstein, *Wharton on Dynamic Competitive Strategy*, John Wiley, New York, 2001.

L. Ferryanto, "DFSS: Lessons Learned." *ASQ Six Sigma Forum*, **4**(2), 24–28 (2005).

N. L. Frigon and D. Matthews, (1996). *Practical Guide to Experimental Design*, John Wiley, New York, 1996.

F. Giudice, G. La Rosa, and A. Risitano, "Indicators for Environmentally Conscious Product Design." IEEE First International Symposium on Environmentally Conscious Design and Inverse Manufacturing, 71–77, 1999.

D. L. Harnett and J. F. Horrell, *Data, Statistics, and Decision Models with Excel*, John Wiley, New York, 1998.

K. Heldman, *PMP: Project management professional: Study guide*, 3rd ed., John Wiley, Hoboken, NJ, 2005.

S.-H. Kim, Y.-H. Yoon, and G.-T. Zeon, "Combine Quality and Speed to Market," *ASQ Six Sigma Forum*, **3**(4), 26–31 (2004).

B. King, *Better Designs in Half the Time: Implementing QFD Quality Function Deployment in America*, 3rd ed., GOAL/QPC, Methuen, MA, 1989.

M. Kutz, *Mechanical Engineers' Handbook*, 3rd ed, John Wiley, Hoboken, NJ, 2005.

C. Loch, A. D. Meyer, and M. T. Pich, *Managing the Unknown: A New Approach to Project Risk Management*, John Wiley, Hoboken, NJ, 2006.

A. Longman and J. Mullins, *The Rational Project Manager: A Thinking Team's Guide to Getting Work Done*, John Wiley, Hoboken, NJ, 2005.

K. Masui and T. Sakao, "Applying Quality Function Deployment to Environmentally Conscious Design," *The International Journal of Quality & Reliability Management*, **20**(1), 90–106 (2003).

C. Nachtsheim and B. Jones, "A Powerful Analytical Tool," *ASQ Six Sigma Forum*, **2**(4), 30–33 (2003).

J. B. ReVelle, *Quality Essentials: A Reference Guide from A to Z*, ASQ Quality Press, Milwaukee, 2004.

J. B. ReVelle, J. W. Moran, and C. A. Cox, *The QFD Handbook*, John Wiley, New York, 1998.

R. K. Roy, *Design of Experiments Using the Taguchi Approach: 16 Steps to Product and Process Improvement*, John Wiley, New York, 2001.

T. T. Soong, *Fundamentals of Probability and Statistics for Engineers*, John Wiley, Hoboken, NJ, 2004.

M. T. Todinov, *Reliability and Risk Models: Setting Reliability Requirements*, John Wiley, Hoboken, NJ, 2005.

REFERENCES CITED

1. National Academies Press, "Future R&D Environments," A Report for the National Institute of Standards and Technology, 2002.
2. B. King, *Better Designs in Half the Time: Implementing QFD Quality Function Deployment in America*, 3rd ed., GOAL/QPC, Methuen, MA, 1989.
3. H. Altov, *And Suddenly the Inventor Appeared: The Art of Inventing*, 2nd ed., Technical Innovation Center, Auburn, MA, 1996.
4. Automotive Industry Action Group (AIAG). Available at www.aiag.org.

Index

D

E

F

G

Q

R

CPSIA information can be obtained at www.ICGtesting.com
Printed in the USA
BVOW09*0016220915

419014BV00016B/425/P